Major topics in primate and human evolution

Major topics in primate and human evolution

Edited by

BERNARD WOOD
Department of Anatomy
The University of Liverpool

LAWRENCE MARTIN
Department of Anthropology
State University of New York

PETER ANDREWS
Department of Palaeontology
British Museum (Natural History), London

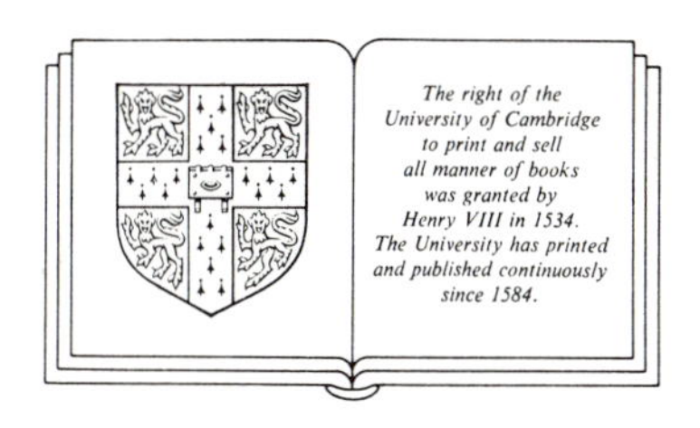

CAMBRIDGE UNIVERSITY PRESS
Cambridge
London New York New Rochelle
Melbourne Sydney

Published by the Press Syndicate of the University of Cambridge
The Pitt Building, Trumpington Street, Cambridge CB2 1RP
32 East 57th Street, New York, NY 10022, USA
10 Stamford Road, Oakleigh, Melbourne 3166, Australia

First published 1986

Printed in Great Britain at the
University Press, Cambridge

British Library cataloguing in publication data

Major topics in primate and human evolution.
1. Human evolution 2. Primates–Evolution
3. Mammals–Evolution
I. Wood, Bernard A. II. Martin, Lawrence
III. Andrews, Peter
573.2 GN281

Library of Congress CIP data available

ISBN 0 521 30700 7

CONTENTS

LIST OF CONTRIBUTORS

L. C. Aiello
Department of Anthropology,
University College London,
London WC1E 6BT

P. Andrews,
Department of Palaeontology
British Museum (Natural History)
Cromwell Road,
London SW7 5BD

A. Bilsborough,
Department of Anthropology,
University of Durham,
43 Old Elvet, Durham, DH1 3HN

M. J. Bishop,
Department of Zoology,
University of Cambridge,
Downing Street,
Cambridge CB2 3EJ

P. M. Butler,
23 Mandeville Court,
Strode Street,
Egham, Surrey TW20 9BU

A. T. Chamberlain,
Department of Anatomy,
The University of Liverpool,
P. O. Box 147, Liverpool L69 3BX

M. H. Day,
Division of Anatomy
United Medical and Dental Schools
of Guy's and St Thomas's Hospitals,
London SE1 7EH

M. C. Dean,
Department of Anatomy and
Embryology,
University College London,
London WC1E 6BT

J. G. Fleagle,
Department of Anatomical Sciences,
School of Medicine,
State University of New York,
Stony Brook, New York 11794

A. E. Friday,
Department of Zoology,
University of Cambridge,
Downing Street,
Cambridge CB2 3EJ

P. D. Gingerich,
Museum of Paleontology
University of Michigan,
Ann Arbor, Michigan 48109

F. E. Grine,
Departments of Anthropology and
Anatomical Sciences,
State University of New York,
Stony Brook, New York 11794

J. S. Jones,
Department of Genetics & Biometry,
University College London,
London NW1 2HE

W. L. Jungers,
Department of Anatomical Sciences,
State University of New York,
Stony Brook, New York 11794

L. Martin,
Departments of Anthropology and
Anatomical Sciences,
State University of New York,
Stony Brook, New York 11794

R. D. Martin,
Department of Anthropology,
University College London,
London WC1E 6BT

D. Pilbeam
Department of Anthropology,
Peabody Museum,
Harvard University
Cambridge, MA 02138

A. L. Rosenberger,
Department of Anthropology,
University of Illinois at Chicago,
Box 4348
Chicago, Illinois 60680

M. Ruvulo,
Department of Biological Chemistry,
Harvard Medical School,
Boston, Mass. 02138

C. B. Stringer,
Department of Palaeontology,
British Museum (Natural History),
Cromwell Road,
London SW7 5BD

B. A. Wood,
Department of Anatomy,
The University of Liverpool,
P.O. Box 147, Liverpool L69 3BX

PREFACE

There have been a number of conferences and publications in recent years which have addressed aspects of primate and human evolution, but none have concentrated solely on two related problems in evolutionary studies, namely the definitions of, and relationships among, taxonomic groups. The Anatomical Society of Great Britain and Northern Ireland invited one of us to organise a symposium on human evolution which stressed these two themes. It was then decided to invite the Primate Society of Great Britain to become a co-sponsor of the symposium, and the three editors were jointly charged with the task of assembling the programme and widening the scope to include all primates.

The subsequent symposium, "Major Topics in Primate and Human Evolution", was designed to try and address research problems associated with the major branching points of primate evolution. We adopted two strategies to try to illuminate some of the thornier topics. First, where appropriate, the data relating to a particular branching point were reviewed by leading proponents of alternative interpretations of the evidence. Secondly, we approached some of the problem areas of hominid evolution by asking experts to contribute papers about fossil evidence, or time periods, which were not necessarily their main research interest. We are grateful to all the contributors for the enthusiastic way they responded to these challenges. Their combined efforts have resulted in what we hope is a fresh and thorough review of the current status of our knowledge of primate evolution. This volume contains the papers that were presented at the symposium, but additionally the Osman Hill Lecture of the Primate Society and the Annual Review Lecture of the Anatomical Society were integrated into the symposium and we are grateful to the respective invited lecturers for agreeing to contribute to this volume.

We are grateful to the Councils of the Anatomical Society of Great Britain and Ireland and of the Primate Society of Great Britain and to the L.S.B. Leakey Trust for their support for the symposium. The meeting was held at The Middlesex Hospital Medical School and we thank the Dean and the staff for their help and cooperation. The assistance of Andrew Chamberlain, Miles Ellis, Susan Evans, Paula Sutton, Graham Wilson, Alan Day and the technical staff of the Department of Anatomy and Biology at that institution was indispensable and the smooth running of the meeting was due to their combined efforts. Many London-based colleagues generously helped us provide hospitality for the speakers from abroad.

With particular regard to this volume, we thank the contributors for adhering to our strict timetable, and we hope that the speed with which this book has been produced will be its own reward. We are grateful to Robin Pellew at Cambridge University Press and to Andrew Chamberlain and Paula Sutton of the Department of Anatomy and Biology for their help with its production. Andrew Chamberlain undertook the onerous task of producing the index, and we are grateful to him for the care he has taken with its preparation. The cover illustration was drawn for us by Anne-Elise Martin.

July 1985

Bernard Wood
Lawrence Martin
Peter Andrews

1 PRIMATES: A DEFINITION

R.D. Martin,
Department of Anthropology,
University College London
London WC1E 6BT.

INTRODUCTION

Interpretations of primate evolution have undergone radical revision over the past fifty years (Cartmill, 1982). Part of this transformation is, of course, attributable to the impressive accumulation of new evidence, but an equally important part is due to increased clarity and precision in the methodology of phylogenetic reconstruction. With respect to the latter, a major contribution was made by Hennig's concepts, first summarized in German, though not made available in English until later (Hennig, 1950, 1966). As part of the ensuing debate, the present author made a very preliminary attempt to re-examine the question of definition of the order Primates (Martin, 1968). Given the further accumulation of new evidence and interpretations since that time, this is perhaps an appropriate time to reconsider some of the issues involved in defining the primates.

Implicitly or explicitly, definition of the order Primates, and of sub-groups thereof, is an essential component of phylogenetic reconstruction. However, there is also a strong element of circularity because of received notions about membership of the order Primates. This being the case, it is probably preferable to begin the process of definition with those forms whose inclusion in the Primates is largely unquestioned and to proceed thereafter to consider marginal cases. There is now a broad consensus that living primates include at least the lemurs, lorises, tarsiers, monkeys, apes and man, whereas the tree-shrews are a marginal case (see Luckett, 1980 for a recent review). Similarly, little doubt has been expressed about the inclusion of an array of fossil forms extending back to Eocene lemuroids (=Adapidae) and tarsioids (=Omomyidae), now commonly termed "primates of modern aspect" (see Simons, 1972; Szalay & Delson, 1979). On the other hand, there is still some doubt about the status of the early Tertiary Plesiadapiformes ("archaic primates", Simons, 1972). Accordingly, it can be argued that the most reliable procedure is to define living primates to the exclusion of tree-shrews and then to assess the affinities of tree-shrews in relation to that definition. Subsequent examination of the fossil "primates of modern aspect" will permit assessment of fossilizable features included in the definition, and will perhaps lead to some refinement of it. Only then is it appropriate to consider the likely relationships of the "archaic primates" (=Plesiadapiformes).

This leads on to the question of the nature of the definition itself, which may be intended purely for descriptive purposes, but which must be framed more precisely for purposes of phylogenetic reconstruction. It is in this context that Hennig's contribution (1950, 1966) has been so important. Definitions must clearly be based on similarities shared by species in a group, and it has long been recognised that similarities acquired by convergent evolution should be excluded from consideration (in so far as it is practically possible to identify such similarities). Hennig, however, made the vital observation that it is also necessary to separate homologous similarities into primitive (plesiomorphic) and derived (apomorphic) categories. For any group of animals under consideration, the common ancestral stock must have been characterized by the possession of a particular set of features that are primitive with respect to all later descendants (primitive character states). Shared retention of primitive character states in any descendants provides no information about subsequent branching-points in the evolutionary tree. The existence and nature of later ancestral stocks within the tree is indicated by shared possession of derived character states. Hence, definitions intended for purposes of phylogenetic reconstruction should include only inferred derived character states, whereas purely descriptive definitions may include both primitive and derived character states (e.g. in relation to a taxonomic key for practical identification of specimens). For instance, one might specify "possession of a rhinarium" as a shared feature of strepsirhine primates (lemurs and lorises), although it is highly probable that possession of a rhinarium is a primitive character state for placental mammals. While it may be useful to cite this feature in a taxonomic key for primates, it does not in itself indicate any special affinity between lemurs and lorises among primates. For the latter purpose, one must cite apparent shared derived features of lemurs and lorises, such as the possession of a tooth-comb, incorporating the canine teeth, in the lower jaw.

This, in fact, ushers in the vexed problem of the relationship between phylogenetic reconstruction and classification (Martin, 1981a). It is, of course, a fundamental tenet of the Hennigian, or cladistic, School that classifications should directly reflect inferred phylogenetic relationships. According to this view, a definition based exclusively on derived character states is required both for phylogenetic reconstruction and for classification (though it still does not follow that taxonomic keys should exclude other kinds of information). But there are numerous reasons, mainly practical, for rejecting the strict Hennigian approach to classification, while accepting the great value of Hennigian concepts for phylogenetic reconstruction. Instead, there is much justification for continuing to use the "classical" approach to primate classification (e.g. see Simpson, 1945; Simons, 1972). This is based on the concept of overall "morphological divergence" and thus depends upon assessments of total homologous similarity (viz. a combination of both primitive and derived character states) and use of the "grade" concept. Regardless of the philosophy of classification adopted, however, it is advisable to maintain a sharp distinction between phylogenetic reconstruction and classification. The former involves inference of the pattern of relationships between species generated by evolutionary divergence over time, while the latter involves construction of a naming system that is compatible with inferred phylogenetic relationships, but requires at least some degree of arbitrary definition. That said, it should be emphasized that the following discussion is concerned exclusively with the inference of phylogenetic relationships.

Hennig's distinction between primitive and derived character states is eminently clear at a theoretical level, but application of that distinction in practice is by no means straightforward. Indeed, it must be emphasized that sorting of similarities between species into convergent, primitive and derived character states depends upon the application of a set of guidelines that permit no more than assessments of probability. Such assessments may vary from highly probable to marginally possible. In practice, therefore, one can only speak of inferred convergent, primitive or derived character states, especially where particular character states exhibit a patchy distribution among the species concerned. This being the case, it is best to take an approach that minimizes the likelihood of erroneous inference. A pragmatic approach that meets this requirement is to concentrate, at least in the first instance, on identifying universal or near-universal features of living primates that separate them from all other living placental mammals. Concentration on such universal or near universal features considerably reduces the problem involved in distinguishing primitive from derived character states (though one cannot rule out *a priori* the possibility that all living primates might universally share certain primitive character states that have been lost in all other placental mammals).

Special Problems of the Fossil Record

Extension of definitions to include fossil forms introduces a set of special problems. There are many reasons for this, most notably the restricted range of characters that can be preserved in the fossil record and the inevitable fact that the common ancestor of a group of living forms must have been more primitive than any of the descendant species. In addition to this, problems of interpretation arise because of a fundamental divergence between the approach taken by those who primarily study living species (neontologists) and the approach taken by those who concentrate mainly on fossil forms (palaeontologists). Although reconstruction of the phylogenetic history of the primates must ultimately rely upon effective fusion of both neontological and palaeontological evidence, there is a marked difference of emphasis that arises from concentrating on either living or fossil primates. The fossil record obviously has unique advantages to offer for reconstruction of phylogenetic relationships (notably with respect to indicating approximate time-scales, adding to the known range of morphology in a group, and permitting tests of inferences based on living forms alone), but it also has unique limitations that may lead to misinterpretation if not explicitly recognised.

The greatest limitation of the fossil record resides in its fragmentary nature, both in terms of preservation of material from individual species and in terms of the sampling of species from the phylogenetic tree. Typically only "hard parts" are preserved in primate fossils and there is a marked disparity in the preservation of individual skeletal elements, with isolated teeth and jaws occurring far more commonly than skulls or associated postcranial elements. It is important here to distinguish fragmentary fossils, known only from teeth and jaw fragments, and substantial fossils, known (ideally) from a combination of dental, cranial and postcranial material. Inclusion of fragmentary fossils in any phylogenetic reconstruction can be particularly misleading if those fossils themselves are made to play a central part in the reconstruction process. In the first place, a fragmentary mammalian fossil only permits the class of judgement: "This dentition resembles the dentitions of a particular sub-group rather than the dentitions of other mammals". Since no other characters are preserved, realistic assessment of likely phylogenetic relationships is impossible and allocation of any such fossil to a given group (e.g. the Primates or a particular division thereof) is typically an act of classification (viz. a general assessment of similarity), rather than an act of phylogenetic reconstruction (viz. assessment of the probability that any shared similarities can only be derived rather than convergent). Secondly, this problem can be compounded in the phenomenon of "classificatory accretion", for example with a series of fragmentary fossils allocated to the primate group as a chain of species exhibiting gradually decreasing resemblance to the original primate model. In this process, fragmentary fossils are recognized as "primates" not because of their conformity with an overall framework of primate evolution, but because of serial resemblances between one fragmentary fossil and another. It should be noted in this context that Simpson (1935) linked *Plesiadapis* to the primates mainly because of dental resemblances to *Pelycodus* (a fragmentary fossil relative of the Eocene lemuroid *Notharctus*, itself a substantial fossil exhibiting many similarities to modern primates - see Gregory, 1920). Other fragmentary fossil species now allocated to the Plesiadapiformes ("archaic primates") because of dental resemblances to *Plesiadapis* bear very little resemblance to the original substantial fossil model (viz. *Notharctus*) and in some cases the link with primates is tenuous in the extreme, as with *Purgatorius* and the Picrodontidae. It is therefore advisable to concentrate on substantial fossil species, at least in the first instance, for phylogenetic reconstruction. Table 1 provides an analysis of currently recognised fossil

Table 1: Analysis of numbers of fossil primate species recognised for different Tertiary epochs. (Data from Szalay and Delson, 1979.)

EPOCH	DURATION (my)	TOTAL NO. OF SPECIES	SPECIES BASED ON SUBSTANTIAL FOSSILS	
Plio-Pleistocene	7	45	26	(58%)
Miocene	19	46	16	(35%)
Oligocene	12	12	3	(25%)
Eocene	16	83	12	(14%)
Palaeocene	10	64*	5*	(8%)
Overall Totals	64	250	62	(25%)

*All "archaic primates" (Plesiadapiformes). No "primates of modern aspect" are known from the Palaeocene as yet.

primate species (based on Szalay & Delson, 1979), showing that there are comparatively few species (see below) and that only about 25% are based on substantial fossils. Not surprisingly, the proportion of substantial fossils known declines (from 58% to 8%) as one passes back through the Tertiary record.

The fragmentary nature of the fossil record is also important in another respect, in that the sample of past primate species is relatively limited and biased in various ways (e.g. with respect to body size and geographical distribution). As a result, there is a danger that certain fossil species - notably those known only from fragmentary remains (e.g. *Purgatorius*) - may be incorporated into the primate evolutionary tree merely because nothing better is available from the known fossil record. There is a common tendency among primate palaeontologists to treat the primate fossil record as a broadly representative sample, reflecting all the key stages in primate evolution. A direct interpretation of the record yields the picture shown in Fig. 1(A). This suggests that there was an early radiation of "archaic primates" during the Palaeocene and early Eocene in the northern continents. Supposedly, an unspecialized survivor from the base of that radiation gave rise to the subsequent, sustained radiation of "primates of modern aspect", beginning in the northern continents and subsequently spreading to certain southern continents. A direct reading of the fossil record indicates that, following a major radiation during the Eocene, "primates of modern aspect" largely disappeared from the northern continents.

There are, however, numerous flaws in such a direct reading from the fossil record. Very little is known about mammals from Cretaceous, Palaeocene or Eocene deposits of the southern continents (see Savage & Russell, 1983). For Africa, fragmentary fossil mammals are now emerging from Palaeocene and Eocene sites at the northernmost extremity of the continent (Cappetta *et al*, 1978), but the earliest fossil site to yield substantial mammal (including primate) remains is the Oligocene Fayum site of Egypt. The recent unexpected discovery of an apparent tarsioid, the fragmentary fossil *Afrotarsius*, from the latter site (Simons & Bown, 1985) clearly shows how long it can take competent palaeontologists to unearth remains of small-bodied prosimians even in the best-documented early Tertiary fossil site of Africa. For South America, it is commonly stated that there is a fairly good record of mammalian evolution from the late Cretaceous onwards. However, the known fossil sites for the Cretaceous, Palaeocene and Eocene are very largely restricted to Patagonia and it cannot be claimed that there is an effective sample of fossil mammals for the entire continent during that critical period of mammalian evolution. Hence, it may not be concluded with any confidence that primates were absent from Africa and South America during the Cretaceous-Eocene.

A very crude guide to the effectiveness of our sampling of the primate fossil record can be obtained by means of a simple calculation beginning with the modern fauna of some 180 primate species and involving two assumptions that should, if anything, minimize the possible numbers of primate species that have existed in the course of primate evolution. The first assumption is that the primates originated only 65 million years ago (it is highly unlikely that they originated more recently than this). The

Figure 1: Two contrasting interpretations of the available primate fossil record:

A. The relative frequencies of past primate species are seen as a representative sample with no major distortions. According to this view the "archaic primates" underwent a significant radiation during the Palaeocene, only to disappear during the Eocene. An Eocene radiation of "primates of modern aspect", presumably derived from an early relative of the "archaic primates", was apparently succeeded by a marked contraction in species numbers during the Oligocene prior to a further major radiation to produce the modern primate fauna.

B. A simple calculation based on the modern primate fauna permits inference of the "expected numbers of species" (ENS) for each geological period. The known numbers of fossil species, relative to the expected number, then yields the "discovery rate" (DR). See Table 1 for data (N.B. "archaic primates" excluded).

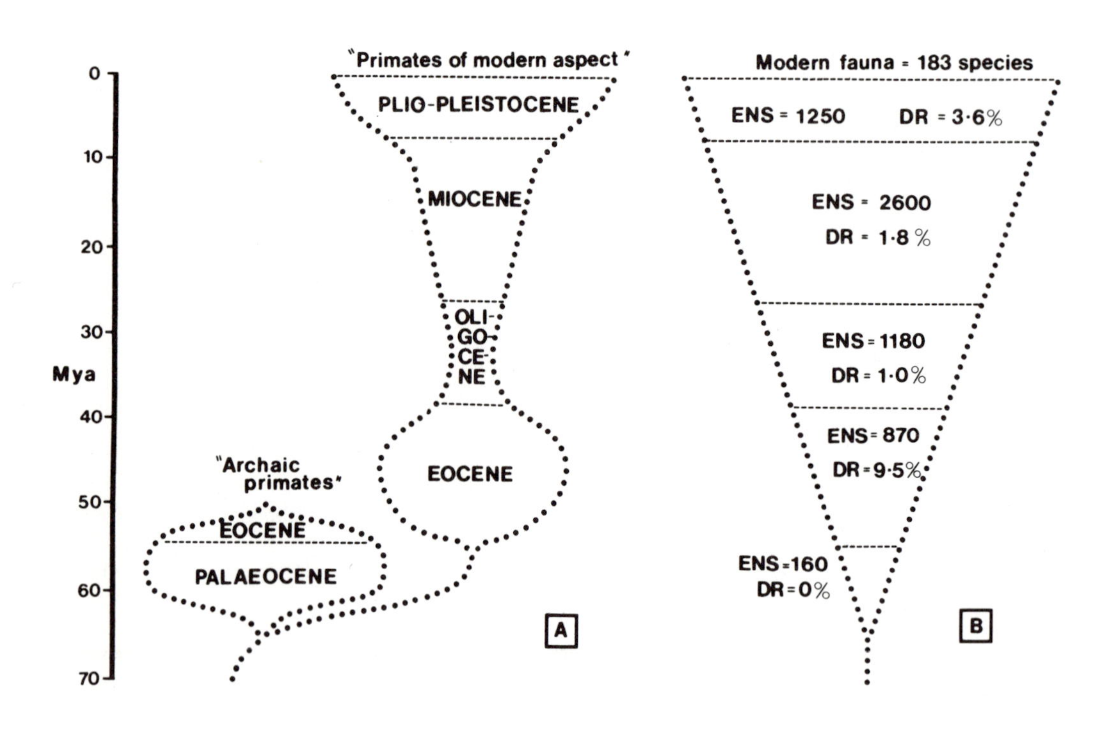

second assumption is that the numbers of primate species have expanded progressively since the time of origin (this is a minimal assumption because most authors have indicated an early rapid diversification followed by slower expansion in numbers of species during the evolution of individual mammal groups). Taking a figure of 1 million years as a reasonable approximation to the average lifetime of each fossil primate species (Stanley, 1976, 1978), it can be calculated that about 6000 primate species have existed (as a minimum) during the course of primate evolution. Table 1 indicates that 186 fossil primate species of "modern aspect" may be recognised, representing a 3% sample of past primate species. If the calculation is restricted to substantial fossils (N=57; Table 1), the sample shrinks to 1%. A similar calculation can be made for placental mammals generally, and a similar result emerges. Hence, it would seem that we have a very poor sample of the primate fossil record and of the mammalian record in general. Furthermore, it is a sample which is also affected by a bias against suitable fossil sites in the southern continents.

The simplistic calculation of numbers of past primate species is illustrated in Figure 1(B) alongside the equally simplistic picture produced by direct reading of the available fossil record Figure 1(A). The former also shows a breakdown of possible species numbers for different periods of the Tertiary, giving the ratio of known fossil species to "expected" numbers of species (ENS); this is designated the "discovery rate" (DR). It is striking that no fossil primates of modern aspect have yet been reported from the Palaeocene, although the common ancestor of Eocene lemuroids and tarsioids and the earliest offshoots thereof must surely have existed prior to the Eocene. By contrast, the inferred discovery rate of 9.5% shown for the Eocene is the highest recorded for any period. Direct reading of the fossil record (Figure 1(A)) would suggest that primates of modern aspect underwent a major radiation during the Eocene, followed by marked contraction. An alternative, reasonable explanation based on the figures in Figure 1(B) is that climatic warming during the Eocene (Wolfe, 1978) led to a northward expansion of primates from the southern continents, which have yet to be effectively sampled. Undoubtedly, such a geographical expansion would have been associated with a temporary global increase in the numbers of primate species during the Eocene, such that the real picture of the evolutionary radiation of the primates would be intermediate between Figure 1(A) and Figure 1(B). Given the alternative interpretation (Figure 1(B)), there is a real danger that the Plesiadapiformes have been inadvertently granted the status of "archaic primates" because we have yet to discover any real "archaic primates" in the fossil record of the southern continents. It should also be noted that with a maximal 3% sample of fossil primate species, any divergence dates that are inferred from known fossils may be much too recent. It is actually quite likely that early "primates of modern aspect" existed prior to 65 million years ago, but have not yet been documented purely because of the very low probability of their discovery. Given all the special difficulties attached to interpretation of the fossil record of primate evolution, there can be little doubt that definition of the living primates provides the correct starting-point for reconstruction of primate evolution history.

DEFINITION OF LIVING PRIMATES

Despite the central importance of providing a comprehensive, but succinct definition of living primates, relatively little attention has been paid to this question in the literature. Indeed, numerous authors have relied fairly heavily on the definition provided by Mivart (1873):

> "Unguiculate, claviculate placental mammals, with orbits encircled by bone; three kinds of teeth, at least at one time of life; brain always with a posterior lobe and calcarine fissure; the innermost digit of at least one pair of extremities opposable; hallux with a flat nail or none; a well-developed caecum; penis pendulous; testes scrotal; always two pectoral mammae".

Even quite recently, it was stated that this definition "has withstood the test of time" (Buettner-Janusch, 1966). Some authors have attempted to expand Mivart's definition. Le Gros Clark (1959) listed the following "evolutionary trends", defining the Primates as a "natural group of mammals".

1. Preservation of generalised limb structure with primitive pentadactyly.
2. Enhancement of free mobility of the digits, especially of the pollex and hallux (both used for grasping).
3. Replacement of sharp, compressed claws by flat nails; development of very sensitive tactile pads on the digits.
4. Progressive shortening of the snout.
5. Elaboration of the visual apparatus, with development of varying degrees of binocular vision.
6. Reduction of the olfactory apparatus.
7. Loss of certain elements of the primitive mammalian dentition. Preservation of a simple molar cusp pattern.
8. Progressive expansion and elaboration of the brain especially of the cerebral cortex.
9. Progressive and increasingly efficient development of gestational processes.

Napier & Napier (1967) cited both Mivart's definition and Le Gros Clark's list and added two additional trends to the latter:

10. Prolongation of postnatal life periods.
11. Progressive development of truncal uprightness leading to a facultative bipedalism.

All these attempted definitions of living primates suffer from the inclusion of likely primitive features of placental mammals and features that have arisen by convergent evolution in other mammal groups. Further, it is of little value to include "trends" in any definition of primates, since these refer not to generally shared features of primates, but to major developments found only in some members of the group. As was noted by Napier & Napier (1967), Wood Jones (1929) examined each of Mivart's criteria in turn and concluded that not one constituted a peculiarity of primates; only the aggregate of the criteria seemed to define them. This is hardly surprising, since Mivart apparently did not believe that the primates are a monophyletic group, descended from a specific common ancestor. Following his much-quoted definition, he considered a list of distinctions between strepsirhine and haplorhine primates and concluded: "...taken together, they render it in the highest degree improbable that the Lemuroids and Apes took origin from any common root-form not equally a progenitor of other Mammalian orders". Clearly, any definition based on the premise that the primates do not form a cohesive phylogenetic unit is unlikely to provide much help in the reconstruction of primate evolution!

It has, in fact, been a common theme throughout the literature on primate evolution that the primates lack any clear-cut diagnostic features of the kind found in other orders of placental mammals. Simpson (1955) specifically noted that primates seem to lack unequivocal identifying features comparable to the wings of bats or the "double-pulley astragalus" of artiodactyls. This view has been widely expressed and was recently repeated in an assessment of the affinities of the plesiadapiforms (MacPhee et al., 1983), "since the order Primates is not clearly definable by unique specialisations, the best grounds for regarding plesiadapiforms as euprimate antecedents are stratigraphic and phenetic". It must be emphasised, however, that the apparent absence of unique, defining features of primates relates to skeletal features identifiable in the fossil record and not to the entire range of features discernible in living primates. Detailed comparisons of living primates reveal that they share numerous universal (or near-universal) features, some of which do seem to be unique. It is possible to define living primates by reference to these apparently unique features, set in the context of other universal features of primates that also occur in other mammals.

Geographical Distribution and Habitat Occupancy

Although it is not customary to include reference to geographical distribution in definitions of mammalian groups, it is significant that living primates are largely confined to tropical and subtropical regions. The prosimian primates (lemurs, lorises and tarsiers) are even more tightly restricted to warmer areas of the world, as a reflection of their generally small body sizes. Since the ancestral primates probably fell within the body size range of modern prosimians, it is quite likely that they were subject to a similar climatic geographical restriction.

Most living primates are essentially arboreal. Only one living prosimian species, Lemur catta, exhibits any significant degree of terrestrial activity and it is only among Old World simians that predominant terrestrial activity has become common, though it is still limited to a minority of species. It is most likely that ancestral primates were essentially arboreal inhabitants of tropical/subtropical forest ecosystems.

Locomotor Adaptations

All living primates except man have a grasping foot with a well-developed, divergent hallux. Further, most primates have at least some prehensile capacity in the hand, though the pollex is in fact completely lacking in a few species. Most living primates have flat nails on digits and all species have at least a flat nail on the hallux, reflecting the typical primate locomotor pattern of grasping arboreal supports rather than clinging to them with claws. The grasping function of the extremities in all living primates is enhanced by possession of tactile pads bearing ridges (dermatoglyphs), which play a dual role (Cartmill, 1974). The ridges resist slipping of the digits along arboreal supports. In addition, they are involved with tactile sensitivity in that dermal counterparts of the ridges enclose Meissner's corpuscles (specialised tactile end-organs), which are apparently unique to primates among placental mammals (Winkelmann, 1963).

The grasping action of the foot, which is dependent upon the divergent hallux, is particularly important for the locomotion of all arboreal primate species. As noted by Morton (1924), the angle between the divergent hallux and the second digit provides the point of thrust during arboreal locomotion (a tarsi-fulcrumating type of foot), in contrast to the typical non-grasping mammalian foot in which the distal ends of the metatarsals of the digits act against the substrate (metatarsi-fulcrumating type). This has direct implications for the dimensions of the tarsal bones in primates exhibiting active arboreal locomotion. One outcome is that in the typical primate foot there is a switch from the probable primitive condition of "alternation" of the tarsus (cuboid in contact with the astragalus) to the apparently unique condition of "reverse alternation" (calcaneus in contact with the navicular; see Lewis, 1980). Although there are numerous features of the primate tarsus that are linked to the requirements of arboreal locomotion, notably with respect to articulation between the astragalus and calcaneus such as to permit inversion and eversion of the foot (Lewis, 1983), most or all of those features are also found in other arboreal mammals (e.g. squirrels). However, living primates do seem to be unique among placental mammals with respect to the ratio between the proximal and distal segments of the calcaneus. It is well known that certain prosimians, notably *Galago* and *Tarsius* species, exhibit pronounced elongation of the distal segment of the calcaneus. Such elongation is also present, though to a lesser degree, in all other primates up to a body weight of about 5 kg (for fairly obvious mechanical reasons, the scope of such elongation of the calcaneus decreases with increasing body size, so any comparisons must take this into account). All living prosimian species and virtually all living simian species weighing 5 kg or less have a longer distal than proximal segment of the calcaneus (viz. a "calcaneal index value" of less than 100%, Walker, 1967), whereas all non-primate placental mammals in that size range have calcaneal index values exceeding 100% (Martin, 1978).

Reliance on the grasping action of the hindfoot in typical primates is associated with *hindlimb domination* (Walker, 1967; Napier & Walker, 1967; Martin, 1972, 1979; Rollinson & Martin, 1981). This is reflected by two special features that are generally characteristic of primates, but not quite unique to them, since they are also found in a few other arboreal mammals (e.g. the kinkajou). The primary feature is that in the body of a typical primate the centre of gravity is located closer to the hindlimbs than to the forelimbs, such that a greater proportion of the body weight is borne by the hindlimbs (Kimura *et al.*, 1979; Rollinson & Martin, 1981). As a direct consequence of this, the symmetrical walking gait of a typical primate follows a *diagonal sequence*, as opposed to the *lateral sequence* typical of non-primate quadrupeds (terms as defined by Hildebrand, 1967).

In other words, the forefoot precedes the hindfoot on each side of the body in the footfall sequence of a walking primate, whereas the converse is the case in typical non-primates. This apparently trivial difference has major implications for the dynamics of locomotion (see Rollinson & Martin, 1981). The grasping foot, tarsi-fulcrumation, elongation of the distal segment of the calcaneus and hindlimb domination together constitute a complex of locomotor features unique to primates among living placental mammals.

Major Sense Organs and Skull Morphology

As has been widely acknowledged, the living primates are characterised by a particular emphasis on vision. However, they are not unique among placental mammals in this; it is only in terms of certain special adaptations of the visual system that primates would seem to be unique. For some of the characters involved (as with other morphological features to be discussed in due course) it is necessary to take body size into account. It has become increasingly obvious from comparisons of living organisms that many characters cited in the assessment of phylogenetic relationships scale in a predictable, but non-linear (viz. allometric) fashion with body size (see Gould, 1966). Allometric scaling has long been taken into account in studies of brain size in mammals (see Jerison, 1973) and several recent books have explicitly reviewed the implications of scaling for morphology, physiology and behavioural ecology (McMahon & Bonner, 1983; Peters, 1983; Calder, 1984; Schmidt-Nielsen, 1984). For present purposes, the essential point is that, in comparisons of species of different body size, it is essential to identify overall scaling trends, such that departures from the general pattern (e.g. in primates in comparison to other placental mammals) can be recognised (Martin, 1980). Thus, in defining primates, instead of simply asking whether a given feature is small or large, it is necessary to conduct a scaling analysis to demonstrate whether it is small or large relative to body size. In this way, many distinctive features of primates that are not immediately apparent, because of the confounding influence of differential body size, can be teased out and incorporated into a definition.

A case in point is provided by the demonstration of Kay & Cartmill (1977) that, in comparison to other placental mammals, primates have large orbits relative to skull length (though they are not unique among mammals in this respect). All living primates also possess a post-orbital bar, formed by the fusion of spurs from the frontal and jugal bones, defining the outer margin of the orbit. But numerous other groups of living placental mammals also possess a postorbital bar (Martin, 1968), so that is not in itself a useful defining feature of living primates. Thus, although all living primates are indeed characterised by large orbits with post-orbital bars, they are not uniquely defined among mammals in this respect. However, as shown by Cartmill (1972), living primates are uniformly characterised by a pronounced degree of forward rotation of the orbits (viz. orbital convergence). Once again, overall skull size must be taken into account, since the scope for forward rotation of enlarged orbits increases with increasing skull size. Relative to skull size, it emerges that living primates, with very few exceptions, are in fact quite distinctive among placental mammals in terms of the degree of orbital convergence. Forward rotation of the orbits is also associated with reduction in the distance between their inner margins (viz. the interorbital breadth) and primates are hence generally outstanding in terms of the narrowness of the inter-orbital distance relative to skull size (Kay & Cartmill, 1977).

As a result of their forward rotation, the orbits of primates encroach on the medial region of the skull to varying degrees. As a direct result of this encroachment, the majority of living primates exhibit exposure of the ethmoid, a nasal component, as a flat plate (os planum) in the medial orbital wall (Cartmill, 1970). Ethmoid exposure is found in all six major groups of living primates (lemurs; lorises; tarsiers; New World monkeys; Old World monkeys; apes and man), and it is only among lemurs that variation is found in this feature of the medial orbital mosaic. It is a moot point whether ethmoid exposure would have been present in the ancestral primates, but Cartmill's analysis (1970) suggests that this would have been the case if orbital convergence was already marked. It is this latter feature, rather than exposure of the ethmoid, which is of major significance.

Forward rotation of the orbits in primates increases the size of the binocular visual field and hence provides the necessary basis for effective stereoscopic vision (see Noback, 1975). But stereoscopic vision also requires the appropriate central nervous apparatus for processing visual information from the binocular field, and it is in this respect that primates seem to be unique among living mammals (Allman, 1982). In the typical mammalian condition, most of the optic fibres from each eye cross over at the optic chiasma and pass to the opposite side of the brain (contralateral projection). Relatively few fibres, generally less than 10%, pass to the same side of the brain (ipsilateral projection). Clearly, stereoscopic vision depends upon matching within the brain of visual inputs from both eyes for the binocular field, and hence upon coordination of contralateral and ipsilateral fibres. It is therefore striking that in all primates investigated ipsilateral fibres account for 40-50% of the retinal projection on each side of the brain. Primates thus have a uniquely well-developed neural basis for stereoscopic vision.

Allman (1982) has noted a further unique feature of the primate visual system with respect to the actual composition of the optic fibre bundles passing to the two sides of the brain:

> "In Primates the optic tectum on each side of the brain contains a systematic representation of the contralateral half of the visual field, whereas in other mammals the optic tectum of each side contains a systematic representation of the visual field that is viewed by the contralateral retina, which is the primitive vertebrate condition found in all non-mammalian vertebrates".

In other words, in all vertebrates other than primates the entire (relatively restricted) binocular field is represented in both sides of the brain, whereas in primates inputs to both eyes from the right half of the binocular field pass exclusively to the left half of the brain and vice-versa. As a result, all of the visual information from one half of the substantial binocular field in primates is analysed by the contralateral half of the brain (Fig. 2). This doubtless provides a more efficient basis for stereoscopic vision. Hence, primates overall have a unique combination of specialisations for effective 3-dimensional vision.

It is commonly maintained in the literature that in all primates an increased emphasis on vision has been accompanied by corresponding reduction in the olfactory system. However, analysis of the size of the olfactory bulbs in relation to body size (Stephan, 1972; Jerison, 1973; Martin, 1979) shows that actual reduction in the olfactory apparatus has probably only occurred in those primates that adopted diurnal habits. Further, there are no firmly identifiable features of the olfactory system that distinguish all living primates (i.e. including forms with nocturnal habits) from other placental mammals. Hence, contrary to past practice, there should be no reference to uniform reduction of the olfactory apparatus in any comprehensive definition of living primates.

With respect to the remaining major special sense organ, the ear, living primates in most respects exhibit a great variety of divergent morphological adaptations rather than uniform defining features that set them apart from other placental mammals. However, it is widely accepted that living primates are universally characterised by a specific pattern of formation of the ventral floor of the auditory bulla. In contrast to other mammals, living primates possess a bulla whose ventral floor is formed largely or exclusively by the petrosal bone (MacPhee et al.,1983).

The Brain

The brain has often figured prominently in discussions of defining characteristics of living primates, both with respect to its overall size and in relation to morphological features. It is, indeed, true that living primates exhibit relatively large brains in comparison to

Figure 2. Comparison of visual projection systems of a typical non-primate (left) and of a primate (right). In the non-primate, there is a relatively small field of binocular overlap and the entire visual field of one eye is projected to the other side of the brain. In the primate, the field of binocular overlap is substantial and projection takes place such that the left half of the total visual field is represented in the right half of the brain and *vice versa*. (Based on illustrations provided by Noback, 1975 and Allman, 1982).

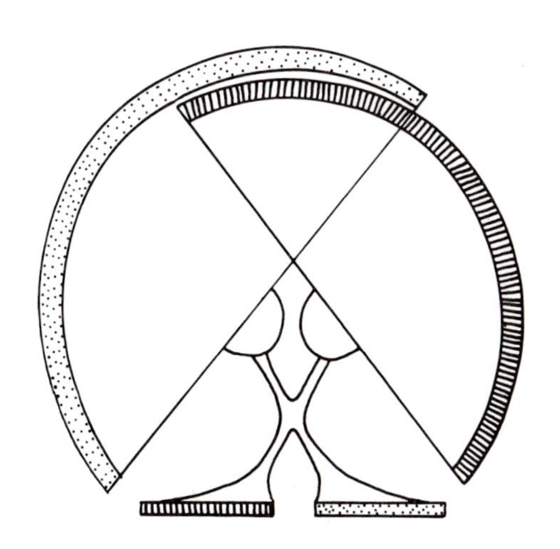

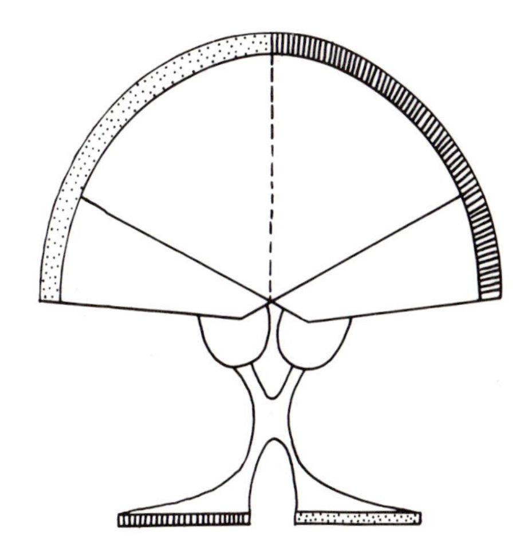

other placental mammals (Jerison, 1973; Martin, 1981b). However, there is considerable overlap between primates and other mammals in this respect. Thus, while living primates as a group tend to exhibit larger brains than other mammals (Martin, 1983), brain size is not a sufficiently reliable criterion to indicate a specific relationship to primates in any individual case. Formation of sulci on the cortical surface of the brain has also been to the fore of discussions of primate characteristics. In fact, folding of the cortical surface - which provides the general basis for the formation of sulci - is a highly predictable concomitant of brain-size alone, since the superficial cortical layer must undergo disproportionate expansion to keep pace with expansion of the volume of white matter in the brain (Prothero & Sundsten, 1984). Nevertheless, there are two specific sulcal features of the living primate brain that do seem to be unique in comparison to other placental mammals (Elliot Smith, 1902). One is the presence of a true Sylvian sulcus, confluent with the rhinal sulcus; the other is the presence of a triradiate calcarine sulcus. Although Mivart's definition (1873) mentions a calcarine sulcus, and although Le Gros Clark (1959) also emphasised this feature, Elliot Smith's extensive comparisons of mammalian brains demonstrated that it is a three-branched sulcus that is universal to (and unique to) primates (see Martin, 1973). Mere possession of a simple sulcus on the internal face of the occipital lobe of the brain is not in itself a defining feature of primates.

It has now emerged that primates also seem to be unique among placental mammals with respect to fetal development of the brain, relative to fetal body size (Holt et al, 1975; Sacher, 1982; Martin, 1983). Primates differ sharply from other placental mammals in the allometric relationship between fetal brain and body size, to the effect that, throughout gestation and for any given body size, a primate fetus possesses a significantly greater proportion of fetal brain tissue. Because this difference applies throughout gestation, it is also found that primate neonates exhibit a greater proportion of brain tissue than non-primate neonates of the same body weight. Since possession of a greater proportion of brain tissue throughout fetal development is apparently a universal feature of primates, it undoubtedly reflects some special adaptation of ancestral primates (Sacher, 1982; Martin, 1983).

Reproductive Biology

Primates also exhibit a number of distinctive features in respect of their reproductive biology. Male primates are universally characterised by permanent descent of the testes into a postpenial scrotum. Although this feature is also found in other placental mammal groups, male primates are unusual (if not unique) in that descent of the testes occurs very early in life, typically close to the time of birth. Female primates are universally characterised by loss of the urogenital sinus, such that the urethra and vagina have more-or-less separate external openings. Although this feature is by no means confined to primates, it does characterise them as a group.

The fetal membranes and placentation of primates have been subjected to intensive comparative study (Hill, 1932; Mossman, 1937, 1953; Luckett, 1975).

However, rather than defining the primates as a group, the major features recorded actually reveal a sharp dichotomy between strepsirhines (with non-invasive, diffuse epitheliochorial placentation) and haplorhines (with invasive, discoidal haemochorial placentation). Numerous accessory characteristics involved in the development of the fetal membranes further underline this marked separation between the two groups of living primates (see Luckett, 1975, for a review). The only universal feature of living primates that may be of some import is early vascularisation of the chorion by allantoic vessels and the reduction of involvement of the yolk-sac in placentation during the latter part of gestation (Hill, 1932).

Living primates do share a universal set of features in their life-history patterns, however. Relative to body size, gestation periods are typically long and this is related to the fact that living primates universally produce small litters of precocial neonates (Portmann, 1962; Martin, 1975, 1983; Martin & MacLarnon, 1985). As noted by Schultz (1948), the number of teats in the female closely matches the litter-size in mammals generally and primates accordingly have a maximum of three pairs of teats. Comparative studies (Payne & Wheeler, 1967a, 1967b, 1968) have shown that the pace of fetal growth is slow in primates, relative to maternal body size (see also Martin & MacLarnon, 1985). The same is true of post-natal growth rates, as is reflected by the constitution of the milk in primates generally (Martin, 1984). In line with the slow pace of their early development, primates attain sexual maturity quite late and also have long life spans relative to their body size. These reproductive characteristics of primates can be summarised as constituting a "precocial complex" associated with low reproductive turnover. Although the possession of a "precocial complex" also characterises other groups of placental mammals (e.g. ungulates, cetaceans), this attribute at least distinguishes these groups from a number of mammal groups exhibiting a contrasting "altricial complex" (e.g. carnivores, most insectivores, most rodents).

Dental Patterns

It is a remarkable fact that, despite the importance of dental features for interpretation of the mammalian fossil record, little mention has been made of the dentition in many definitions of living primates (Mivart, 1873; Le Gros Clark, 1959; Napier & Napier, 1967). Students of modern primates have been struck more by the general primitiveness of their dentitions rather than by shared distinctive features.

It is widely accepted that the ancestral placental mammals had a dental formula of $\frac{3.1.4.3}{3.1.4.3}$ and that the molars exhibited a simple tritubercular pattern. It is also generally accepted that the dental evolution of placental mammals has been marked by two major trends: reduction in the number of teeth and an increasingly complex crown morphology of the cheek teeth. Overall, the primates stand out among placental mammals only to the extent that there has been less change in these two directions than in most other groups (with a few notable exceptions, such as the insectivores). However, there has been some reduction in the dental formula among living primates. The maximum formula observed is $\frac{2.1.3.3}{2.1.3.3}$ and it follows that living primates are universally characterised by the loss of at least one incisor and one premolar from each tooth-row. Reduction in the number of incisors in living primates is universally associated with a considerable limitation of the length of the premaxilla, such that the upper incisors are arranged more transversely than longitudinally. It may be argued that in primates there has been a reduction in the size of the premaxilla and in the number of incisors because of a switch of prehensile functions (e.g. grasping of arthropod prey or fruits) from the snout to the forelimb.

Although there has clearly been some change in the organisation of the cheek teeth among living primates, as reflected by a general reduction in the number of premolars, little attempt has been made in the literature to establish that living primates exhibit a common pattern of shared derived features of molar morphology. Indeed, it is noteworthy that living tarsiers possess molar teeth that are very close to the hypothetical ancestral condition for placental mammals generally. It can be argued that living primates share a number of subtle modifications of the molars, such as a relatively low, rounded, cusp morphology and relatively large, raised talonid basins on the lower molars, but such modifications do not uniquely define primates in contrast to all other living placental mammals.

A Definition of Living Primates

Taking into account the points discussed above, it is in fact quite feasible to produce a new definition of living primates, summarising universal or near-universal features that together clearly demarcate primates from all other placental mammals:

Primates are typically arboreal inhabitants of tropical and sub-tropical forest ecosystems. Their extremities are essentially adapted for prehension, rather than grappling arboreal supports. A widely divergent hallux provides the basis for a powerful grasping action of the foot in all genera except Homo, while the hand usually exhibits at least some prehensile capacity. The digits typically bear flat nails rather than bilaterally compressed claws; the hallux always bears a nail. The ventral surfaces of the extremities bear tactile pads with cutaneous ridges (dermatoglyphs) that reduce slippage on arboreal supports and provide for enhanced tactile sensitivity in association with dermal Meissner's corpuscles. Locomotion is hindlimb-dominated, with the centre of gravity of the body located closer to the hindlimbs, such that the typical walking gait follows a diagonal sequence (forefoot preceding hindfoot on each side). The foot is typically adapted for tarsi-fulcrumation, with at least some degree of relative elongation of the distal segment of the calcaneus, commonly resulting in reverse alternation of the tarsus (calcaneo-navicular articulation).

The visual sense is greatly emphasised. The eyes are relatively large and the orbits possess (at least) a postorbital bar. Forward rotation of the eyes ensures a large degree of binocular overlap. Ipsilateral and contra-lateral retinofugal fibres are approximately balanced in numbers on each side of the brain and organised in such a way that the contralateral half of the visual field is represented. Enlargement and medial approximation of the orbits is typically associated with ethmoid exposure in the medial orbital wall (though there are several exceptions among the lemurs). The ventral floor of the well-developed auditory bulla is formed predominantly by the petrosal. The olfactory system is unspecialized in most nocturnal forms and reduced in diurnal forms. Partly because of the increased emphasis on vision, the brain is typically moderately enlarged, relative to body size, in comparison to other living mammals. The brain of living primates always possesses a true Sylvian sulcus (confluent with the rhinal sulcus) and a triradiate calcarine sulcus. Primates are unique among living mammals in that the brain constitutes a significantly larger proportion of body weight at all stages of gestation.

Male primates are characterised by permanent precocial descent of the testes into a postpenial scrotum; female primates are characterised by the absence of a urogenital sinus. In all primates, involvement of the yolk-sac in placentation is suppressed, at least during the latter half of gestation. Primates have long gestation periods, relative to maternal body size, and produce small litters of precocial neonates. Fetal growth and postnatal growth are characteristically slow in relation to maternal size. Sexual maturity is attained late and life-spans are correspondingly long relative to body size. Primates are, in short, adapted for slow reproductive turnover.

The dental formula exhibits a maximum of $\frac{2.1.3.3}{2.1.3.3}$ *. The size of the premaxilla is very limited, in association with the reduced number of incisors, which are arranged more transversely than longitudinally. The cheek teeth are typically relatively unspecialized, though cusps are generally low and rounded and the lower molars possess raised, enlarged talonids.*

ARE THE TREE-SHREWS PRIMATES?

It is, of course, a complex matter to decide whether the tree-shrews (Tupaiidae) are more closely related to primates than to any other group of placental mammals. However, a very good starting-point is provided by the definition of living primates set out above. If it were to emerge that tree-shrews are covered, at least partially, by that list of universal or near-universal special features, then there would obviously be a good case for suggesting an evolutionary relationship between tree-shrews and primates.

Like many living primates, tree-shrews are restricted to tropical forest regions and they do exhibit varying degrees of arboreality. However, the arboreal locomotion of tree-shrews is very different from that of primates. All digits bear bilaterally compressed claws and there is no definite grasping adaptation of the hallux. Tree-shrews rely upon the grappling action of their claws to negotiate relatively broad arboreal supports and they lack the typical primate propensity for locomotion among relatively fine branches. While it is true that the digits of tree-shrews bear moderately developed tactile pads possessing dermatoglyphs, Meissner's corpuscles appear to be absent (Winkelman, 1963). Hence, it would seem that tree-shrews lack the special tactile sensitivity of the digits that characterises primates. In contrast to the typical primate condition, tree-shrew locomotion is not hindlimb-dominated. The centre of gravity is closer to the forelimbs than to the hindlimbs and the walking gait follows the typical non-primate lateral sequence (Jenkins, 1974; Rollinson & Martin, 1981). The foot is adapted for metatarsi-fulcrumation, as is to be expected from the absence of a grasping adaptation of the hallux, and there is no elongation of the distal segment of the calcaneus. Calcaneal index values for tree-shrews fall clearly into the range of non-primates (Martin, 1979).

Tree-shrews do, in fact, have comparatively well developed eyes, but relative to skull size their orbits are significantly smaller than in living primates. Further, although there is some degree of forward rotation of the orbits, providing for a limited amount of binocular overlap, that is far less marked than in the typical primate condition (Cartmill, 1970; Kay & Cartmill, 1977). Thus, given the fact that the postorbital bar occurs quite frequently in a variety of different mammal groups, the presence of a postorbital bar in the tree-shrew skull is not associated with a typical primate-like degree of enlargement and forward rotation of the orbits. In agreement with the relatively limited amount of forward rotation of the orbits, tree-shrews still exhibit a large interorbital distance (relative to skull size) and there is accordingly no exposure of the ethmoid in the orbit. In line with these differences in the size, orientation and composition of the orbit, tree-shrews lack the special adaptations of the nervous system for stereoscopic vision found in primates. In tree-shrews, less than 10% of the retinofugal fibres from each eye remain ipsilateral and the tectal projection doubtless follows the typical non-primate pattern (see Allman, 1982). Tree-shrews do possess a well developed auditory bulla, but in contrast to the universal condition among living primates the ventral floor of the bulla is formed by an entotympanic and not by an extension of the petrosal (Val Valen, 1965; Spatz, 1966).

Moderate development of the visual system in tree-shrews has been matched by moderate expansion of the brain. As a result, relative brain size in tree-shrews is found to lie within the primate range (Stephan, 1972). However, since many other non-primate mammals also lie within the primate range this can hardly be taken as an adequate criterion for suggesting a phylogenetic link between tree-shrews and primates. In fact, tree-shrews lack a true Sylvian sulcus and the triradiate calcarine sulcus, both universal attributes of the brain in living primates (Martin, 1973), and fetal development of the brain in tree-shrews follows the typical non-primate pattern (see Martin, 1983).

In male tree-shrews, as in male primates, the testes descend into a scrotum. However, the scrotum is prepenial and descent does not take place until the attainment of sexual maturity and remains reversible throughout life and testis retraction can occur as a response to stress (Martin, 1968; von Holst, 1969). Female tree-shrews are also distinguished from female primates by retention of a substantial urogenital sinus and there is a significant difference in gestational processes in that the yolk-sac remains functional until term (Hill, 1965). There is an even greater contrast between tree-shrews and primates in terms of life-history patterns, since tree-shrews uniformly exhibit the "altricial complex". The gestation period is short, relative to body size, and the neonates are poorly developed at birth

(Martin, 1968). Fetal growth and postnatal growth proceed relatively rapidly, in relation to maternal body size, while sexual maturity is achieved very early and there is only a moderate life-span. Tree-shrews are therefore adapted for a more rapid reproductive turnover than are living primates.

Although the dental formula in tree-shrews has undergone some reduction relative to the presumed ancestral placental condition, from $\frac{3.1.4.3}{3.1.4.3}$ to $\frac{2.1.3.3}{3.1.3.3}$ there remain features which contrast with the typical primate condition. Unlike the primates, tree-shrews have retained a full complement of three incisors in the lower jaw. Further, despite the fact that the number of incisors has been reduced to two in the upper jaw, as in primates, tree-shrews are unusual in that the two remaining incisors are separated by a noticeable gap and they are arranged longitudinally, rather than transversely, on a relatively well developed premaxilla. It is a curious fact that there seems to have been no suggestion anywhere in the literature that the molar teeth of tree-shrews resemble those of primates in any significant features. The cusps are high and sharply defined and the molars exhibit none of the features commonly discussed with respect to the primate fossil record (see later). Given the general emphasis on dental features in reconstruction of mammalian evolution (albeit largely dictated by the predominance of dental remains in the fossil record) it is unusual and puzzling that relatively little has been written about the virtually complete absence of dental resemblances between tree-shrews and primates.

It is apparent from the above that tree-shrews do not fit at all well with the suggested definition of living primates. Tree-shrews lack all of the unique, universal features of living primates and they differ in a number of respects, notably in possessing cursorial locomotor features and in exhibiting the "altricial complex". In fact, most or all of the less specific resemblances between tree-shrews and primates are also found in arboreal squirrels, and this greatly increases the probability that any similarities between tree-shrews and primates may be due to general adaptation for arboreal habits rather than to a common ancestry. Of course, it remains possible that tree-shrews are closer to the primates than any other group of mammals, having branched off before the special dental characteristics of the living primates had become established. However, for those who would suggest a phylogenetic link between tree-shrews and primates, there remains an obligation to demonstrate that any residual similarities they share can be explained only as derived character states, rather than as primitive retentions or as convergent developments. Since residual similarities between tree-shrews and primates do not involve character states that are unique among placental mammals, a great deal depends upon the accuracy with which the derived condition of any character state can be inferred.

FOSSIL PRIMATES

Because of the fragmentary and incomplete nature of the fossil record, any attempt to apply the above definition of living primates to the primate fossil record is beset with problems. In particular, since fossil specimens are usually restricted to "hard parts", the definition contracts considerably with respect to the scope for its practical application. Only the following defining features remain:

i. Well-developed, divergent hallux with a flat terminal phalanx in the foot.
ii. Elongated distal segment of the calcaneus (viz. low calcaneal index value).
iii. Relatively large, convergent orbits with restricted interorbital distance. Postorbital bar present; ethmoid exposure in the orbit possible (depending on interorbital distance relative to skull size).
iv. Petrosal bulla.
v. Relatively large braincase. Sylvian sulcus on endocast.
vi. Dental formula maximally $\frac{2.1.3.3}{2.1.3.3}$. Premaxilla short; upper incisors arranged more transversely than longitudinally. Molars with low, rounded cusps. Lower molars with raised, enlarged talonids.

Despite this marked contraction of the neontological definition, the recognised fossil "primates of modern aspect" documented from the basal Eocene onwards (Simons, 1972; Szalay & Delson, 1979) accord extremely well with the residual features. The known postcranial elements from Eocene lemuroids (Adapidae) and Eocene tarsioids (Omomyidae) clearly indicate the presence of a divergent grasping hallux, and this is true of all more recent fossil primates of modern aspect for which appropriate postcranial evidence is available. Eocene primates of modern aspect also typically exhibit elongation of the distal segment of the calcaneus and there are correspondingly low values for the calcaneal index (Martin, 1979). All known Eocene tarsioids exhibit low calcaneal index values and the same is true of North American Eocene lemuroids (*Notharctus*; *Smilodectes*). However, a problem arises in the case of the European Adapinae. Isolated calcanei attributed to *Adapis parisiensis* and to *Adapis* (*Leptadapis*) *magnus* exhibit high calcaneal index values markedly exceeding those of living primates and other fossil primates of modern aspect in the same body size range (Martin, 1979). Recently, Dagosto (1983) has conducted a major review of isolated postcranial bones attributable to *Adapis* and reconstructions of the feet indicate that a grasping hallux was indeed present, but that the distal segment of the calcaneus lacked the typical primate features discussed above. Dagosto interprets this as a specialised feature of *Adapis*, associated with a slow-climbing (possibly potto-like) mode of locomotion and representing a departure from the likely ancestral primate condition. This latter interpretation is supported by the fact that von Koenigswald (1979) has described the rear end of a primate skeleton, from the Eocene deposits of Messel, as exhibiting the typical primate pattern of a grasping foot combined with elongation of the distal segment of the calcaneus. Unfortunately, there are no cranial or dental remains to permit proper identification of this specimen, but on size grounds it would seem to be an adapine (probably with a body weight in the region of 1 kg).

The problem is that this would indicate a sharp difference in tarsal morphology between species within the subfamily Adapinae, and it is difficult to imagine which of the known genera other than *Adapis*, or its close relatives, would be likely to accommodate the Messel specimen. It should be noted that although the modern lorises and pottos (Lorisinae) are specially adapted for show climbing, they still exhibit relative elongation of the distal segment of the calcaneus, even though they lie in the upper range of calcaneal index values for their body size within the primate distribution (Martin, 1979). Overall, the available evidence indicates that one should accept Dagosto's interpretation that *Adapis* exhibited an unusual specialisation of the tarsus away from the typical primate condition, but we must really await the discovery of properly associated postcranial remains before drawing any firm conclusions. The fact remains that the calcanei attributed to *Adapis* are quite unusual in comparison with those from other fossil primates of modern aspect and from living primates. Any definition of the likely ancestral primate condition should therefore exclude *Adapis* as an exceptional case.

As with the postcranial evidence, known fossil primates of modern aspect also resemble modern primates with respect to the orbital region of the skull. All species are known to possess a postorbital bar, where appropriate evidence is available. Further, as has been shown by Kay & Cartmill (1977), adapids and tarsioids broadly overlap the modern primates with respect to the size of the orbits, the degree of forward rotation of the orbits and the resulting limitation on the interorbital breadth, relative to skull size. *Adapis* species do tend to have rather small orbits compared to other primates, but in other respects they exhibit the typical primate pattern, doubtless reflecting a shared emphasis on stereoscopic vision.

Much has been written about the presence or absence of ethmoid exposure in the skulls of fossil primates of modern aspect. However, it must be emphasised that this character can only be assessed reliably for living forms, on the basis of developmental evidence. Because of fusion of sutures in the adult skull, it is actually impossible to identify the components of the orbital mosaic with any confidence in adult skulls of many extant primate species. As was pointed out by Cartmill (1970, 1971), the medial orbital mosaic of the adult skull of the Eocene tarsioid *Necrolemur* does not seem to show ethmoid exposure (see Simons & Russell, 1960), but the pattern of bones resembles the condition in adult *Cheirogaleus*, in which the ethmoid suture with the frontal is no longer visible. A particularly well preserved skull of *Adapis parisiensis* has now been shown to exhibit a small ethmoid component in the medial orbital mosaic (Gingerich & Martin, 1981).It is therefore likely that at least some of the Eocene primates of modern aspect exhibited ethmoid exposure, though this cannot usually be detected in the available adult skulls. In any case, it should be emphasised that ethmoid exposure is probably a secondary reflection of orbital convergence and is therefore only a symptom of the significant primate feature of emphasis on stereoscopic vision.

It is generally accepted that primates of modern aspect all possess a petrosal bulla. As with the composition of the medial orbital mosaic, however, it is impossible to identify with certainty the composition of the ventral bulla wall in fossil primates, since fusion of component bones may not be apparent in adult skulls. In modern tree-shrews, only developmental evidence demonstrates beyond doubt that the ventral floor of the bulla is formed by an entotympanic. In the adult skull, partial fusion of the entotympanic with the petrosal has in the past led to misidentification of the tree-shrew bulla as a petrosal bulla. Since many other features of the auditory region of early primates of modern aspect resemble those in modern primates (e.g. in comparisons between _Notharctus_ and _Lemur_ - see Gregory, 1920), it is reasonable to infer that fossil primates of modern aspect are generally characterised by a petrosal bulla. This inference must, however, remain somewhat speculative in the absence of evidence from immature skulls.

In comparison to living primates, Eocene fossil primates of modern aspect, notably the Adapidae, have relatively small braincases. Adapids have brain sizes less than half those found in modern lemurs of comparable body size (Jerison, 1973; Martin, 1973), whereas tarsioids tend to have brain sizes closer to those of the modern prosimians. This, indeed, underlines the unreliability of brain size as such as a criterion for identification of primates. Relative to body size, brain size has probably progressively expanded in the evolution of most, if not all, mammalian orders since their departure from the common placental stock. Accordingly, early representatives of any given order are likely to have small relative brain sizes in comparison to their modern relatives. Nevertheless, it is likely that Eocene primates of modern aspect possessed relatively large brains in comparison to contemporary Eocene representatives of other mammalian orders, and it seems likely that primates have been typified by relatively large brains throughout their evolution, though this is masked both by overlap with other mammals and by progressive increase in brain size (relative to body size) in mammals generally.

Apart from providing an indication of overall size of the brain, endocasts from fossil skulls also provide some information regarding brain morphology. As is to be expected from their relatively small brains, in comparison to living primates, fossil primates of modern aspect exhibit a relatively primitive pattern, notably with respect to exposure of the cerebellum, associated with limited expansion of the occipital pole of the brain. Nevertheless, a survey of early primate brains conducted by Radinsky (1974) showed that Eocene lemuroids and tarsioids typically exhibit a true Sylvian sulcus. However, _Smilodectes_ was an exception to this and it is a moot point whether the absence of a Sylvian sulcus on the endocast indicates that this sulcus was not in fact an ancestral primate feature, or whether in this particular case the endocast did not accurately reflect actual brain morphology.

With respect to dental patterns, early primates of modern aspect (Eocene lemuroids and tarsioids) partially confirm the expectation from a definition based on living primates and partially require modification thereof. The maximum dental formula for the Eocene primates of modern aspect is $\frac{2.1.4.3}{2.1.4.3}$, suggesting that reduction in the number of incisors, but not reduction in the number of premolars, had already taken place early in primate evolution. In association with reduction in incisor numbers, Eocene primates of modern aspect exhibit limited development of the premaxilla, with the incisors arranged more transversely than longitudinally. This suggests that transfer of prehensile functions from the anterior dentition to the forelimb, associated with reduction in the size of the premaxilla and in the number of incisors, was an early development in the primates, whereas changes in the cheek teeth (e.g. reduction in the number of premolars) have to a large extent, at least, developed in parallel in separate lineages derived from the ancestral primates. In common with all living primates, Eocene primates of modern aspect (along with all more recent fossil primates) exhibit low, rounded cusps on the molars and molar talonids are raised and enlarged.

Whereas fossil primates of modern aspect, from the Eocene onwards, conform quite well with a definition based on living primates, this is not at all the case with the Plesiadapiformes (the "archaic primates"). This is, of course, generally recognised and even those who would argue most strongly that the Plesiadapiformes should be included with the order Primates accept that they must represent a separate early Tertiary radiation linked to the radiation of the primates of modern aspect (the "Euprimates" of Szalay & Delson, 1979) only by a remote common ancestry (see Figure 1(A)). But for this very reason one must be very cautious in interpreting the similarities (e.g. in molar morphology) shared by Plesiadapiformes and certain early fossil primates of modern aspect (e.g. *Pelycodus*).

At first sight, there would seem to be little to link Plesiadapiformes with primates of modern aspect with regard to locomotor morphology. The locomotor skeleton is known only from *Plesiadapis*, and even then only incompletely. Reconstructions of the *Plesiadapis* skeleton indicate that the digits bore well-developed, bilaterally compressed claws (Simons, 1972; Gingerich, 1976). There is no evidence that the hallux was divergent, but unfortunately the hallux has not been preserved in the known fossil specimens (Szalay *et al*, 1975). Szalay (1977) has suggested, on the basis of certain fine details of calcaneal morphology, that *Plesiadapis* might have possessed a divergent, grasping hallux. However, Gingerich (1976) saw no basis for such a suggestion and, following a re-examination of certain specimens (see Gingerich, this volume), now believes that the hallux might have been totally lacking. This, of course, would represent a complete departure from the typical primate condition. It is also noteworthy that *Plesiadapis* lacks the typical primate feature of elongation of the distal segment of the calcaneus and exhibits a calcaneal index well in excess of 100% (Szalay & Decker, 1974; Martin, 1979). At first sight, this might not seem to present much of a problem, since (as explained above) calcanei attributed to *Adapis* exhibit a similar condition. However, Dagosto (1983) has shown that the condition indicated for *Adapis* must be derived, relative to the ancestral condition for primates of modern aspect, a process which most probably involved relative elongation of the distal segment of the calcaneus. Hence, any similarity with respect to a high calcaneal index value provides no grounds for allying *Plesiadapis* to *Adapis*. Szalay & Decker (1974) provide a detailed comparison of tarsal morphology in *Plesiadapis* with that in early primates of modern aspect (see also Decker

& Szalay, 1974) and suggest several points of similarity. However, it should be noted that Szalay & Decker took isolated calcaneal bones from the Late Cretaceous, allocated to *Procerberus* and *Protungulatum*, as representative of the primitive mammalian condition. An integral part of their reasoning was the assumption that the ancestral mammals were terrestrial (as supposedly exemplified by *Procerberus* and *Protungulatum*) and that locomotion in *Plesiadapis* was intermediate between this presumed ancestral condition and the typical arboreality of primates. Lewis (1980) and Novacek (1980) have since shown that many arboreal mammals exhibit the same "primate-like" features of the tarsus as identified for *Plesiadapis* by Szalay & Decker (1974). Further, it is by no means certain that the ancestral placental mammals were terrestrial. The available fossil evidence from the Cretaceous is extremely limited and probably unrepresentative (e.g. the best-preserved postcranial remains from Mongolia, described by Kielan-Jaworowska (1977, 1978) come from mammals that were apparently adapted to a marginal semi-desert habitat). There remains a distinct possibility that the ancestral placental mammals were at least semi-arboreal and any similarities shared by later groups (e.g. Plesiadapiformes, tree-shrews and primates) might therefore represent primitive retentions rather than derived conditions indicative of specific common ancestry. In fact, Gingerich (1976, this volume) presents a number of compelling arguments to suggest that at least some Plesiadapiformes (notably *Plesiadapis*) were terrestrial in habits. Hence, any apparent "arboreal" adaptations in the tarsus of these forms must surely represent retention of features from a more arboreal ancestor of some sort. It is as yet unknown whether there were any arboreal (or semi-arboreal) members of the Plesiadapiformes, as the postcranial skeleton is only known from the apparently terrestrial *Plesiadapis*.

Plesiadapiformes also lack the typical primate features of the orbital region of the skull. Kay & Cartmill (1977) have shown that, relative to skull size, their orbits are remarkably small in comparison to *all* other placental mammals. Further, there is very little orbital convergence and the interorbital distance is relatively great in the skulls of plesiadapiforms. Accordingly it is only to be expected that ethmoid exposure would not have occurred in the orbits of plesiadapiforms, and it is generally agreed that the ethmoid is not involved in the medial orbital mosaic (though the evidence for this is subject to the problem of sutural fusion discussed above). No plesiadapiform is thought to have possessed even a postorbital bar, so the Plesiadapiformes clearly lacked every single defining feature of the orbital complex of primates of modern aspect.

Much has been made of the fact that Plesiadapiformes seemed, at one time, to possess a petrosal bulla. As has been noted above in the discussion of fossil primates of modern aspect, the evidence for this could never have been secure because of the problem of suture closure in adult skulls. It has now emerged, however, that if Microsyopidae are included, the Plesiadapiformes exhibited some variability in bullar morphology, and that possession of a petrosal bulla, if it occurred at all, was not universal (McPhee *et al.*, 1983). Present evidence does not provide any firm grounds for proposing that ancestral plesiadapiforms shared with ancestral primates the unique, derived, condition of the ventral floor of the bulla being formed from the petrosal.

Because of the limited cranial material available for the Plesiadapiformes, little can be said about relative brain size other than that the braincase was fairly small in relation to overall skull size. Gingerich (1976) calculated that the maximum relative brain size for Plesiadapis could have fallen just within the range for modern primates, but Radinsky (1977) provides estimates indicating a relative brain size smaller than for any living primate and possibly inferior to any known primate of modern aspect. It seems likely that plesiadapiforms were quite small-brained creatures and there is no evidence that they exhibited any primate-like expansion of the brain in comparison to contemporary mammals of other groups. Since no endocasts are available for plesiadapiforms, it is not known whether a Sylvian sulcus was present on the brain.

Dental characteristics have, of course, been to the fore in discussions of the possible affinities between the plesiadapiforms and the primates (Gingerich, 1974, 1976, this volume). Such discussions, though, have tended to concentrate on molar morphology. Since the earliest known plesiadapiform, Purgatorius, apparently had a dental formula of $\frac{3.1.4.3}{3.1.4.3}$, there is no evidence that Plesiadapiformes exhibited a shared ancestral condition of reduction in tooth numbers with primates, even though many later plesiadapiforms have formulae of $\frac{2.1.3.3}{2.1.3.3}$ or even less. Further, although some plesiadapiforms seem to have undergone some shortening of the premaxilla, this was by no means a universal condition. Plesiadapis, for example, has a substantial premaxilla. Further, the incisors were not typically arranged transversely rather than longitudinally on the premaxilla in plesiadapiforms. Thus, we are left with similarity of molar morphology between certain plesiadapiforms and certain early primates of modern aspect (notably Pelycodus) as the major reason for suggesting an ancestral connection between Plesiadapiformes and Primates. Although the resemblance in molar morphology between, say, Plesiadapis and Pelycodus is striking, it is not impossible that convergent evolution may have taken place.

CONCLUSIONS

It has been shown that - contrary to the prevailing view - living primates can be defined, to the exclusion of other mammals, on the basis of several complexes of universal (or near-universal) features, each including certain unique and presumably derived traits. Since living tree-shrews share none of these distinctive features with living primates, there is no compelling case for suggesting that they share a specific common ancestry with primates. Although various special arguments have been produced to support a hypothesis of common ancestry of tree-shrews and primates, it has yet to be shown convincingly that any similarities between them can only be derived, rather than primitive or convergent.

Although a definition based on living primates contracts considerably when confronted with the real limitations of the fossil record, enough of the definition remains to accommodate all known fossil primates of modern aspect quite comfortably. By contrast, Plesiadapiformes do not fit such a contracted definition at all well and the case for including them in the Primates rests essentially on resemblances in molar morphology between certain plesiadapiforms and certain early primates of modern aspect. Given the fragmentary nature of the known fossil record, the hypothesis that plesiadapiforms are early relatives of the primates must be treated with considerable scepticism.

In closing, it should be noted that inclusion of tree-shrews in the Primates was linked quite early to inclusion of plesiadapiforms. Simpson (1935), argued as follows with respect to *Plesiadapis* and its relatives:

> "The question of primate or insectivore relationships has become one purely of primate or tupaioid affinities, as there is no question of close resemblance to any insectivores other than the tupaioids. Since the tupaioids are now universally recognised as a conservative offshoot of the primate ancestry, the plesiadapids are related to the primates in either case, and the question as to their inclusion in that order or in a protoprimate division of the Insectivora is purely verbal...If the tupaioids are primates, then the plesiadapids are necessarily primates also."

On the one hand, this indicates that denial of an evolutionary relationship between tree-shrews and primates might weaken the case for accepting a relationship between the plesiadapiforms and primates. On the other, it emphasises the consensus view that the plesiadapiforms are somewhat more likely to be related to primates of modern aspect than are the tree-shrews. Whereas the former relationship remains a marginal possibility in view of certain dental evidence, the latter now seems to be most unlikely. In both cases, however, opinions have far too often been influenced by the naive expectation that the common ancestor of living primates was derived from an animal intermediate between a hedgehog and a lemur (hence the frequent references to a hypothetical insectivore/primate boundary). In fact, comparison of living primates (excluding tree-shrews) indicates that they are all derived from an ancestral form combining basic features of mouse-lemurs, bushbabies and tarsiers. There is no need to invoke, at any stage, an ancestral form that looked even remotely like any modern insectivore. Living tree-shrews may seem to be intermediate between *modern* hedgehogs and lemurs, but they do not share with primates any of the key characteristics that define them and therefore reflect their evolutionary origins.

ACKNOWLEDGEMENTS

Thanks go to Ann MacLarnon and to Lawrence Martin for their helpful comments on the draft manuscript. Figure 2 was prepared by Anne-Elise Martin and her help is gratefully acknowledged. Thanks also go to Paula Sutton for typing the manuscript.

REFERENCES

Key References

Allman, J. 1982 Reconstructing the evolution of the brain in primates through the use of comparative neurophysiological and neuro-anatomical data. *In* Primate Brain Evolution, eds. E.Armstrong and D. Falk, pp. 13-28. New York: Plenum Press.

Cartmill, M. 1972 Arboreal adaptations and the origin of the order Primates. *In* The Functional and Evolutionary Biology of Primates, ed. R. Tuttle, pp. 97-122. Chicago: Aldine-Atherton.

Cartmill, M. 1974 Rethinking primate origins. Science, *184*, 436-443.

Kay, R.F. and Cartmill, M. 1977 Cranial morphology and adaptations of Palaechthon nacimienti and other Paromomyidae (Plesiadapoidea, ?Primates), with a description of a new genus and species. J. Hum. Evol., *6*, 19-53.

MacPhee, R.D.E., Cartmill, M. and Gingerich, P.D. 1983 New Palaeogene primate basicrania and the definition of the order Primates. Nature, Lond., *301*, 509-511.

Martin, R.D. 1968 Towards a new definition of primates. Man, n.s., , 377-401.

Rollinson, J. and Martin, R.D. 1981 Comparative aspects of primate locomotion, with special reference to arboreal cercopithecines. Symp. zool. Soc. Lond., *48*, 377-427.

Main References

Allman, J. 1982 Reconstructing the evolution of the brain in primates through the use of comparative neurophysiological and neuro-anatomical data. *In* Primate Brain Evolution, eds. E. Armstrong and D. Falk, pp. 13-28. New York: Plenum Press.

Buettner-Janusch, J. 1966 Origins of Man. New York: John Wiley.

Calder, W.A. 1984 Size, Function and Life History. Cambridge, Mass: Harvard University Press.

Cappetta, H., Jaeger, J-J., Sabatier, M., Sudre, J. and Vianey-Liaud, M. 1978 Decouverte dans le Paléocene du Maroc des plus anciens mammifères euthériens d'Afrique. Geobios, *11*, 257-263.

Cartmill, M. 1970 The Orbits of Arboreal Mammals: A Reassessment of the Arboreal Theory of Primate Evolution. Ph.D. Thesis: University of Chicago.

Cartmill, M. 1971 Ethmoid component in the orbit of primates. Nature, Lond., *232*, 556-567.

Cartmill, M. 1972 Arboreal adaptations and the origin of the order Primates. *In* The Functional and Evolutionary Biology of Primates, ed. R. Tuttle, pp. 97-122. Chicago: Aldine-Atherton.

Cartmill, M. 1974 Rethinking primate origins. Science, *184,* 436-443.

Cartmill, M. 1982 Basic primatology and prosimian evolution. *In* A History of American Physical Anthropology, 1930-1980, ed. F. Spencer, pp. 147-186. New York: Academic Press.

Dagosto, M. 1983 Postcranium of Adapis parisiensis and Leptadapis magnus (Adapiformes, Primates). Folia primatol., *41*, 49-101.

Decker, R.L. and Szalay, F.S. 1974 Origins and functions of the pes in the Eocene Adapidae (Lemuriformes, Primates). *In* Primate Locomotion, ed. F.A. Jenkins, pp. 261-291. New York: Academic Press.

Elliot Smith, G. 1902 On the morphology of the brain in the Mammalia, with special reference to that of the lemurs, recent and extinct. Trans. Linn. Soc. Lond. (Zool.), *8*, 319-432.

Gingerich, P.D. 1974 Dental function in the Paleocene primate Plesiadapis. *In* Prosimian Biology, eds. R.D. Martin, G.A. Doyle and A.C. Walker, pp. 531-541. London: Duckworth.

Gingerich, P.D. 1976 Cranial anatomy and evolution of early Tertiary Plesiadapidae (Mammalia, Primates). Mus. Paleontol. Univ. Michigan. Pap. Paleontol., *15*, 1-140.

Gingerich, P.D. and Martin, R.D. 1981 Cranial morphology and adaptation in Eocene Adapidae. II. The Cambridge skull of Adapis parisiensis. Amer. J. Phys. Anthrop. *56*, 235-257.

Gould, S.J. 1966 Allometry and size in ontogeny and phylogeny. Biol. Rev., *41*, 487-640.

Gregory, W.K. 1920 On the structure and relations of Notharctus, an American Eocene primate. Mem. Amer. Mus. Nat. Hist. n.s., *3*, 49-243.

Hennig, W. 1950 Grundzüge einer Theorie der Phylogenetischen Systematik. Berlin: Deutscher Zentralverlag.

Hennig, W. 1966 Phylogenetic Systematics. Chicago: University of Illinois Press (Reprinted 1979).

Hildebrand, M. 1967 Symmetrical gaits of primates. Amer. J. Phys. Anthrop., *26*, 119-130.

Hill, J.P. 1932 The developmental history of the primates. Phil. Trans. Roy. Soc. Lond., B, *221*, 45-178.

Hill, J.P. 1965 On the placentation of Tupaia. J. Zool. Lond., *146*, 278-304.

Holt, A.B., Cheek, D.B., Mellits, E.D. and Hill, D.E. 1975 Brain size and the relation of the primate to the nonprimate. *In* Fetal and Postnatal Cellular Growth: Hormones and Nutrition, ed. D.B.Cheek pp. 23-44. New York: John Wiley.

Jenkins, F.A. 1974 Tree shrew locomotion and primate arborealism. *In* Primate Locomotion, ed. F.A. Jenkins, pp. 85-115. New York: Academic Press.

Jerison, H.J. 1973 Evolution of the Brain and Intelligence. New York: Academic Press.

Kay, R.F. and Cartmill, M. 1977 Cranial morphology and adaptations of Palaechthon nacimienti and other Paromomyidae (Plesiadapoidea, ?Primates), with a description of a new genus and species. J. Hum. Evol., *6*, 19-53.

Kielan-Jaworowska, Z. 1977 Evolution of therian mammals in the Late Cretaceous of Asia. Part II. Postcranial skeleton in Kennalestes and Asioryctes. Palaeont. Polonica, *37*, 65-83.

Kielan-Jaworowska, Z. 1978 Evolution of therian mammals in the Late Cretaceous of Asia. Part III. Postcranial skeleton in the Zalambdalestidae. Palaeont. Polonica, *38*, 5-41.

Kimura, T., Okada, M. and Ishida, H. 1979 Kinesiological characteristics of primate walking: its significance in human walking. *In* Environment, Behaviour and Morphology: Dynamic Interactions in Primates, eds. M.E. Morbeck, H. Preuschoft and N. Gomberg, pp. 297-311. New York: Gustav Fischer.

Le Gros Clark, W.E. 1959 The Antecedents of Man. (First Edition) Edinburgh: Edinburgh University Press.

Lewis, O.J. 1980 The joints of the evolving foot. Part II. The intrinsic joints. J. Anat., *130*, 833-857.

Luckett, W.P. 1975 Ontogeny of the fetal membranes and placenta: their bearing on primate phylogeny. *In* Phylogeny of the Primates: a Multidisciplinary Approach, eds. W.P. Luckett and F.S. Szalay, pp. 157-182. New York: Plenum Press.

Luckett, W.P. (ed.) 1980 Comparative Biology and Evolutionary Relationships of Tree-Shrews. New York: Plenum Press.

MacPhee, R.D.E., Cartmill, M. and Gingerich, P.D. 1983 New Palaeogene primate basicrania and the definition of the order Primates. Nature, Lond., *301*, 509-511.

Martin, R.D. 1968 Towards a new definition of primates. Man, n.s., *3*, 377-401.

Martin, R.D. 1972 Adaptive radiation and behaviour of the Malagasy lemurs. Phil. Trans. Roy. Soc. Lond., B, *264*, 295-352.

Martin, R.D. 1973 Comparative anatomy and primate systematics. Symp. Zool. Soc. Lond., *33*, 301-337.

Martin, R.D. 1975 The bearing of reproductive behavior and ontogeny on strepsirhine phylogeny. *In* Phylogeny of the Primates: a Multidisciplinary Approach, eds. W.P. Luckett and F.S. Szalay, pp. 265-297. New York: Plenum Press.

Martin, R.D. 1979 Phylogenetic aspects of prosimian behavior. *In* The Study of Prosimian Behavior, eds. G.A. Doyle and R.D. Martin, pp.45-77. New York: Academic Press.

Martin, R.D. 1980 Adaptation and body size in primates. Z. Morph. Anthrop., *71*, 115-124.

Martin, R.D. 1981a Phylogenetic reconstruction versus classification: the case for clear demarcation. Biologist, *28*, 127-132.

Martin, R.D. 1981b Relative brain size and basal metabolic rate in terrestrial vertebrates. Nature, Lond., *293*, 57-60.

Martin, R.D. 1983 Human Brain Evolution in an Ecological Context. (52nd James Arthur Lecture) New York: American Museum of Natural History.

Martin, R.D. 1984 Scaling effects and adaptive strategies in mammalian lactation. Symp. Zool. Soc. Lond., *51*, 87-117.

Martin, R.D. and MacLarnon, A.M. 1985 Gestation period, neonatal size and maternal investment in placental mammals. Nature, Lond., *313*, 220-223.

McMahon, T.A. and Bonner, J.T. 1983 On Size and Life. New York: Scientific American Books..

Mivart, St. G. 1873 On Lepilemur and Cheirogaleus, and on the zoological rank of the Lemuroidea. Proc. Zool. Soc. Lond., *1873*, 484-510.

Morton, D.J. 1924 Evolution of the human foot. Amer. J. Phys. Anthrop., *7*, 1-51.

Mossman, H.W. 1937 Comparative morphogenesis of the fetal membranes and accessory uterine structures. Contrib. Embryol. Carnegie Inst. Wash., *26*, 129-246.

Mossman, H.W. 1953 The genital system and the fetal membranes as criteria for mammalian phylogeny and taxonomy. J. Mammal., *34*, 289-298.

Napier, J.R. and Napier, P.H. 1967 A Handbook of Living Primates. London: Academic Press.

Napier, J.R. and Walker, A.C. 1967 Vertical clinging and leaping - a newly recognised category of locomotor behaviour of primates. Folia primatol., *6*, 204-219.

Noback, C.R. 1975 The visual system of primates in phylogenetic studies. *In* Phylogeny of the Primates: a Multidisciplinary Approach, eds. W.P. Luckett and F.S. Szalay, pp. 199-218. New York: Plenum Press.

Novacek, M.J. 1980 Cranioskeletal features in tupaiids and selected Eutheria as phylogenetic evidence. *In* Comparative Biology and Evolutionary Relationships of Tree-Shrews, ed. W.P. Luckett, pp. 35-93.

Payne, P.R. and Wheeler, E.F. 1967a Growth of the foetus. Nature, Lond., *215*, 849-850.

Payne, P.R. and Wheeler, E.F. 1967b Comparative nutrition in pregnancy. Nature, Lond., *215*, 1134-1136.

Payne, P.R. and Wheeler, E.F. 1968 Comparative nutrition in pregnancy and lactation. Proc. Nutr. Soc., *27*, 129-138.

Peters, R.H. 1983 The Ecological Implications of Body Size. Cambridge: Cambridge University Press.

Portmann, A. 1962 Cerebralisation und Ontogenese. Medizin. Grundlagenforsch *4*, 1-62.

Prothero, J.W. and Sundsten, J.W. 1984 Folding of the cerebral cortex in mammals: a scaling model. Brain Behav. Evol., *24*, 152-167.

Radinsky, L. 1974 Prosimian brain morphology: functional and phylogenetic implications. *In* Prosimian Biology, eds. R.D. Martin, G.A. Doyle and A.C. Walker, pp. 781-798. London: Duckworth.

Radinsky, L. 1977 Early primate brains: facts and fiction. J. Hum. Evol., *6*, 79-86.

Rollinson, J. and Martin, R.D. 1981 Comparative aspects of primate locomotion, with special reference to arboreal cercopithecines. Symp. Zool. Soc. Lond., *48*, 377-427.

Sacher, G.A. 1982 The role of brain maturation in the evolution of the primates. *In* Primate Brain Evolution: Methods and Concepts, eds. E. Armstrong and D. Falk, pp. 97-112. New York: Plenum Press.

Savage, D.E. and Russell, D.E. 1983 Mammalian Paleofaunas of the World. Reading, Mass: Addison-Weslely.

Schmidt-Nielsen, K. 1984 Scaling: Why is Animal Size so Important? Cambridge: Cambridge University Press.

Schultz, A.H. 1948 The number of young at birth and the number of nipples in primates. Amer. J. Phys. Anthrop., *6*, 1-23.

Simons, E.L. 1972 Primate Evolution: An Introduction to Man's Place in Nature. New York: MacMillan.

Simons, E.L. & Bown, T.M. 1985 *Afrotarsius chatrathi*, first tarsiiform primate (?Tarsiidae) from Africa. Nature, Lond. *313*, 475-477.

Simons, E.L. and Russell, D.E. (1960) Notes on the cranial anatomy of Necrolemur. Breviora, No. 127, pp. 1-14.
Simpson, G.G. 1935 The Tiffany fauna, upper Paleocene. 2. Structure and relationships of Plesiadapis. Amer. Mus. Novitates, *816*, 1-30.
Simpson, G.G. 1945 The principles of classification and a classification of mammals. Bull. Amer. Mus. Nat. Hist., *85*, 1-350.
Simpson, G.G. 1955 The Phenacolemuridae, new family of early Primates. Bull. Amer. Mus. Nat. Hist., *105*, 411-442.
Spatz, W.B. 1966 Zur Ontogenese der Bulla tympanica von Tupaia glis Diard 1820 (Prosimiae, Tupaiiformes), Folia primatol., *4*, 26-50.
Stanley, S.M. 1976 Stability of species in geologic time. Science, *192*, 267-269.
Stanley, S.M. 1978 Chronospecies' longevities, the origin of genera, and the punctuational model of evolution. Paleobiology, *4*, 26-40.
Stephan, H. 1972 Evolution of primate brains; a comparative anatomical investigation. *In* The Functional and Evolutionary Biology of Primates, ed. R. Tuttle, pp. 155-174. Chicago: Aldine-Atherton.
Szalay, F.S. 1977 Phylogenetic relationships and a classification of the eutherian Mammalia. *In* Major Patterns of Vertebrate Evolution, eds. M.K. Hecht, P.C. Goody and B.M. Hecht, pp. 315-374. New York: Plenum Press.
Szalay, F.S. and Decker, R.L. 1974 Origins, evolution and function of the tarsus in the Late Cretaceous eutherians and Paleocene primates. *In* Primate Locomotion, ed. F.A. Jenkins, pp. 223-259. New York: Academic Press.
Szalay, F.S. and Delson, E. 1979 Evolutionary History of the Primates. New York: Academic Press.
Szalay, F.S., Tattersall, I., and Decker, R.L. 1975 Phylogenetic relationships of Plesiadapis - postcranial evidence. Contrib. Primatol., *5*, 136-166.
Van Valen, L. 1965 Tree-shrews, primates and fossils. Evolution, *19*, 137-151.
von Holst, D. 1969 Sozialer Stress bei Tupajas (Tupaia belangeri). Z. vergl. Physiol., *63*, 1-58.
von Koenigswald, W. 1979 Ein Lemurenrest aus dem eozanen Ölschliefer der Grube Messel bei Darmstadt. Paläont. Zeitschr., *53*, 63-76.
Walker, A.C. 1967 Locomotor Adaptations in Recent and Fossil Madagascan Lemurs. Ph.D. Thesis: University of London.
Winkelman, R.L. 1963 Nerve endings in the skin of primates. *In* Evolutionary and Genetic Biology of Primates, vol. 1, ed. Buettner-Janusch, J., pp. 229-259.
Wolfe, J.A. 1978 A paleobotanical interpretation of Tertiary climates in the northern hemisphere, Amer. Sci., *66*, 694-703.
Wood Jones, F. 1929 Man's Place Among the Mammals. London: Edward Arnold.

2 PLESIADAPIS AND THE DELINEATION OF THE ORDER PRIMATES

Philip D. Gingerich,
Museum of Paleontology
University of Michigan,
Ann Arbor,
Michigan 48109, U.S.A.

INTRODUCTION

Biologists sometimes speak or write as if life in the past was pretty much like it is today, there just wasn't as much of it! Living species are most familiar to all of us. The false impression that species living in the past were little different from those living today may result from palaeontologists' constant comparison of extinct plants and animals with living models, or it may simply reflect poor communication between palaeontologists and neontologists. Fossils provide the only direct evidence of life in the past, and the fossil record indicates that enormous change has taken place since living organisms became abundant and diverse. Phanerozoic time, the past 600 million years, is conventionally divided into three eras, Palaeozoic, Mesozoic, and Cainozoic, based on the evolutionary grade of plants and animals living during each era. Mammals are known only from the last one-third of the Phanerozoic, and primates are known from the last one-third of this one-third. It is not surprising that primates should come on the scene so late - the major diversification of placental mammals has taken place only within the past 65 million years, that is, within the Cainozoic era.

Cainozoic time can be subdivided into two periods, Tertiary and Quaternary (or alternatively Palaeogene and Neogene), and these in turn can be subdivided into six epochs (Figure 1). Each of these epochs contains distinctive mammalian faunas, but for our purposes here it is useful to recognize four intervals of primate history comprising the Palaeocene, the Eocene, the Oligo-Miocene, and the Plio-Pleistocene. Each of these intervals, comprising an episode of climatic warming and cooling, contains a progressively more advanced radiation of primates. Members of the first radiation, in the Palaeocene, are so primitive relative to later primates that they may not be primates at all. Members of the second radiation, the first "euprimates" in the Eocene, are so primitive relative to later primates that they may not be primates either. Members of the third radiation, "eu-euprimates" in the Oligo-Miocene, are so primitive relative to later primates that they too may not be primates at all. Members of the fourth radiation, "eu-eu-euprimates" (Hominidae) in the Plio-Pleistocene, are the only primates sufficiently advanced in ways we all deem important to rank unequivocally as members of our own esteemed mammalian order.

The order Primates could reasonably be restricted to Hominidae, and it could as well be restricted to Hominoidea, to Anthropoidea, to Prosimii

+ Anthropoidea, or to Plesiadapiformes (Praesimii) + Prosimii + Anthropoidea. What one includes in an order or any other taxonomic category matters far less than what one sees in examining the evolutionary history of a particular group. Here I shall claim that the question of whether *Plesiadapis* is a primate or not can only be answered arbitrarily, and hence it is not a question of real interest or importance. I would hope that this perspective might temper our natural enthusiasm for boundary arguments. The more important questions concern what *Plesiadapis* and its allies tell us about primate evolutionary history.

Figure 1. Overview of primate evolution in the northern hemisphere viewed in context of changing climates affecting their biogeographic distribution (climatic curve from Buchardt, 1978). *A*, appearance of *Purgatorius* following terminal Cretaceous cooling. *B*, radiation of plesiadapiform primates during cool phase of middle and late Palaeocene. *C*, appearance of primates of modern aspect in Europe and North America during phase of climatic warming (primates of modern aspect probably evolved in more equatorial areas, possibly in Africa), followed by final opening of North Atlantic. *D*, acme of primate diversity in Europe and North America, coinciding with warmest climates. *E*, decline and extinction of Eocene primate lineages in Europe and North America, coinciding with climatic cooling; *Mahgarita* apparently migrated to North America from Eurasia just before the Grande Coupure in the latest Eocene or early Oligocene. *F*, nadir of primate diversity in northern continents; known primate faunas are all in more equatorial areas (Bolivia, Egypt). *G*, appearance of Hominoidea in Europe and Asia during climatic warming of middle Miocene. *H*, disappearance of Hominoidea in northern continental faunas coinciding with cooler, dryer climates during Messinian crisis; major diversification of savanna- and woodland-adapted Cercopithecoidea; emergence of Hominidae. *I*, rapid evolution and dispersal of humans during Pleistocene climatic fluctuations.

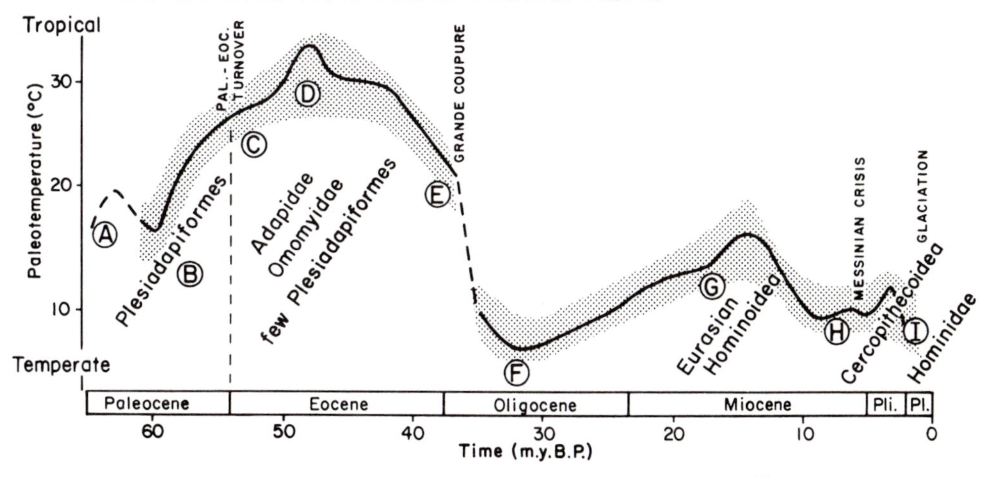

SYSTEMATIC POSITION OF PLESIADAPIS

Plesiadapis has always been problematical. Etymologically Plesiadapis means "near-adapis", clear reference to the common Eocene lemuriform Adapis. The first specimens of Plesiadapis were collected by the French physician Victor Lemoine of Reims. When Paul Gervais named and described Plesiadapis, he considered it to be "un petit mammifere de classification encore douteuse" (Gervais, 1877). Gervais was himself an authority on Adapis, and it is difficult to imagine what resemblance he saw to Adapis in the very limited fragments of Plesiadapis available in 1877. It seems likely that Gervais had other specimens in mind when he compared Plesiadapis to Adapis. These other specimens were described by Lemoine the following year as Plesiadapis, but later placed in Protoadapis. Adapis was (and is) widely recognized as lemuriform in grade, if not in class. Thus it is natural that Plesiadapis, given a name compounded from Adapis, should be compared with lemurs.

Many genera closely related to Plesiadapis are now known from Europe and North America. These are conventionally classified in five (or six) families: the microsyopoid Microsyopidae (and possibly Apatemyidae), and the plesiadapoid Paromomyidae, Picrodontidae, Carpolestidae, and Plesiadapidae. For reference, the evolutionary diversification of Plesiadapidae is shown in outline in Figure 2. Pronothodectes and Nannodectes are known only from

Figure 2. Outline of evolutionary diversification of European and North American Plesiadapidae, all drawn to same scale. Plesiadapidae range in age from middle Paleocene through early Eocene. Reproduced from Gingerich (1976).

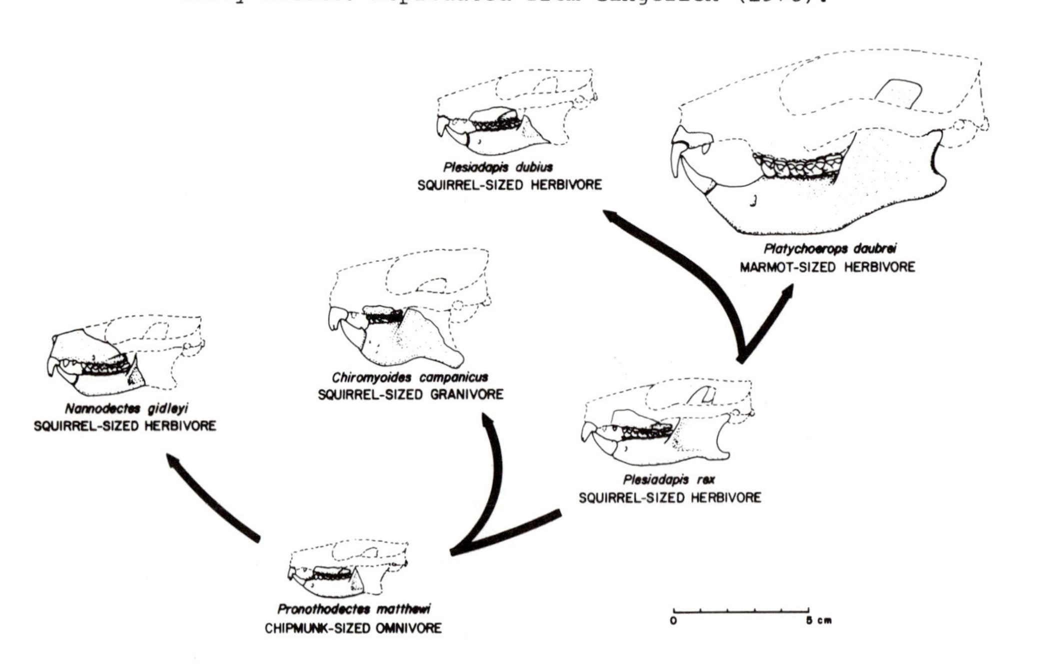

North America, and Platychoerops is known only from Europe. Chiromyoides and Plesiadapis are known from both Europe and North America.

Lemoine (1887) clearly recognized that Plesiadapis was not in fact a lemur, and he expressed this in attributing to Plesiadapis "des charactères lémuriens avec un facies marsupial." Plesiadapis is not a marsupial either, but comparison with marsupials serves to highlight the great number of primitively generalized and also divergently specialized morphological characteristics of Plesiadapis. Stehlin (1916) regarded Plesiadapis and its allies as primates, in part because of their enlarged superficially Daubentonia-like ("Chiromys"-like) central incisors, and in part because of similarity in the structure of their molars to those of Eocene primates. Matthew (1917) was more cautious, reserving judgement until cranial and skeletal characteristics of plesiadapids could be more fully studied. Gidley (1923:16-17) discussed at length the dental evidence for primate affinities of Plesiadapis and its allies, writing:-

> "I know of no species, certainly not primate, in which is found the peculiar combination of modifications described...as characteristic of Paromomys, and which is observed also in the Notharctidae [now a subfamily of Adapidae] and in some at least of the Eocene Tarsiidae [Omomyidae]. These modifications include for the lower molars a broad basined heel and narrower, more or less forwardly sloping trigonid in which the paraconid is progressively diminishing, or absent, its function being taken over by the anterior cingulum ridge continuing with the anterior flank of the protoconid to form a trigonid basin similar to but smaller and more elevated in position than the talonid depression; while the correlated modifications of the upper molars are a shallow anterior basin external to the protocone, which basin and cone function with the talonid portion of the corresponding lower molar, and a posterior basin, somewhat higher in position, formed by the posterior cingulum ridge continuing with the backwardly expanded border of the protocone to the summit of that cusp. The latter basin and ridge function with the trigonid of the lower molar next behind. This peculiar structure of the upper molars at least, while apparently distinctively primate, is, however, not observed in all groups of the order. But the stage just described, which in the Notharctidae is followed by the budding off of a hypocone from the posterior flank of the protocone, seems to have been an important basis of modification in all the anthropoids, including man, and in some but not all groups of Lemurs. The molar pattern of modern anthropoids seems to have been built on this model."

Simpson (1935) concluded his study of the dentition, limited cranial remains, and partial postcranial skeleton of *Plesiadapis* by ranking Plesiadapidae within the superfamily Lemuroidea. In discussing the dentition of *Plesiadapis*, Simpson wrote (1935:26):-

> "As regards molar pattern, *Plesiadapis* resembles the primitive Notharctinae more closely than any other group. The resemblance to the Adapinae is more distant, but still striking in many respects. There is also considerable resemblance to *Necrolemur*, a later tarsioid, and to *Paromomys*, a middle Paleocene genus of doubtful position, perhaps tarsioid. Resemblance to the other main groups of early primates is more distant.
>
> The resemblance to *Pelycodus* [*Cantius*], most primitive known notharctine, is really amazing and extends to the apparently most insignificant details. The upper molars are of almost identical structure throughout, differing only in details of the cingula and proportions such as may characterize species of one genus. In the lower molars, *Pelycodus* [*Cantius*] has the paraconids slightly more distinct, but the resemblance is equally striking and includes even such features as the minute grooving of the trigonid face of the metaconid and the exact structure of the complex grooving of the talonid face of the hypoconid and of the whole heel of M3. Matthew and others have noted this resemblance, although hardly recognizing its very complete character, but have tended to distrust or even reject it because of the well-known fact that early tuberculosectorial dentitions are all more or less alike and that erroneous allocations have frequently resulted from comparisons of molar teeth alone. This is, of course, true, but it is also true, as Gidley pointed out, that such complete convergence in a really complicated pattern as occurs between *Plesiadapis* and *Pelycodus* *Cantius* has rarely or never been found in mammals not truly related, and that the characteristic structure of *Plesiadapis* molars is encountered in no order other than the primates. The evidence of molar pattern is decidedly in favor of rather close relationship to the Notharctinae. As this happens to be correlated with other resemblances to that subfamily or, more broadly, to the general division of primates which it represents, there is every reason to accept this evidence as valid."

It is important to note, in evaluating Simpson's (1935) appraisal of the postcranial remains of *Plesiadapis*, that he accepted tree shrews as primates on the experience and authority of W.E. Le Gros Clark. Simpson wrote (p.25): "If the tupaioids are primates, then the plesiadapids are necessarily primates also." The converse would not necessarily follow, that is, inclusion of plesiadapids in Primates would not require that tupaioids be classified here as well.

The first reasonably complete skull of *Plesiadapis* was described by Russell (1959), who reported that the auditory bullae were completely ossified and "dérivent seulement de l'os pétreux." He also noted the presence of an ossified tubular ectotympanic in *Plesiadapis*. Russell later (1964:105) concluded from the cranial anatomy of *Plesiadapis*:-

> "Comme on l'a montré, les Adapidés qui ont vécu de l'Éocène moyen à l'Éocène tardif, présentaient, en common avec *Plesiadapis*, un certain numbre de caractères craniens fondamentaux; le ressemblance va presque aussi loin avec les Notharctidés de l'Éocène ancien et moyen. Ces faits suggèrent que les trois groupes étaient apparentés de trés près au début du Paléocène, mais que les Notharctidés se sont développés plus ou moins isolément en Amérique du Nord. Les Adapidés d'Europe, plus conservateurs, ont gardé des traces plus nettes de leur héritage commun, comme en témoignent certains traits du *Plesiadapis* paléocène. Toutefois les spécialisations présentées par celui-ci montrent que les trois lignes évolutives étaient déjà séparées dès le milieu ou peut-être même dès le début des temps paléocènes. Même en admettant que les faunes paléocènes d'Europe soient moins bien connues que les faunes contemporaines d' Amérique du Nord, il est vraisemblable que ce dernier pays fut le centre évolutif des Plésiadapidés."

Russell (1964) reported little on the postcranial skeleton of *Plesiadapis*, but he did describe several features of the clawed ungual phalanges present in newly collected material, noting that these were strongly curved, laterally compressed, and fissured dorsally over the distal one-third of their length, with a large basal tuberosity for insertion of a strong flexor tendon. R ssell's postcranial material of *Plesiadapis* was described by Szalay *et al*. (1975), who argued forceably that the mode of articulation of the radius and humerus, and a complex of characteristics of the tarsus of *Plesiadapis*, are evolutionarily advanced and shared only with primates.

My own research on *Plesiadapis* has been concerned primarily with evolutionary patterns and systematic relationships within Plesiadapidae, and to some extent with the systematic relationships of Plesiadapidae to other families within Plesiadapiformes. My comparisons of the abundant dental remains of Palaeocene *Plesiadapis* with those of later Eocene primates led me to conclude, as Gidley (1923) and Simpson (1935) had before, that *Plesiadapis* shares detailed resemblances of upper and lower molar morphology with *Cantius* of similar size. As noted by Gidley (1923), molars of smaller plesiadapiform genera, e.g. *Elphidotarsius*, resemble those of smaller tarsioids like *Tetonius* or *Tarsius* itself. Thus it seemed to me that molar resemblances might link *Plesiadapis* and its allies phyletically to all later primates and not just to lemuroids as Simpson (1935) had concluded.

Other dental and cranial features suggested special affinity with tarsioids. These included enlarged, pointed central incisors, an ossified auditory bulla continuous with (if not actually comprised of) the petrosal, an ectotympanic anulus anchored to the auditory bulla by ossified *Necrolemur*-type struts, and an ossified tubular extrabullar extension continuous with (if not actually comprised of) the ectotympanic (Gingerich, 1975, 1976). Hence I proposed (1976) that primates be divided into two principal clades, Plesitarsiiformes, uniting Plesiadapiformes and Tarsiiformes, and Simiolemuriformes, uniting Anthropoidea (Simiiformes) and Lemuriformes. I still think this arrangement has as much cladistic merit as the popular actualistic alternative which groups tarsiers with anthropoids as Haplorhini, retains lemuroids and lorisoids in the primitive Strepsirhini, and removes *Plesiadapis* from Primates entirely.

Plesiadapis, and other plesiadapiform genera known from the Palaeocene, are clearly more primitive than any Eocene primates of modern aspect. All plesiadapiform genera lack a postorbital bar. *Plesiadapis* not only retained pedal grooming claws, as tarsioids, lemuroids, and lorisoids do today, but evidently retained claws on all phalanges. Parts of two lower limbs of a late Palaeocene specimen from Menat, preserved in articulation, are illustrated in Figure 3. The posterior part of the left foot is preserved in articulation with the tibia and fibula. Immediately below this, one can see the anterior part of a right foot that has become disarticulated and rotated through 180 degrees. The hallux of the rotated right foot now lies just below and parallel to metatarsals of the left foot. Judging from the preserved impression, the hallux (digit I) bears a sharply pointed claw. Pedal digits II through V are longer than digit I, and these too bear claws. The pollex is not preserved in the counterpart of this specimen, but digits II through V of the hand all bear claws (Gingerich, 1976, Pl.12).

Primitive mammalian characteristics retained by *Plesiadapis* and its allies, combined with many distinctive specializations, render any attempt to link Plesiadapiformes definitively with Tarsiiformes or Lemuriformes difficult. *Plesiadapis* is older geologically than any known tarsiiform or lemuriform primate, and it represents an evolutionary grade more primitive than these primates of modern aspect. Detailed, stratigraphically documented, studies of mammalian evolution across the Palaeocene-Eocene boundary in western North America, carried out intensively for the past ten years, suggest that major faunal turnovers may take place rapidly (on a geological time scale). The stratigraphic record and fossil record are rarely complete enough to document such turnover events in detail, and major worldwide episodes of climatic change sometimes facilitate intercontinental faunal dispersal in any case. Consequently, one might predict *a priori* that it is unlikely that such important faunal transitions will be preserved in the fossil record as a continuous series of intermediate genera and species linking early ancestors to later more advanced descendants. Our search for the origin of the tarsiiform Omomyidae and the lemuriform Adapidae in North America has permitted us to say that both groups appeared in the early part of the Wasatchian land-mammal age (early Eocene) as immigrants from elsewhere, with no likely ancestors evident in North America in older geological strata. Such negative evidence is never fully satisfying, but in this case the first appearance of Omomyidae and Adapidae coincides with the first appearance of hyaenodontid creodonts and primitive equid perissodactyls likely to be African in oridin. Hyaenodontids dominate Oligocene faunas of carnivorous mammals in Africa as nowhere else, and the dawn horse *Hyracotherium* resembles closely the hyracoid genus *Microhyrax* from the Eocene of Africa, a likely centre of origin for primates of modern aspect.

Fig. 3. Posterior part of skeleton of *Plesiadapis insignis* from French locality of Menat. Specimen is preserved as a negative (and thus reversed) impression in carbonaceous shale. Note preserved impression of fur on bushy tail, and impressions of nearly complete left and right posterior limbs. What appears here as a left foot remains articulated with the tibia and fibula. What appears here as a right foot has become disarticulated and rotated. Note clawed hallux (digit I) of right foot lying just below and parallel to metatarsals of left foot and clawed digits II-V as well. Scale in cm. Specimen is better illustrated with stereophotographs on Plate 12 in Gingerich (1976).

Intermediates linking primitive Plesiadapiformes to Tarsiiformes and/or Lemuriformes are unknown and unlikely to be discovered unless the African Palaeocene begins to yield a larger mammalian fauna. As a result of phylogenetic uncertainty regarding the origin of Tarsiiformes and Lemuriformes, any formal classification of Plesiadapiformes with one or another of these groups cannot readily be justified. A grade classification appears to express what we know (and do not know) about the phylogeny of early primates more accurately than a cladistic classification (Gingerich 1981, MacPhee et al, 1983).

PLESIADAPIS AND PRIMATE EVOLUTION

Plesiadapis is one of the most common mammals in late Palaeocene faunas of Europe and North America. This genus is represented by tens or hundreds of specimens in quarry samples representing the principal faunal zones of continental Thanetian (late Palaeocene) and Ypresian (early Eocene) ages in Europe, and continental Tiffanian (late Palaeocene) and Clarkfordian (transitional Palaeocene-Eocene) land-mammal ages in North America. Plesiadapis is important in providing a remarkably detailed record of mammalian evolution through an interval of 4-5 million years of geological time. The outline of changing tooth size and, by inference, body size shown in Figure 4 illustrates one of many evolutionary patterns recorded for this genus. The continuity of evolutionary change over time seen in Plesiadapis is evidence of the continuity of evolution as a general process (discontinuities too are common in the fossil record, as shown in the same figure by the absence of forms connecting Carpodaptes and Carpolestes). Figure 4 illustrates how phylogenetic hypotheses can be constructed on the basis of available stratigraphic and morphological information, and integrated phenetically without a priori assignment of primitiveness or derivedness to morphological characteristics to produce a temporally ordered minimum spanning tree. Figure 4 also illustrates how such a phylogenetic hypothesis can be tested when it serves as a target for future discoveries (closed circles and accompanying numerals represent new discoveries made after the phylogenetic pattern itself was drawn - these generally fall within the predicted target area).

The pattern linking successive species of Plesiadapidae together in Fig. 4 reflects a traditional approach to the study of evolutionary history that differs in several important ways from much current practice in systematic biology. For reasons that are not completely clear, systematics has become in recent years implicitly (and sometimes explicitly) actualistic. Available fossils are either ignored or they are treated as imperfect specimens coeval in time but secondary in importance to anatomically and biochemically complete living animals. The ages of fossils are often ignored lest such an independent perspective on time and evolutionary history bias an intuitive sense for how life in the past must have been, based on life in the present (hence the sentiment expressed in opening this essay - life in the past was "pretty much like today, there just wasn't as much of it."). The focus of systematic research today, as in Richard Owen's day, seems to be on archaetypes rather than ancestors.

Figure 4 is explicitly about ancestors evidenced in the fossil record.

Given present knowledge of North American Plesiadapidae, Plesiadapis praecursor is the oldest known species of Plesiadapis, and it appears to be a plausible ancestor for all later North American species of Plesiadapis. P. praecursor is real, it is not a hypothetical archaetype. This is not to say that detailed study of morphological patterns in stratigraphic context will always identify plausible ancestors for all species under study. Carpolestes jepseni stands out clearly as a species whose ancestry at the species level is unknown (although it is likely to have been derived from some species of Carpodaptes, perhaps even one of the known species).

PLESIADAPIS AND DELINEATION OF THE ORDER PRIMATES

The same approach to the study of evolutionary history illustrated by Plesiadapis can be applied on a broader scale to the study of all primate relationships (see Figure 5). In Figure 4 Plesiadapis dubius was linked to P. fodinatus, which was linked in turn to P. churchilli, P. rex, P. anceps, and P. praecursor, which was linked finally to Pronothodectes jepi. Each of these successive species is distinctive, and together they

Figure 4. Outline of the species-level phylogeny of North American Palaeocene Carpolestidae and Plesiadapidae, showing pattern of evolution in tooth size, and by inference body size, in stratigraphic column 1400 metres thick representing 4-5 million years on geological time. History of diversification of Plesiadapis might serve as a reasonable model for diversification of Plio-Pleistocene Hominidae. Reproduced with additions from Gingerich (1976, 1980).

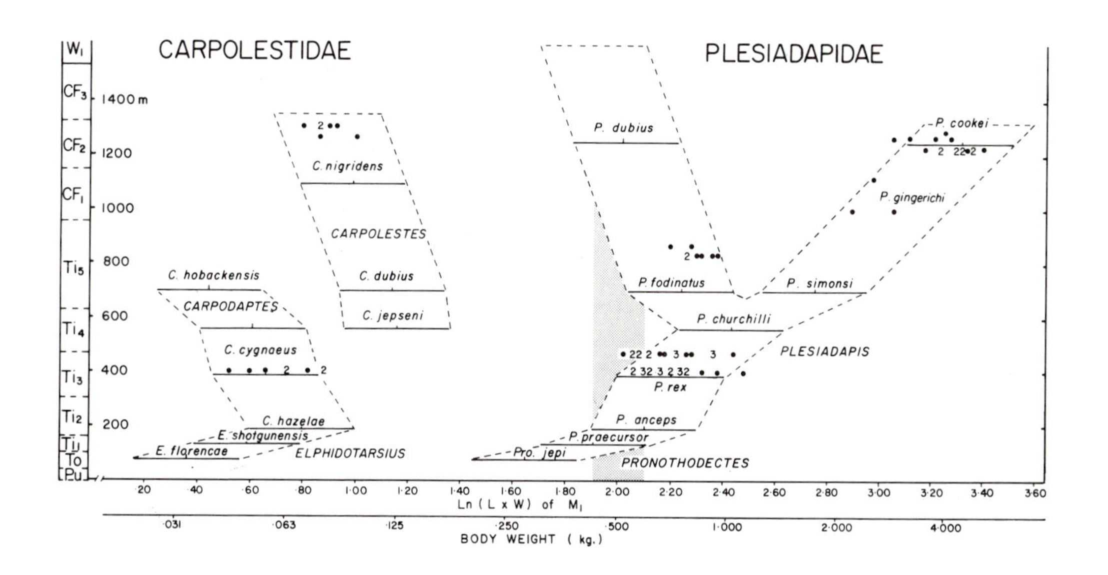

record change over time. Species of *Plesiadapis* are included in the same genus because they exhibit such a high degree of evolutionary continuity. Given present evidence, there is little discontinuity between *Pronothodectes* and *Plesiadapis* and two genera are recognized as distinct, in part, because they were named and defined before a connected series of intermediates became known. Similarly, in Figure 5, anthropoid primates can be traced back in time and linked to older and more primitive prosimians, and prosimians can be traced back in time and linked to older and more primitive Plesiadapiformes or praesimians. As discussed above, discontinuities separate praesimians, prosimians, and anthropoids, and for this reason it

Figure 5. Outline of the history of primates in three grades: Praesimii (Plesiadapiformes, possibly including Apatemyidae, together with Dermoptera and Tupaiidae), Prosimii (shaded, including Tarsiiformes and Lemuriformes - Lorisiformes could as well be included as a superfamily of the latter), and Anthropoidea (monkeys, apes, and humans). Modified from MacPhee *et al.* (1983).

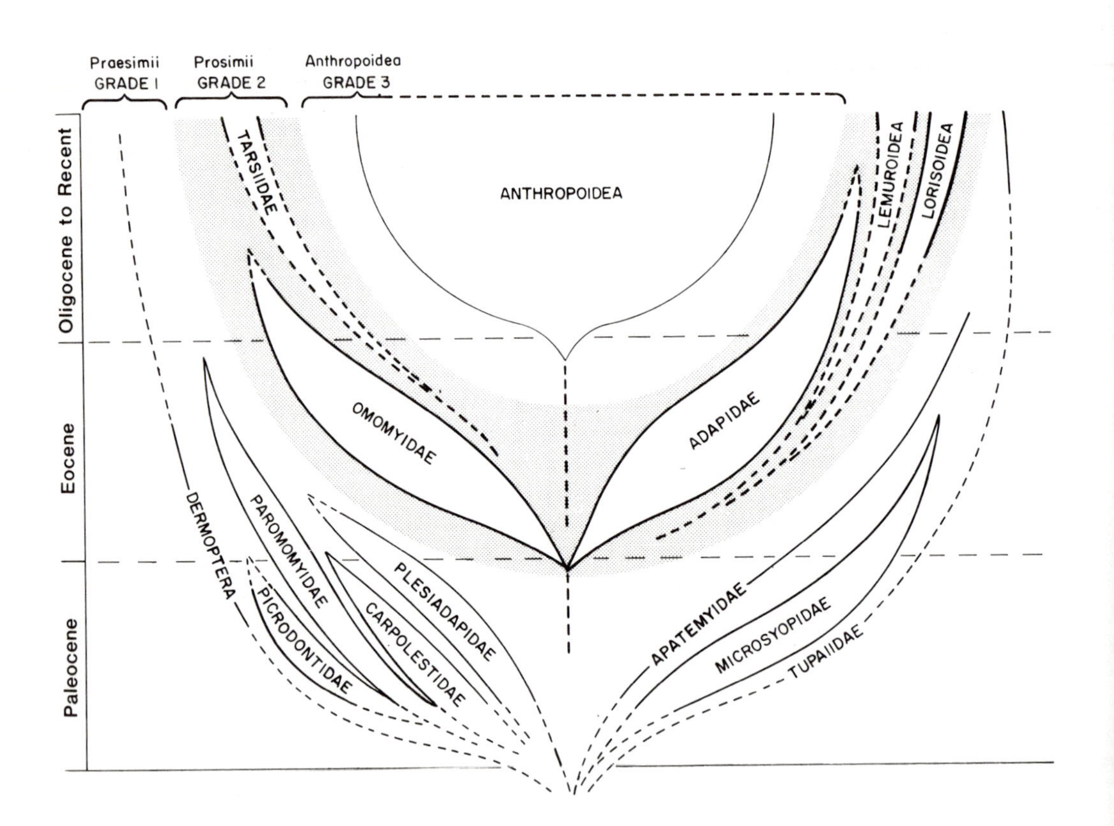

is appropriate at present to visualize the history of primates in terms of three major grades.

Is Plesiadapis a primate? This question depends entirely on how one defines Primates. The very fact that primatologists cannot discover and hence agree on any true definition of primates (but see Martin, this volume) is evidence of the evolutionary continuity by which our order is connected to other mammalian orders. We will probably never all agree on the placement of such an arbitrary boundary. Indeed science is by nature so democratic as to encourage disagreement on such unanswerable questions (there will probably always be a niche for devil's advocates on questions such as these). It might be possible to delimit or "define" higher taxonomic groups like Primates if the lower taxa that compose them were discrete, but transitions, real or imagined, between lower taxonomic groups preclude a priori definition (even in terms of "total morphological pattern"). In evolutionary context, taxonomic groups are delineated in time, not delimited morphologically. One consequence of transition is that taxonomic categories, developed in, and for, an actualistic world, become arbitrary because names are discrete while the underlying phenomenon, evolution, is continuous.

Biology is the study of life, most of which is now extinct. Contemporary organisms can be observed and compared in greater detail than is possible for those living in the past. However, biology is more than "neontology", it is more than the study of living plants and animals. Palaeontology, the study of life in the geological past is an integral component of biology providing unique information about the history of life. The difference between neontological and palaeontological perspectives on life and its history can be illustrated by considering how differently trees or branching bushes (common metaphors of phylogeny) appear when viewed from the top and from the side (Figure 6). Viewed from the top, the tip of each branch of a "phyletic" bush is discrete. The species and higher taxonomic categories of a neontologist are discrete because they are all compared at one instant of time. Change takes place through time. Hence the same species and higher taxonomic categories that appear to be discrete at an instant of time may intergrade through time. Palaeontologists view phyletic bushes as neontologists do, in sequential slices from the top, but palaeontologists must also attempt to characterize phyletic bushes as seen from the side. The view from the side, necessary for documentation of transitions through time, is in fact the goal of any evolutionary science.

If Plesiadapis and its allies, along with the best living models for primitive praesimians, the tree shrews, are removed from Primates, what more can we then say about the origin and early evolution of the order? How satisfied are we to trace our ancestry back to the beginning of the Eocene and say nothing of how similar or different mammals were in the Palaeocene? One can, of course, speak of Palaeocene mammals in any case, whether Plesiadapis is a primate or not. However, I doubt that we will ever be satisfied with an empty Palaeocene chapter in the history of Primates. The evolutionary importance of Plesiadapis and its allies, like that of tree shrews, would be little diminished if they were to be excluded from Primates, but I think in the case of Plesiadapis, as in the case of tree shrews, we would diminish our understanding of primate evolution by excluding them and the perspective they provide on the form and grade of early primates.

SUMMARY AND CONCLUSIONS

Plesiadapis is one of the best known Palaeocene mammals. It is primate-like in molar form and some features of the basicranium and ankle. All these areas are conservative anatomically and are acknowledged to be of special utility in mammalian systematics. *Plesiadapis* differs from later primates of modern aspect in retaining claws on all digits, including the hallux, and in lacking a postorbital bar and a promontory branch of the internal carotid artery. Like other Palaeocene mammals, *Plesiadapis* and its allies retain more primitive characteristics than do Eocene and later forms, while exhibiting divergent specializations as well. Palaeocene mammals are all of distinctly primitive grade. In evolutionary perspective, orders are taxonomic groups to be delineated rather than

Figure 6. Comparison of a hypothetical phyletic tree or bush viewed from the top (A) and from the side (B). Note that the same tree has a very different appearance depending on how it is viewed. Terminal branches (species) and aggregations of these (higher taxonomic categories) appear to be discrete and "real" when viewed from the top (at one instant of time), while the same entities appear as arbitrary parts of a continuously growing bush when viewed through time. Viewing a phyletic bush from above as neontologists are constrained to do reduces its complexity while obscuring its essential continuity.

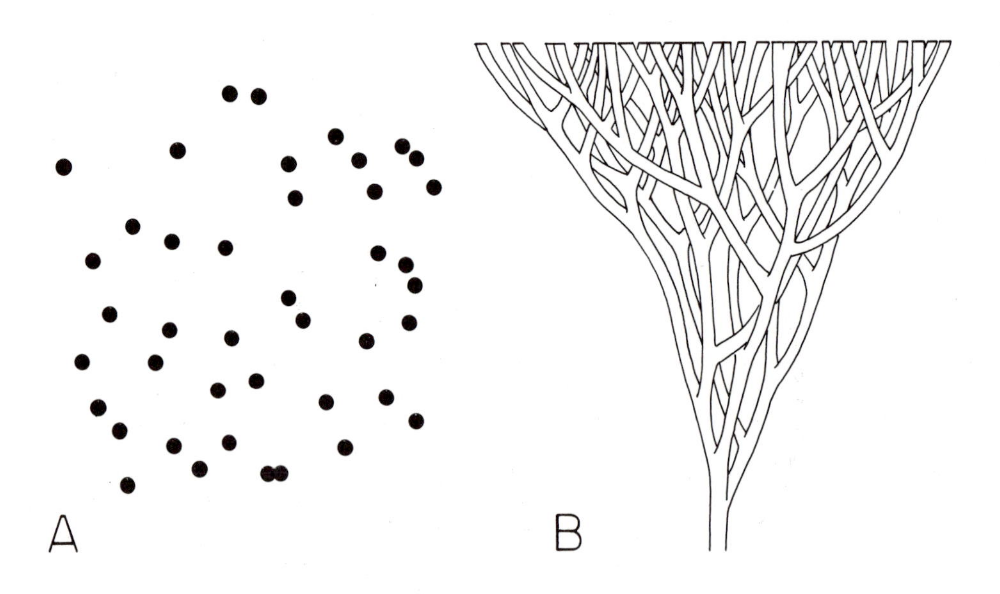

delimited. Any "definition" of primates is necessarily arbitrary. The question of whether Plesiadapis is classified with primates is far less interesting than how it relates to primate phylogeny.

ACKNOWLEDGEMENTS

Figures 1,4,5 and 6 were drawn by Karen Klitz while supported by U.S. National Science Foundation grant BNS 80-16742.

REFERENCES

Key references

Gingerich, P.D. (1976). Cranial anatomy and evolution of early Tertiary Plesiadapidae. Univ. Mich. Pap. Paleont., *15*, 1-140.

MacPhee, R.D.E., Cartmill, M., & Gingerich, P.D. (1983). New Palaeogene primate basicrania and definition of the order Primates. Nature, *301*, 509-511.

Russell, D.E. (1964). Les mammifères Paléocènes d'Europe. Mém. Mus. Nat. d'Hist. Natur. (C), *13*, 1-324.

Szalay, F.S., Tattersall, I., and Decker, R.L. (1975). Phylogenetic relationships of Plesiadapis - postcranial evidence. Contrib. Primatol., *5*, 136-166.

Main References

Buchardt, B. (1978). Oxygen isotope palaeotemperatures from the Tertiary period in the North Sea area. Nature, *275*, 121-123.

Gervais, P. (1877). Enumération de quelques ossements d'animaux vertébrés recueillis aus environs de Reims par M. Lemoine. J. Zool. (Paris), *6*, 74-79.

Gidley, J.W. (1923). Paleocene primates of the Fort Union, with discussion of relationships of Eocene primates. Proc. U.S. Nat. Mus., *63*, 1-38.

Gingerich, P.D. (1975). Systematic position of Plesiadapis. Nature, *253*, 111-113.

Gingerich, P.D. (1976). Cranial anatomy and evolution of early Tertiary Plesiadapidae. Univ. Mich. Pap. Paleont., *15*, 1-140.

Gingerich, P.D. (1980). Evolutionary patterns in early Cenozoic mammals. Ann. Rev. Earth Planet. Sci., *8*, 407-424.

Gingerich, P.D. (1981). Early Cenozoic Omomyidae and the evolutionary history of tarsiiform primates. J. Human Evol., *10*, 345-374.

Lemoine, V. (1887). Sur le genre Plesiadapis, mammifère fossile de l'éocène inférieur des environs de Reims. C.R. Acad. Sci., *1887*, 190-193.

MacPhee, R.D.E., Cartmill, M. & Gingerich, P.D. (1983). New Palaeogene primate basicrania and definition of the order Primates. Nature, *301*, 509-511.

Matthew, W.D. (1917). The dentition of Nothodectes. Bull. Am. Mus. Nat. Hist., *37*, 831-839.

Russell, D.E. (1959). Le crâne de Plesiadapis. Bull Soc. Géol. France (7th sér.), *1*, 312-314.

Russell, D.E. (1964). Les mammifères Paléocènes d'Europe. Mém. Mus. Nat. d'Hist. Natur. (C), *13*, 1-324.

Simpson, G.G. (1935). The Tiffany fauna, upper Paleocene. II. - Structure and relationships of Plesiadapis. Am. Mus. Novitates, *816*, p. 1-30.

Stehlin, H.G. (1916). Die Saugetiere des schweizerischen Eocaens. Caenopithecus usw. Abh. Schweiz. Paläont. Gesell., *41*, 1299-1552.

Szalay, F.S., Tattersall, I., and Decker, R.L. (1975). Phylogenetic relationships of Plesiadapis - postcranial evidence. Contrib. Primatol., *5*, 136-166.

3 THE RELATIONSHIPS OF THE TARSIIFORMES: A REVIEW OF THE CASE FOR THE HAPLORHINI

Leslie C. Aiello,
Department of Anthropology,
University College London,
London WC1E 6BT.

INTRODUCTION

There are three species of the genus Tarsius (T. syrichta, T. bancanus, T. spectrum). All three species are extant and the genus has a range which encompasses Sumatra, the southern Philippines and the Celebes. The tarsiers are among the smallest of living primates with a body size range between approximately 100 and 200 grams and they are totally nocturnal in activity pattern. They are among the most specialized of living primates in some aspects of their anatomy while, at the same time, among the most generalized, or primitive, in other aspects. Their specialized features include their extremely large eyes in relation to their body sizes and their radically elongated tarsal bones and fused tibia and fibula associated with their highly developed Vertical Clinging and Leaping locomotor pattern. Their generalized features include the simple arrangement of the peritoneum and alimentary canal and the retention of the flexores breves digitorum manus, a humeral entepicondylar foramen and a three-branched aorta (Hill, 1955).

The phylogenetic and taxonomic relationships of the genus Tarsius have been a source of controversy since these primates were first reported in the literature in the early years of the eighteenth century (Petiver, 1702-04). Until 1777, when Erxleben recognized their primate affinities and placed them in the genus Lemur, they were considered to be most closely related to jerboas or opossums (Buffon, 1765; Linnaeus, 1767-1770). By the end of the eighteenth century they were recognized by some to be distinct from the lemur (Link, 1795; Illiger, 1811; Gervais, 1854-55). However, it was not until the turn of the century that Hubrecht recognized that the tarsiers were characterized by a haemochorial placenta and, in this respect, resembled the higher primates (see Hubrecht, 1908). At this time, Cope (1885, 1896) also pointed out the similarities between the Eocene fossil Anaptomorphus and the extant tarsiers and considered the tarsiers and their fossil antecedents to represent a grade intermediate between the lemurs and the monkeys. In 1898 Gadow created the suborder Tarsii for both the fossil and extant tarsiers and later Elliot Smith (1907) renamed the suborder Tarsioidea. In 1915 Gregory replaced this suborder with the 'series' Tarsiiformes.

Since this time there has been considerable debate over the fossil taxa which should be classified together with the genus Tarsius in the higher order taxon Tarsiiformes (see Szalay, 1976) as well as considerable controversy over the classification of the Tarsiiformes within the Order

Primates. In relation to the issue of the taxonomic position of the Tarsiiformes, there are three major and current schools of thought. The first school is consistent with the work of Gregory (1915) who placed his 'series' Tarsiiformes together with the 'series' Lorisiiformes and the 'series' Lemuriiformes in the suborder Lemuroidea. The modern expression of this classification is the system proposed in 1945 by G.G. Simpson (Fig. 1). This classification differs from the remaining two, in being both a vertical and a horizontal classification. Under this system overall morphological similarity between taxa is used as a criterion for classification and, in some cases such as that of the Tarsiiformes, can override the evidence of inferred phylogenetic relationship.

The other two schools of thought are based on the principle that classification should accurately reflect inferred phylogenetic relationships. The first of these is based on the work of Pocock (1918). Pocock emphasized the fact that the tarsier lacks a naked rhinarium and is thereby similar to the higher primates and distinct from the lemurs and lorises. This fact, together with the shared presence of the haemochorial placenta and the postorbital partition in the tarsiers and the higher primates, provided grounds for Pocock to establish two grades within the Order Primates which emphasized the putative relationship of the extant tarsiers and the higher primates. Pocock erected the grade Haplorhini to include the suborder Tarsioidea and the suborder Pithecoidea (monkeys, apes and humans) while at the same time placing the lemurs and lorises in a second grade, the Strepsirhini. While Pocock was concerned primarily with the relationships

Figure 1. Three current systems of classification of the Order Primates. See text for discussion.

Simpson (1945)

ORDER	Primates
SUBORDER	Prosimii
INFRAORDER	Lemuriformes
INFRAORDER	Lorisiformes
INFRAORDER	Tarsiiformes
SUBORDER	Anthropoidea
SUPERFAMILY	Ceboidea
SUPERFAMILY	Cercopithecoidea
SUPERFAMILY	Hominoidea

Szalay and Delson (1979)

ORDER	Primates
SUBORDER	Strepsirhini
INFRAORDER	Lemuriformes
INFRAORDER	Adapiformes
SUBORDER	Haplorhini
INFRAORDER	Tarsiiformes
INFRAORDER	Platyrrhini
INFRAORDER	Catarrhini

Gingerich (1981)

ORDER	Primates
SUBORDER	Tarsiiformes
INFRAORDER	Tarsiiformes
SUBORDER	Simiolemuriformes
INFRAORDER	Lemuriformes
INFRAORDER	Simiiformes

of the extant tarsier and the extant higher primates, later authors have applied this classification to fossil, as well as living, primates (Hill, 1955; Szalay & Delson, 1979).

The remaining school of classification is consistent with the work of Wortman (1903) and Gregory (1920) who suggested that the higher primates (particularly the platyrrhine primates) are derived from the Eocene Lemuriformes (adapids) rather than from the Eocene Omomyidea (Tarsiiform primates). Gingerich (1973, 1975, 1978, 1980, 1981; Gingerich & Schoeninger, 1977) has provided modern support for the hypothesis that the lemuriform primates have a closer relationship to the higher primates (simiiforms) than do the tarsiiform primates. In order to reflect this inferred relationship, he has proposed that the order Primates be divided into the suborder Tarsiiformes on the one hand and the suborder Simiolemuriformes, including both the infraorder Lemuriformes and the infraorder Simiiformes, on the other (Gingerich, 1981).

In the context of these three conflicting models of primate classification, the case for the Haplorhini is the case for Pocock's (1918) classification of the Primates into the two grades Haplorhini and Strepsirhini. At the most basic level this case rests on two points. Firstly, it rests on the proposal that the living tarsiers are more closely related to the higher primates than they are to the lemurs and lorises. Secondly, it rests on the point that related taxa should be classified together and that these taxa should be monophyletic. Rejection of the first point led directly to the establishment of the Tarsiiformes-Simiolemuriformes classification of the primates (Gingerich, 1981), while objections to the second point, particularly among palaeontologists, has given preference to the Prosimii-Anthropoidea classification (Simons, 1974; Cartmill & Kay, 1978). These objections primarily concern the practicalities of applying an evolutionary based system of classification to fossil primates.

The following assessment of the case for the Haplorhini is divided into two parts. The first part addresses the evidence for and against relating the extant tarsiers with the extant higher primates. The second section addresses the problems and implications of applying the Strepsirhini-Haplorhini classification in the context of the fossil record. This section gives particular attention to the still active debate over the evolutionary antecedents of the living tarsier as well as to the controversies involved in the definition of the Haplorhini in the light of the fragmentary nature of the fossil record.

TARSIUS AND THE EXTANT PRIMATES

In the first instance, the case for the Haplorhini is the case for the hypothesis that among living primates the tarsiers are more similar to the anthropoid primates (monkeys, apes and humans) than are the lemurs and lorises, or in other words, that among living primates the tarsiers are the sister group of the anthropoid primates. In this context, similarity or sister group status, is defined by the common possession of synapomorphic, or uniquely derived, morphological features that must meet two criteria. Firstly, they must not occur as the result of parallel evolution in the groups concerned. Secondly, they must not be symplesiomorphic, or primitive features. Symplesiomorphic features are features that would have characterised a remote ancestor of the groups involved and would, therefore, be expected to be found not only in the anthropoid primates and the tarsiers but also in more remotely related primates. Therefore, to falsify the hypothesis that the tarsiers are the sister group of the anthropoid primates, it is necessary to establish beyond a reasonable doubt that the morphological features upon which this hypothesis is based are either symplesiomorphic features or parallelisms.

Table 1 lists 23 morphological features that are shared in common by the living tarsiers and the anthropoid primates and which have been used to support the sister group status of these primates. There is little concern over the problem of symplesiomorphy in relation to these features. (The reader is referred to the references given in Table 1 for detailed justifications for each feature.) The only feature that has been suggested possibly to be primitive is the posteromedial position of the carotid foramen (Cartmill *et al*, 1981). This suggestion is based on the fact that the posterior medial position of the foramen is found in lorisiformes as well as most of the possible outgroups of primates, including tree shrews, elephant shrews and lipotyphlous Insectivora. In addition, ontogenetic evidence of Malagasy lemurs suggests that the lateral position of the foramen in these primates is most probably associated with differences in the extent and pattern of pneumatization in the rear part of the petrosal plate (Cartmill *et al*, 1981).

There is, however, room for concern that the shared presence of some of the features by the tarsiers and anthropoid primates may well have occurred as the result of parallel evolution. There are three specific arguments that can be made in favour of parallel evolution for some of these characteristics. The first centres on the effect of the tarsier's extraordinarily large eyes on the surrounding cranial morphology, and particularly on the nasal architecture and on the partial postorbital septum. It is possible to explain the presence of the apical interorbital septum (feature 20: see Table 1) and the absence of an olfactory recess (feature 21) in *Tarsius* through the necessary approximation of the huge orbital cavities. The resulting reduction in the size of the nasal cavity could then be correlated with an overall reduction in the olfactory sense including a reduction in the size of the olfactory bulbs (feature 10) and loss of the naked rhinarium (feature 12) and resulting in the short, deep, low hafted facial skull (feature 23). In addition, Rosenberger & Szalay (1980) believe that the alisphenoid element in the tarsier postorbital septum (feature 22) is also correlated with the hypertrophic eyeballs and cite in support the similar occurrence of postorbital flanges in large-eyed strepsirhines such as *Loris*.

There is no obvious neontological argument against the possibility of parallelism in these characteristics. However an argument against the parallel evolution of the apical interorbital septum can be made from palaeontological evidence. The apical septum is present in Eocene omomyids whose eyes are the same size relative to body size as are the eyes of the extant lemurs and lorises, which lack an apical interorbital septum (Kay & Cartmill, 1978; Cartmill, 1975). The fact that this feature is not correlated

Table 1. Morphological features shared in common by *Tarsius* and the anthropoid primates. Cited references support the synapomorphic status of these features.

	Feature	Reference
Soft Tissue Morphology		
1.	No choriovitteline placenta, rudimentary allantois, well developed body stalk, ovarian bursa reduced/absent, primordial amniotic cavity transitory, blastocyst attachment invasive, monodiscoidal haemochorial placenta	Luckett, 1980
2.	Relatively high ratio neonatal weight to maternal weight	Leutengger, 1973
3.	Biochemical similarity	Dene *et al.*, 1976
4.	Karyotypic similarity	Martin, 1978
5.	Absence of a tapetum lucidum	Martin, 1973
6.	Presence of a fovea centralis	Wolin & Massoupust, 1970
7.	Well developed promontory artery	Rosenberger & Szalay, 1980
8.	Prenatal loss of a functional stapedial artery	Cartmill *et al.*, 1981
9.	Loss of the contralateral sulcus	Rosenberger & Szalay, 1980
10.	Reduced olfactory bulbs	Rosenberger & Szalay, 1980
11.	Absence of well developed vibrissae	Cartmill, 1980
12.	Absence of a naked rhinarium	Pocock, 1918
13.	Absence of a sublingua	Hill, 1955
Hard Tissue Morphology		
14.	Perbullar path of the internal carotid artery	Cartmill *et al.*, 1981
15.	Loss of the subtympanic recess beneath the eototympanic	Cartmill *et al.*, 1981
16.	Presence of an intrabullar transverse septum partioning the hypotympanic sinus	Cartmill *et al.*, 1981
17.	Anteromedially enlarged hypotympanic sinus	Rosenberger & Szalay, 1980
18.	Ossified auditory tube derived from the ectotympanic rather than the petrosal	Conroy, 1980
19.	Posteromedial position of the carotid foramen	Rosenberger & Szalay, 1980
20.	Apical interorbital septum	Cave, 1973
21.	Lack of an olfactory recess	Cave, 1973
22.	Presence of an alisphenoid contribution to the postorbital septum	Cartmill, 1980
23.	Short, deep low hafted facial skull	Rosenberger & Szalay, 1980

with enlarged eyes in a putative ancestor of Tarsius suggests that it is highly unlikely that it occurred as the result of enlarged eyes in the extant genus. In addition, among living mammals only Tarsius and the anthropoid primates possess an apical interorbital septum (Cave, 1967, 1973). The unique occurrence of this feature in these primates also suggests that it is most probably a robust synapomorphic characteristic.

The second argument for parallelism of some of the features in common to the tarsiers and the anthropoid primates comes from the possibility that the tarsiers and the anthropoid primates underwent a parallel adaptation to a diurnal activity pattern. The basis for this lies in the fact that, although the extant tarsiers are nocturnal animals, the presence of a fovea centralis (feature 6) in their retinas provides near conclusive evidence that they stem from a diurnal ancestor (Cartmill, 1980; Martin, 1978). In addition, features 5 to 12 in Table 1 can all be related to this diurnal adaptation in the context of an emphasis on the sense of vision and a reduction in emphasis of olfaction. This is even true in relation to the well developed promontory artery and the prenatal loss of the stapedial artery. This arterial pattern has been argued to be expected in any line undergoing forebrain enlargement (Cartmill & Kay, 1978), a feature which has been correlated with a diurnal environment (Martin, 1978).

An argument against parallel evolution of a diurnal activity pattern in the tarsier line and in the anthropoid primates can be made from the detailed structure of the retinas in these primates. Not only do both groups possess a fovea centralis but they also possess a yellow spot, or macula lutea (Polyak, 1957). The function of the yellow spot is to reduce violet light and thereby protect the rods in the retina from high levels of light intensity. The combination of a fovea and yellow spot is unique only to Tarsius and the anthropoid primates among living primates. The unique occurrence of the fovea and the yellow spot in these primates suggests not only that they are robust synapomorphic features, but also that the diurnal activity pattern in the ancestors of Tarsius and in the anthropoid primates did not occur in parallel.

The third argument for parallelism centres on the interpretation of the ear region of the tarsiers and the anthropoid primates. Cartmill & Kay (1978) and Cartmill et al. (1981) have argued that the ear of the extant tarsiers and anthropoid primates differs from the ear of lemurs and lorises as well as all known Palaeocene and Eocene primates in the absence of the subtympanic recess beneath the tympanic (feature 15), the presence of an intrabullar transverse septum partioning the hypotympanic sinus (feature 16) and a perbullar as opposed to transpromontorial pathway of the internal carotid artery (feature 14). However, in support of parallelism, Rosenberger & Szalay (1980), argue that the morphology of the transverse septum delimiting the anterior accessory cavity and the route of the carotid artery are different in the platyrrhines on the one hand, and the living tarsiers and the catarrhine primates on the other. They argue that the primitive anthropoid condition would be most probably represented by platyrrhines such as Cebus where there is a restricted communication between the hypotympanic and tympanic cavities due to the approximation of the promontorium and the promontory canal with the lateral wall and ventral floor of the bulla rather than to the interposition of a transverse septum. Citing the ontogenetic work of van Kampen (1905), they argue that the development of the septum may be related to verticalization of the carotid canal as the bullar chamber increases in height; the canal dragging behind it tissues that later ossify to become the transverse septum. If this interpretation is correct, the presence of an intrabullar transverse septum partioning the hypotympanic sinus (feature 16) and the perbullar path of the internal carotid artery (feature 14) would have to have arisen through parallel evolution in the tarsiers and the catarrhine primates.

However, this argument does not incorporate an explanation for parallel evolution of the third feature, the loss of the subtympanic recess beneath the ectotympanic. Although Rosenberger & Szalay (1980) suggest that the absence of the subtympanic recess in the modern Tarsius and in the anthropoid primates could have been derived in parallel from an ancestor with a reduced recess (such as Necrolemur), it is difficult at present to accept such an explanation. Szalay (1975) has argued that an increasingly extrabullar position of the ectotympanic (associated with the reduction of the subtympanic recess) could be achieved through the ossification of the annulus membrane. However MacPhee (1977) has demonstrated that this could not be possible because the annulus membrane consists only of ectodermally-derived meatal skin and intertympanic mucous membrane, and, therefore, could not ossify. Based on this analysis, Conroy (1980) has also argued that the auditory tubes in any primate with a subtympanic recess would, by necessity, be derived from the petrosal, and not from the ectotympanic, as they are in Tarsius and the anthropoid primates which lack the recess. In addition, Cartmill (1975) has argued that the loss of the subtympanic recess (combined with the presence of an extrabullar position of the ectotympanic) would be the expected morphology in a very small primate with an enclosed bullar chamber. His argument is based on the negative allometry shown by the ratio of eardrum area to bullar volume which is explained by the mechanical necessity of large bullar volume in small animals in order to facilitate acute auditory reception. Therefore parallel evolution of the loss of the subtympanic recess would necessarily require independent derivation of both the tarsiers and the anthropoid primates from ancestors which independently achieved smaller size from a primate similar to Necrolemur (in body size and bullar morphology). Although not impossible, when coupled with the previous argument in favour of a diurnal adaptation for both the tarsiers and anthropoid primates at a larger body size, this scenario suffers from problems of parsimony. A much simpler explanation would be to consider the loss of the subtympanic recess as a robust synapomorphic feature characteristic of a small-bodied ancestor of the tarsiers and the anthropoid primates, an ancestor which then increased in size and became diurnal in activity pattern before the separation of the line leading to the modern tarsiers on the one hand and the anthropoid primates on the other.

Therefore, although the perbullar path of the internal carotid artery (Table 1, feature 14) and the presence of an intrabullar transverse septum partitioning the hypotympanic sinus (feature 16) may have arisen through parallel evolution in the tarsiers and the anthropoid primates, evidence suggests that the loss of the subtympanic recess (feature 15) is best considered as a robust synapomorphic feature in common to the tarsiers and the anthropoid primates. The same interpretation is most probably true for the anteromedially enlarged hypotympanic sinus (feature 18, see Rosenberger & Szalay, 1980), the fetal membrane evidence (feature 1, see Luckett, 1980), the neonatal body weight/maternal body weight evidence (feature 2, see Leutenegger, 1973) and the karyotypic evidence (feature 4, see Martin, 1978). The sister group status of Tarsius and the anthropoid primates is also supported by biochemical evidence (Dene et al., 1976; Beard et al., 1976; Goodman et al., 1974, Goodman, 1973).

In the context of the falsification of the hypothesis that, among living primates, the tarsiers are the sister group of the anthropoid primates there is room for greater doubt in relation to the synapomorphic status of some of the features shared in common by these primates than there is in relation to other characters. Therefore, on the basis of the interpretation given above, the hypothesis that, among living primates, the tarsiers are the sister group of the anthropoid primates cannot be rejected. However it must be remembered that decisions in relation to the synapomorphic status of individual morphological features are based on largely subjective assessments of the probability of parallel evolution. It is in this context that the alternative hypothesis for tarsier relationships has been proposed and supported by Gingerich (1981). Gingerich argues that, among living primates, the lemurs and lorises, and not the tarsiers, are the sister group of the anthropoid primates. He bases this hypothesis not on the similarities between living lemurs and lorises and the anthropoid primates, but on the strength of the morphological similarities of the putative ancestors of these primates. Because the anthropoid primates cannot be both the sister group of the lemurs and lorises and the sister group of the tarsiers, Gingerich argues that the features shared in common by the living tarsiers and the anthropoid primates by necessity have to be the result of parallel evolution.

Most recently, Gingerich (1981) has supported this hypothesis of anthropoid relationships by an analysis of a series of characteristics of primitive Oligocene anthropoid primates in comparison to those of the Eocene Omomyidae (putative tarsier ancestors) and the Eocene Adapidae (putative lemur and loris ancestors) (Table 2). He argues that because the great majority of these features are shared in common by the Adapidae and the Oligocene anthropoid primates, the anthropoids had their ancestry within the Adapidae and, among living primates, the lemurs and lorises would be the sister group of the anthropoid primates. There are three points which must be taken into consideration in the evaluation of this analysis. Firstly Gingerich has not included among his characters the morphological features that can reasonably be considered to be derived features in common to the omomyids on the one hand and the tarsiers and the anthropoid primates on the other (see Table 3 and the accompanying discussion). Secondly, some of the features used by Gingerich are probably correlated. Therefore, the sheer number of features in common to the adapids and the Oligocene anthropoids are not as impressive as they might otherwise seem. And thirdly, if Gingerich's features are assessed by the same criteria used to assess the features in common to the tarsiers and the anthropoid primates, strong arguments can be made against their synapomorphic status.

Of the twelve features shared in common by the adapids and the Oligocene anthropoid primates (Table 2), ten (features 2-9, 11-12) are most probably directly related to body size (feature 1). Not only is body size of limited genealogical value (Delson & Rosenberger, 1980) but also it places definite constraints on behaviour and associated morphology. Features 2-9 are gnathic features and are directly related to dietary adaptations. Kay & Simons (1980) have demonstrated a close relationship between body size and

Table 2: Morphological features of primitive Oligocene anthropoids compared to those of Eocene Omomyidae and Adapidae after Gingerich (1981). AAA = features shared by Oligocene anthropoids and Eocene adapids. AOA = features of the Oligocene anthropoids shared equally by Eocene adapids and Eocene omomyids. OOO = features shared by Oligocene anthropoids and Eocene omomyids. xxx = features of minimal phylogenetic significance (nontubular (partially free) ectotympanic is based on a specimen which is unlikely to be a primate, Cartmill <u>et al.</u>, 1981). 111, 222 = features which are most probably correlated. 3p3 = features which are most probably primitive and correlated. ppp = features which are most probably primitive. (After Delson & Rosenberger, 1980). bbb = features most probably correlated with body weight.

1.	Body size over 500 grams	AAA			+	xxx	xxx	+
2.	Tendency to fuse mandibular symphysis	AAA			+	111	bbb	+
3.	Vertical spatulate incisors	AAA			+	111	bbb	+
4.	Lower I1 smaller than lower I2	AAA			+	111	bbb	+
5.	Interlocking canines	AAA			+	222	bbb	+
6.	Canines sexually dimorphic	AAA			+	222	bbb	+
7.	Canine/premolar honing complex	AAA			+	222	bbb	+
8.	Molarized lower P4	AAA			+		bbb	+
9.	Tendency toward quadrate lower molars	AAA			+	ppp	bbb	+
10.	Nontubular (partially free) ectotympanic	AAA			+	xxx	xxx	+
11.	Relatively short calcaneum	AAA			+	3p3	bbb	+
12.	Unfused tibia-fibula	AAA			+	3p3	bbb	+
13.	2:1:3:3 dental formula		AOA		+			+
14.	Position of hypocone on basal cingulum		AOA		+			+
15.	Absence of postorbital closure		AOA		+			+
16.	Presence of stapedial artery		AOA		+			+
17.	Encephalization quotient: .85 anthropoids,			OOO	+			+
	.42-.97 omomyids, .39-.41 adapids				+			+

dietary adaptation, while Kay (1977, 1981) has demonstrated the relationship between molar tooth form, diet and body size, and Hylander (1975) has shown the relationship between incisor size and diet in anthropoids. The fact that the adapids converge on the living and fossil anthropoid body size range (Fig. 2) provides sufficient grounds upon which to argue that the dental similarities are also convergences resulting from a parallel adaptation to a frugivorous, or folivorous, dietary adaptation required by their relatively large body sizes. In addition, Rosenberger & Szalay (1980) have argued that some of the omomyids (e.g. *Ourayia* at a relatively large body size, Fig. 2) have incisor teeth similar in shape and proportion to those of anthropoids, while another large sized omomyid, *Rooneyia*, has molar teeth which are similar to those of the Oligocene anthropoids and indicate a frugivorous dietary adaptation (Szalay & Delson, 1979).

Figure 2. Weight distribution of the Eocene adapids in relation to the Eocene omomyids and the Oligocene anthropoids. Weight data after Gingerich 1981, 1982. Horizontal axes = logged body weight (base 10). Omomyids (1) = weight estimates based on length of lower M2 regressed against all living primates (Gingerich 1981). Omomyids (2) = weight estimates based on length of lower M2 regressed against living tarsiers (Gingerich 1981). r = *Rooneyia*, o = *Ourayia*, h = *Hemiacodon*

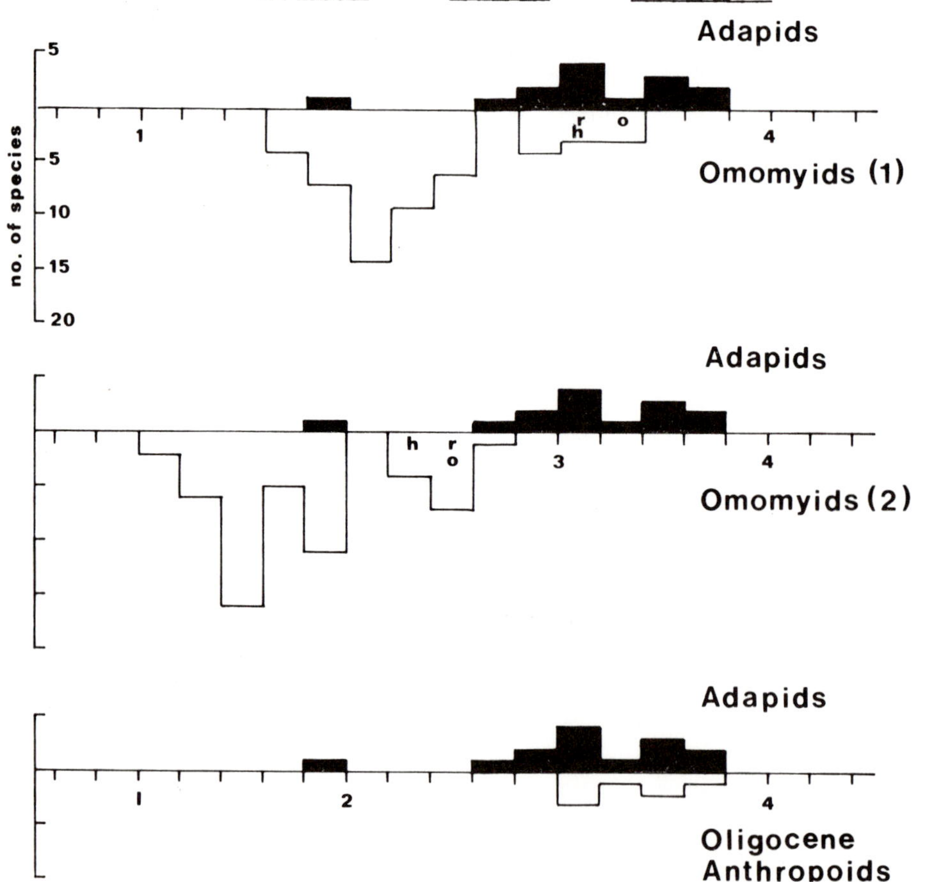

In relation to the postcranial features (nos. 11 & 12), body size also plays a significant role in their interpretation. Not only has it been argued that both a relatively short calcaneum and an unfused tibia and fibula are part of the primitive morphotype of primates of modern form (including the omomyids and adapids as well as living primates)(Delson & Rosenberger, 1980), but also that the body size range of the adapids would prohibit, on mechanical grounds, the elongated calcaneum and fused tibia and fibula which are known to occur in parallel in the small bodied tarsiers and some of the lemurs and lorises (Fleagle & Simons, 1983).

It follows, then, that there is room for serious doubt in relation to the synapomorphic status of all of the features upon which the hypothesis of sister group status for the living lemurs and lorises and the anthropoid primates is based. Therefore, this hypothesis provides insufficient evidence to force the explanation of parallel evolution for the features, which otherwise underpin the alternative hypothesis that among living primates the tarsiers are the sister group of the anthropoid primates.

These analyses, therefore, support the case for a Haplorhini-Strepsirhini classification rather than the alternative Tarsiiformes-Simiolemuriformes classification.

DEFINITION OF THE HAPLORHINI: PROBLEMS AND IMPLICATIONS

If we were dealing only with extant primates, the grade Haplorhini could be defined on the basis of the characteristics listed in Table 1, and particularly on the basis of those characteristics that have been interpreted in the foregoing discussion to be most probable robust derived features shared in common by Tarsius and the anthropoid primates. However many of these features are not observable on the fossil material and the others that can be either directly observed on the fossils or inferred from the hard tissue anatomy are either not present in the omomyids or have a variable distribution among these fossils (Table 3). Because of this situation, the definition of the Haplorhini as well as the manner in which it should be applied in relation to fossil primates, is problematical. The classification of fossil primates is also complicated by the fact that the great majority of Palaeocene and Eocene primate fossils are gnathic fossils that do not preserve the areas of hard tissue morphology that are relevant to the putative derived characteristics common to Tarsius and the anthropoid primates. Among the largely Eocene Omomyidae, which have been traditionally included with Tarsius in the taxon Tarsiiformes, there are only three genera that preserve cranial remains upon which the majority of the hard tissue features given in Table 1 can be assessed (Tetonius, Early Eocene, North America; Rooneyia, Late Eocene/ Early Oligocene, North America; Necrolemur, Late Eocene, Europe).

The distribution of characteristics in Table 3 allows three possible definitions for the Haplorhini that are all consistent with the basic premise that, among living primates, Tarsius is the sister group of the anthropoid primates. Firstly, the Haplorhini could be defined on the basis of the features shared in common by all of the known omomyids for which there is available evidence, Tarsius and the anthropoid primates (features 20-25, Table 3). Secondly, it could be defined on the basis of the inferred presence of a fovea centralis, absence of a tapetum lucidum and reduction of the olfactory bulbs to include some of the omomyids, *Tarsius* and the anthropoid primates. And thirdly, it could be defined on the absence of the subtympanic recess to include only *Tarsius* and the anthropoid primates. All of these possibilities have been suggested in the literature, however in some cases they have been based on alternative assessments of the

Table 3. Occurrence of morphological features in the Eocene Omomyidae that are shared in common by *Tarsius* and the anthropoid primates. YES, NO or INFERRED in the first column = whether or not the features can be directly observed on fossils specimens. YES or NO in the second column = whether or not the features are present in the omomyid fossils. VAR = present in some omomyids but not in others, ? = problems involved with the interpretation of the features. Numbers = references discussing the features: (1) Conroy 1980, (2) Cartmill 1980, (3) Radinsky 1970, (4) Cartmill & Kay 1978, (5) Rosenberger & Szalay 1980, (6) Cartmill *et al*. 1981.

	Feature	Observed	Present	?	Ref.
1.	No choriovitteline placenta, rudimentary allantois, well developed body stalk, ovarian bursa reduced/absent, primordial amniotic cavity transitory, blastocyst attachment invasive, monodiscoidal haemochorial placents	NO			
2.	Absence of a sublingua	NO			
3.	Relatively high ratio neonatal wt./maternal wt.	NO			
4.	Biochemical similarity	NO			
5.	Karyotypic similarity	NO			
6.	Perbullar path of the internal carotid artery	YES	NO		
7.	Loss of the subtympanic recess beneath the ectotympanic	Yes	NO		
8.	Presence of an intrabullar transverse septum partioning the hypotympanic sinus	YES	NO		
9.	Presence of an alisphenoid contribution to the postorbital septum	YES	NO		
10.	Auditory tube derived from the ectotympanic rather than the petrosal	INFERRED	NO	?	1
11.	Absence of a naked rhinarium	INFERRED	NO	?	2
12.	Absence of well developed vibrissae	INFERRED	NO	?	2
13.	Absence of a tapetum lucidum	INFERRED	VAR	?	2
14.	Presence of a fovea centralis	INFERRED	VAR	?	2
15.	Reduced olfactory bulbs	INFERRED	VAR	?	3
16.	Well developed promontory artery	INFERRED	YES	?	4
17.	Prenatal loss of a functional stapedial artery	INFERRED	YES	?	4
18.	Lack of an olfactory recess	INFERRED	YES	?	5
19.	Short, deep low hafted facial skull	YES	YES	?	5
20.	Posteromedial position of the carotid foramen	YES	YES	?	6
21.	Apical interorbital septum	YES	YES		
22.	Loss of the contralateral sulcus	YES	YES		
23.	Anteromedially enlarged hypotympanic sinus	YES	YES		

relevant synapomorphic features (see particularly Cartmill & Kay, 1978, Cartmill, 1980, Rosenberger & Szalay, 1980).

Conclusive rejection of the second and third definitions of the Haplorhini, in favour of the first definition, would require that among both living and fossil primates, Tarsius was the sister group of a subset of the omomyids which were not characterised by the features used in support of these alternative definitions of the Haplorhini. The Microchoerinae are such a subset of omomyids that frequently has been suggested to be the sister group, or evolutionary ancestor, of Tarsius (Simons, 1961, 1974; Simpson, 1940). If this interpretation is correct, it would necessarily imply that the absence of a naked rhinarium, of well developed vibrissae, of a tapetum lucidum, and of a subtympanic recess together with the presence of a fovea centralis and reduction of the olfactory bulbs would have been the result of parallel evolution in Tarsius on the one hand and the anthropoid primates on the other (see references in Tables 1 & 3 for a detailed discussion of the distribution of these features in the Omomyidae). In particular relation to the presence of a naked rhinarium in the Microchoerinae, Martin (1973) has argued that a naked rhinarium is necessarily associated with separation of the upper central incisors to accommodate the backward passage of the ventral rhinarium. This bears a central groove leading to the apertures of the Jacobson's organ which lie anteriorly in the roof of the mouth. Such a gap is clearly present in extant lemurs and lorises which possess a naked rhinarium and is absent in Tarsius and the anthropoid primates which lack a naked rhinarium. The gap between the upper central incisors is also clearly present in the two microchoerine species for which we have the relevant fossil evidence (Microchoerus edwardsi Szalay & Delson, 1979, Fig. 133D; Pseudoloris parvulus Szalay & Delson, 1979, Fig. 134C).

As with the previous assessment of sister group status for Tarsius and the anthropoid primates, a similar status for Tarsius and the microchoerines is dependent on the synapomorphic nature of the morphological features shared in common by these taxa. However the assessment of the dental, cranial and postcranial features shared by these taxa (Simons, 1961, Table 1) suggests that there are reasonable grounds to believe that they are either symplesiomorphic or features which have resulted from parallel evolution.

Szalay (1976, Szalay & Delson, 1979) has drawn attention to the marked differences in the form and function of the anterior dentition in the microchoerines and Tarsius as well as to the differences in the proportions of the mandibular lower anterior teeth and to the symplesiomorphic nature of the similarities in the postcanine teeth. In addition, Cartmill et al. (1981) and Szalay (1976) note the extreme differences in petromastoid inflation in Necrolemur on the one hand and Tarsius on the other which would preclude tarsier ancestry within the microchoerines in spite of any arguments based on the similarity between the reduced subtympanic space in Necrolemur and its absence in Tarsius (see previous discussion). In relation to the postcrania, Godinot & Dagosto (1983) argue that although the lower limb morphology of Necrolemur is consistent with some kind of vertical clinging and leaping locomotion, differences in the morphology of the talus and of the humerus (Szalay & Dagosto, 1980) suggest that the exact mode of locomotion of *Necrolemur* differed from that of *Tarsius* (as well as galagos and other omomyids) and evolved independently. In this context they suggest that any similarities in the foot of *Necrolemur* and *Tarsius* are the result of parallel evolution and therefore are of no phylogenetic significance.

It follows from the above that in spite of certain phenetic similarities between *Tarsius* and the microchoerines, there is little, if any, evidence at present to support the sister group status of these primates. As a result it is not possible on this evidence alone to argue that the Haplorhini should necessarily be defined to include all of the omomyids, *Tarsius* and the anthropoid primates, over the other allowable definitions for the Haplorhini. Nor is it possible to argue on this evidence alone that the features shared in common by *Tarsius* and the anthropoid primates, which either show a variable distribution among the omomyids or are not found in these fossils, by necessity would have had to have been the result of parallel evolution in *Tarsius* and the anthropoid primates. The truth of the situation is that there is no consensus over the sister group status of *Tarsius* in relation to the omomyid Tarsiiform primates. This situation results from the difficulties involved in interpreting sister group status from the fragmentary and primarily gnathic fossil record where very few, and often non-diagnostic, features are the only ones available for analysis.

It is the fragmentary fossil record that has caused Simons (1974) to favour the grade based Prosimii-Anthropoidea classification of the primates over the phylogenetically-based Strepsirhini-Haplorhini classification. He argues that the Strepsirhini-Haplorhini classification is unworkable for the classification of fossil primates on the basis that the fossil record does not preserve the bony features that are diagnostic for each suborder. In particular, he argues that 'Our quite inadequate knowledge of many Early Tertiary primate species and even genera would leave a very large number of them in an *incertae sedis* position between the two suborders.' (Simons 1974:416).

Cartmill & Kay (1978) are in full agreement with Simons (1974) and introduce an additional problem with the Haplorhini-Strepsirhini classification in relation to fossil primates. They argue that, among living primates, the Strepsirhini are defined only by the shared presence of a tooth comb in the lemurs and lorises. The problem arises with the subordinal classification of the Adapidae (the second Eocene primate family) which possess neither a tooth comb, nor the features which would allow their inclusion under any of the possible definitions of the Haplorhini. They argue that if the Strepsirhini were defined on the basis of the shared presence of the tooth comb, the Haplorhini would, by the necessary inclusion of the Adapidae, become a 'waste basket' taxon. Alternatively, if the Haplorhini were defined by any of the three alternatives suggested in this paper, it is the Strepsirhini that would, by necessity, become a 'waste basket' taxon.

Therefore, in spite of the probable sister group status of *Tarsius* and the anthropoid primates, there are a number of serious problems in applying the Strepsirhini-Haplorhini classification to all primates, living and fossil. Firstly, the variable distribution in the omomyids of the features supporting the sister group status of *Tarsius* and the anthropoid primates (Table 3), together with the uncertain sister group relationship of *Tarsius* and any of the omomyids, allow three possible definitions of the Haplorhini that are all consistent with the basic premise that among living primates *Tarsius* is the sister group of the anthropoids. Secondly, the fragmentary nature of the primate fossil record would leave many known primate specimens in an *incertae sedis* position as to their subordinal affinities. And thirdly, any subordinal definition of the Haplorhini or the Strepsirhini would leave one of the two suborders as a 'waste basket' taxon.

These are real problems in relation to the utility of the Haplorhini-Strepsirhini classification that are not often recognized, or appreciated, by neontologists. However, the neontologists should recognize that the case for the Haplorhini is more than the case for the sister group status of *Tarsius* and the anthropoid primates. The evidence discussed in this paper gives strong palaeontological reasons to agree, at least for the moment, with Cartmill & Kay (1978) when they say that '...there is something to be said for retaining a subordinal division of Primates into Prosimii and Anthropoidea...while recognizing that some prosimians are more closely related to anthropoid primates than others.'

SUMMARY

The case for the Haplorhini rests on two basic points. Firstly, it rests on the hypothesis that the living tarsiers are more closely related to the anthropoid primates than they are to the lemurs and lorises. Secondly, it rests on the point that related taxa should be classified together and that these taxa should be monophyletic.

The foregoing discussion has supported the case for the Haplorhini in so far as the hypothesis of sister group status for *Tarsius* and the anthropoid primates cannot be rejected. This conclusion is based on two lines of evidence. Firstly, there are insufficient grounds upon which to argue that all of the morphological features shared in common by the tarsiers and anthropoid primates are parallelisms or symplesiomorphies, rather than synapomorphies. Secondly, there is room for serious doubt in relation to the synapomorphic status of *all* of the features upon which the alternative hypothesis of sister group status for the living lemurs and lorises and the anthropoid primates is based.

However, it has also been argued that the case for the Haplorhini falters on the second point, the utility of a system of classification based on the principle that related taxa should be classified together. This conclusion rests on the fact that there are serious problems in applying the phylogenetically based Strepsirhini-Haplorhini classification to all primates, living and fossil. These problems lie in three areas. Firstly, the variable distribution in the omomyids of features supporting the sister group status of *Tarsius* and the anthropoid primates, together with the uncertain sister group relation of *Tarsius* and any of the omomyids, allow three possible definitions of the Haplorhini that are all consistent with the basic premise that, among living primates, *Tarsius* is the sister group of the anthropoids. Secondly, the fragmentary nature of the primate fossil record would leave many known primate specimens in an incertae sedis position as to their subordinal affinities. And thirdly, any subordinal definition of the Haplorhini or of the Strepsirhini would leave one of the two suborders as a 'waste basket' taxon.

In view of these problems, it is argued that the best course for primatologists at the moment is to separate the issue of phylogenetic reconstruction from the issue of classification. There are sufficient grounds upon which to accept the phylogenetic hypothesis that the Tarsiiform primates have a closer evolutionary relationship with the anthropoid primates than they do with the lemurs and lorises. However, the practical problems associated with the application of the phylogenetically-based Strepsirhini-Haplorhini classification in the context of the fossil record, argue for the retention, at least for the present, of the grade-based subordinal classification of the Order Primates into Prosimii and Anthropoidea.

REFERENCES

Key References

Cartmill, M. (1980). Morphology, function, and evolution of the post-orbital septum. *In* Evolutionary Biology of the New World Monkeys and Continental Drift, eds. R.L. Ciochon & A.B. Chiarelli pp. 243-274. New York: Plenum Press.

Cartmill, M. & Kay, R.F. (1978). Craniodental morphology, tarsier affinities and primate suborders. *In* Recent Advances in Primatology, eds. D.J. Chivers & K.A. Joysey, Vol. *3* (Evolution) pp. 205-14. London-Academic Press.

Gingerich, P.D. (1981). Early Cenozoic Omomyidae and the Evolutionary History of Tarsiiform Primates. Journal of Human Evolution *10*, 345-74.

Luckett, W.P. & Szalay, F.S. (1978). Clades versus grades in primate phylogeny. *In* Recent Advances in Primatology, eds. D.J. Chivers & K.A. Joysey, Vol. *3* (Evolution) pp. 227-35. London: Academic Press.

Rosenberger, A. & Szalay, F.S. (1980). On the tarsiiform origins of the Anthropoidea. *In* Evolutionary Biology of the New World Monkeys and Continental Drift, eds. R.L. Ciochon & A.B. Chiarelli, pp. 139-57. New York: Plenum Press.

Main References

Beard, J.M., N.A. Barnicot & D. Hewett-Emmett (1976). Alpha and beta chains of the major haemoglobin and a note on the minor component of *Tarsius*. Nature *259*, 338-40.

Buffon, G.L.L., Comte de (1765). *Histoire Naturelle General et Particuliere xiii* pp. 87-91. Paris: L'imprimerie du Roi.

Cave, A.J.E. (1967). Observations on the platyrrhine nasal fossa. American Journal of Physical Anthropology *26*, 277-88.

Cave, A.J.E. (1973). The primate nasal fossa. Biological Journal of the Linnean Society *5*, 377-87.

Conroy, G.C. (1980). Ontongeny, auditory structures and primate evolution. American Journal of Physical Anthropology *52*, 443-51.

Cope, E.D. (1885). The Vertebrata of the Tertiary formations of the West. Report of the United States Geological Survey of the Territories, (F.V. Hayden, Geologist), Washington *3*, 1-1009.

Cope, E.D. (1896). The Primary Factors of Organic Evolution. Chicago: Open Court Publishing Co.

Delson, E. & Rosenberger, A.L. (1980). Phyletic perspectives on platyrrhine origins and anthropoid relationships. *In* Evolutionary Biology of the New World Monkeys and Continental Drift, eds. R.L. Ciochon & A.B. Chiarelli, pp. 445-58. New York: Plenum Press.

Dene, H., Goodman, M., Prychodoko, W. & Moore, G.W. (1976). Immunodiffusion Systematics of the Primates III. The Strepsirhini. Folia Primatologia *25*, 35-61.

Elliot Smith, G. (1907). On the Relationship of Lemurs and Apes. Nature, London *lxxvi*, 7-8.

Erxleben, C.P. (1777). Systema Regni Animalis. Leipzig.

Fleagle, J.G. & Simons, E.L. (1983). The tibio-fibular articulation in Apidium phiomense, an Oligocene anthropoid. Nature *301*, 238-9.

Gadow, H. (1898). A Classification of Vertebrata Recent and Extinct. London: Adam and Charles Black.

Gervais, F.L.P. (1854-1855). Histoire Naturelle des Mammiferes. Paris.

Gingerich, P.D. (1973). Anatomy of the temporal bone in the Oligocene anthropoid Apidium and the origin of Anthropoidea. Folia Primatologica *19*, 329-37.

Gingerich, P.D. (1975). A new genus of Adapidae (Mammalia, Primates) from the Late Eocene of southern France, and its significance for the origin of higher primates. Contrib. Mus. Paleont. Univ. Mich. *24*, 163-70.

Gingerich, P.D. (1978). Phylogeny, reconstruction and the phylogenetic position of Tarsius. *In* Recent Advances in Primatology, eds. D.J. Chivers & K.A. Joysey, Vol. *3* (Evolution) pp. 249-55. London: Academic Press.

Gingerich, P.D. (1980). Eocene Adapidae, paleobiogeography, and the origin of South American Platyrrhini. *In* Evolutionary Biology of the New World Monkeys and Continental Drift, eds. R.L. Ciochon & A.B. Chiarelli, pp. 123-38. New York: Plenum Press.

Gingerich, P.D. (1981). Early Cenozoic Omomyidae and the Evolutionary History of Tarsiiform Primates. Journal of Human Evolution *10* , 345-74.

Gingerich, P.D. & Schoeninger, M. (1977). The fossil record and primate phylogeny. Journal of Human Evolution *6*, 483-505.

Godinot, M. & Dagosto, M. (1983). The astragalus of Necrolemur (Primates, Microchoerinae). Paleontology *57*, 1321-24.

Goodman, M. (1973). The chronicle of primate phylogeny contained in proteins. Symposia of the Zoological Society of London *33*, 339-75.

Goodman, M., Moore, G.W., Prychodko, W. & Sorenson, M.W. (1974). Immunodiffusion systematics of the primates. II. Findings on Tarsius, Lorisidae, and Tupaiidae. *In* Prosimian Biology, eds. R.D. Martin, G.A. Doyle & A.C. Walker, pp. 881-90. London: Duckworth.

Gregory, W.K. (1915). I. On the relationship of the Eocene Lemur Notharctus to the Adapidae and to other primates, II. On the classification and phylogeny of the Lemuroidea. Bulletin of the Geological Society of America *26*, 419-46.

Gregory, W.K. (1920). On the structure and relations of Notharctus, an American Eocene primate. Memoirs of the American Museum of Natural History, n.s., *3*, 51-243.

Hill, W.C.O. (1955). Primates, Comparative Anatomy and Taxonomy: 2. Haplorhini: Tarsiodea. Edinburgh.

Hubrecht, A.A.W. (1908). Early ontogenetic phenomena in mammals and their bearing on our interpretation of the phylogeny of the vertebrates. Quarterly Journal of the Microscopical Society *53*, 1-181.

Hylander, W.L. (1975). Incisor size and diet in anthropoids with special reference to Cercopithecidae. Science *189*, 1095-98.

Illiger, J.K.W. (1811). Prodromus systemates mammalium et avium. Berolini.

Kampen, P.N. van (1905). Die Tympanalgegend des Säugetierschädels. Morph. Jb. *34*, 321-722.

Kay, R.F. (1977). Diets of early Miocene African hominoids. Nature *268*, 628-30.

Kay, R.F. (1981). The nut-crackers - a new theory of the adaptations of the Ramapithecinae. American Journal of Physical Anthropology *55*, 141-51.

Kay, R.F. & Cartmill, M. (1978). Cranial morphology and adaptations of Palaechthon nacimienti and other Paromomydae (Plesiadapoidea, ? Primates), with a description of a new genus and species. Journal of Human Evolution *6*, 19-53.

Kay, R.F. & Simons, E.L. (1980). The ecology of Oligocene African Anthropoidea. International Journal of Primatology *1*, 21-37.

Leutenegger, W. (1973). Maternal-fetal weight relationships in Primates. Folia Primatol. *20*, 280-93.

Link, H.F. (1795). Beytrage zur Naturgeschichte. Rostock & Leipzig 1794-1801, Bd. i, Stck 2.

Linnaeus (1767-1770). Systema Natura, 13th ed. Vienna: J.T. Trattner.

Luckett, W.P. (1980). Monophyletic or diphyletic origins of Anthropoidea and Hystricognathi: Evidence of the fetal membranes. *In* Evolutionary Biology of the New World Monkeys and Continental Drift, eds. R.L. Ciochon & A.B. Chiarelli, pp. 347-68. New York: Plenum Press.

Luckett, W.P. & Szalay, F.S. (1978). Clades versus grades in primate phylogeny. *In* Recent Advances in Primatology, Vol. *3* (Evolution), eds. D.J. Chivers & K.A. Joysey, pp. 227-35. London: Academic Press.

Martin, R.D. (1973). Comparative anatomy and primate systematics. Symp. zool. soc. London *33*, 301-37.

Martin, R.D. (1978). Major features of prosimian evolution: A discussion in the light of chromosomal evidence. *In* Recent Advances in Primatology, Vol. *3* (Evolution) eds. D.J. Chivers & K.A. Joysey, pp. 3-26. London: Academic Press.

MacPhee, R.D.E. (1977). Ontogeny of the ectotympanic-petrosal plate relationship in strepsirhine prosimians. Folia primatologia *27*, 245-83.

Petiver, J. (1702-1704). Gazophylacii Naturae et Artis. London.

Pocock, R.I. (1918). On the external characters of the Lemurs and of Tarsius. Proc. zool. Soc. London *1918*, 19-53.

Polyak, S. (1957). The Vertebrate Visual System. Chicago: University of Chicago Press.

Rosenberger, A. & Szalay, F.S. (1980). On the Tarsiiform origins of the Anthropoidea. *In* Evolutionary Biology of the New World Monkeys and Continental Drift, eds. R.C. Ciochon & A.B. Chiarelli, pp. 139-57. New York: Plenum Press.

Simpson, G.G. (1940). Studies on the earliest primates. Bulletin of the American Museum of Natural History *77*, 185-212.

Simpson, G.G. (1945). The principles of classification and a classification of mammals. Bulletin of the American Museum of Natural History *85*, 1-350.

Simons, E.L. (1961). Notes on Eocene Tarsioids and a revision of some Necrolemurinae. Bulletin of the British Museum (Natural History) (Geology) *5*, 45-69.

Simons, E.L. (1974). Notes on Early Tertiary prosimians. *In* Prosimian Biology, eds. R.D. Martin, G.A. Doyle & A.C. Walker, pp. 415-33. London: Duckworth.

Szalay, F.S. (1975). Phylogeny of higher primate taxa: The basicranial evidence. *In* Phylogeny of the Primates, eds. P. Luckett & F.S. Szalay, pp. 91-126. New York: Plenum Press.

Szalay, F.S. (1976). Systematics of the Omomyidae (Tarsiiformes, Primates): Taxonomy, phylogeny and adaptations. Bulletin of the American Museum of Natural History *156*, 157-450.

Szalay, F.S. & Dagosto, M. (1980). Locomotor adaptations as reflected on the humerus of Paleogene Primates. Folia primatologia *34*, 1-45.

Szalay, F.S. & Delson, E. (1979). Evolutionary History of the Primates. New York: Academic Press.

Wolin, L.R. & Massopust, L.C. (1970). Morphology of the primate retina. *In* The Primate Brain, eds. C.R. Noback & W. Montagna, pp. 1-28. New York: Appelton-Century Crofts.

Wortman, J.L. (1903). Studies of Eocene Mammalia in the Marsh collection, Peabody Museum, Part II. Primates. American Journal of Science, Series 4 *16*, 345-368.

4 PLATYRRHINES, CATARRHINES AND THE ANTHROPOID TRANSITION

Alfred L. Rosenberger,
Dept. of Anthropology,
University of Illinois at Chicago,
Box 4348,
Chicago, Illinois 60680, USA.

INTRODUCTION

Four basic questions are fundamental to our inquiries into the origins and early history of the anthropoid primates: (1) Are platyrrhines and catarrhines monophyletically or diphyletically related? (2) How did their geographical division into New and Old World radiations come about? (3) Who were the ancestors of the anthropoids? (4) What adaptive breakthroughs, if any, were achieved during the anthropoid transition? These large, much debated subjects will be the topic of this paper, scaled down to a size and formulation that will perhaps suggest more questions than it answers. Another equally basic matter, predicated on the others and therefore more interesting in many ways, is: What are the differences between the platyrrhine and catarrhine adaptive radiations, and how can they be explained? That this more comparative concern has received attention only recently (Delson & Rosenberger, 1984) is in large part due to the imbalance of fossil evidence for the Old and New World anthropoids. Despite its interest, it is a question that will not be considered in this paper.

Monophyly or diphyly - how are platyrrhines and catarrhines related?

The obvious distinctions between living platyrrhines, catarrhines and other primates were well known to the authors of the earliest higher level primate classifications (e.g. E. Geoffroy, 1812; Gray, 1821). More detailed work on skeletal morphology (e.g. Mivart, 1874; Flower, 1866) expanded the range of their differences, and it was this body of evidence that formed the backdrop for theories of the affinities of platyrrhines and catarrhines. Two schools of thought emerged. One suggested that anthropoids were the monophyletic descendants of a single protoanthropoid ancestor that was genealogically linked with a non-anthropoid (variously specified as adapid or omomyid). The other argued that each group arose in parallel from distinct 'lower primate' stock. Wood Jones (1929) attributed the origins of the parallelism hypothesis to St. George Mivart (1874), who was an accomplished primate anatomist, but decidedly aphylogenetic in his thinking. Mivart seemed fascinated by cases of adaptive similarity in disparate taxonomic groups, such as the long armed Ateles and Hylobates, the thumbless Ateles and Colobus and the long faced Alouatta and Papio. The origins of the monophyly theory can probably be traced to primatology's phylogenetically orientated thinkers, such as Elliot Smith (1924), although other prominent phylogeneticists of the period, like Haeckel (1899) and Wood Jones (1929), were convinced that platyrrhines and catarrhines arose independently. The complexity of the problem is evident when one realizes that Le Gros Clark, one of the masters of comparative primate morphology, opted for monophyly in his first classic synthesis of primate evolution (1934), but seemed swayed by diphyly in his heavily updated revision (1963).

Although the majority of today's workers have rejected the diphyly hypothesis, it was the overwhelming favourite of researchers until the last decade (e.g. Gregory, 1920; Schultz, 1969), and there is lingering support for it (e.g. Groves, 1972; Cachel, 1979, 1981). Since this debate shaped so much of modern primatology, it would seem fitting to attempt a brief, but by no means exhaustive, historical synopsis and critique of the major propositions of the diphyly theory (Fig. 1).

Let us examine three propositions pertaining to taxonomy, evolutionary theory and geography, in sequence:

1. Platyrrhines and catarrhines are markedly different in form, suggesting a lack of affinity and warranting their separation at higher taxonomic levels (e.g. Flower, 1866; Mivart, 1874; Keith, 1934).
2. Primate higher taxa should be ordered and interpreted as successive grades of organization, reflecting the existence of directed evolutionary trends, and suggesting that advanced grades of organization could be attained, or traversed, independently by separate lineages (e.g. Huxley, 1863; Le Gros Clark, 1963).
3. The disjunct distribution of New and Old World faunas were products of parallel evolution, anthropoid primates being just one example (e.g. Wallace, 1876; Matthew, 1915; Simpson, 1961).

As Martin (1973) explained, using primate examples, classifications were an inspirational source for evolutionary hypotheses, rather than the other way around. Additionally, early classifications were usually devised as aids to zoological identification, rather than evolutionary statements. To maintain such a typological structure and convey diagnostic messages, such classifications were exclusive in design, that is, not built upon the inclusive notions akin to the monophyly concept, or genealogy (Mayr, 1982). Consequently, when applied to primates, the differences between platyrrhines and catarrhines were exaggerated; when evolutionary questions were posed, the answers were often correspondingly awry. Thus platyrrhines and catarrhines were assumed to be genetically far removed from one another.
be genetically far removed from one another.

The gradistic perspective, which implied an inherently progressive ordering of groups not too different from the *scala naturae* paradigm which preceded it (e.g. Mayr, 1982), offered an evolutionary explanation for the set morphological dichotomies that distinguished taxa. The differences between platyrrhines and catarrhines became explicable if each had attained somewhat different levels of evolutionary rank along the trajectory exhibited by living primates (Fig. 1). The similarities of platyrrhines and catarrhines were also thus explained, since the evolutionary trends that guided primate diversification operated in parallel in all groups, enabling each to reach comparable higher grades of organization.

Fig. 1. The gradistic view of anthropoid evolution as a parallel transition. Based upon the ordering of "types" along a scale of progress (see insert) in the human direction, allegedly common evolutionary trends in groups across the order and geographical division of possible ancestral stocks (Diphyly I), later replaced by a less definite ancestral-descendant scheme (Diphyly II).

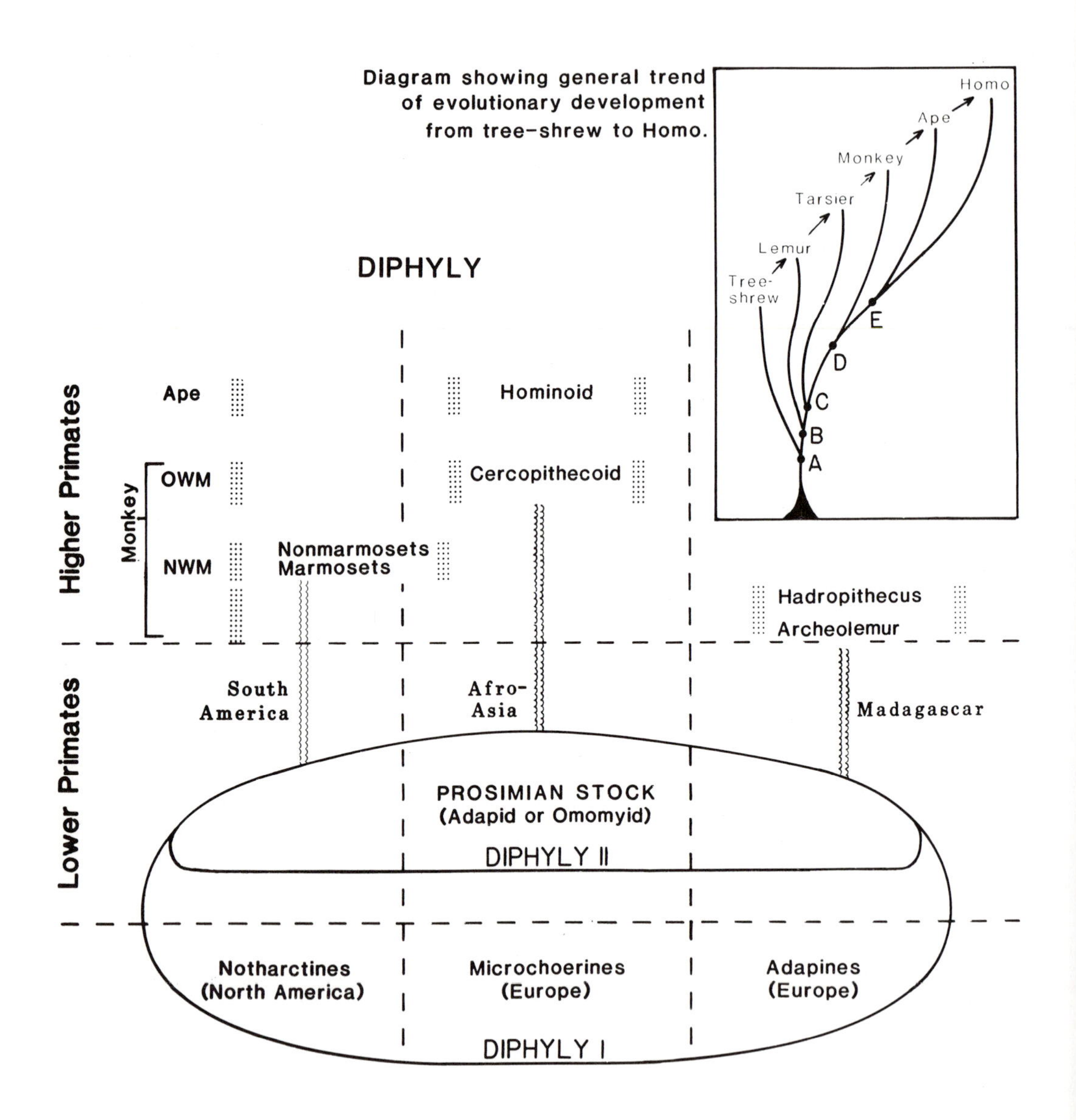

Grades, like taxa, were typologically defined. The suggestion that higher taxa did, in fact, arise in parallel was but a simple extension of the accepted principle of convergent adaptive evolution at the genus level. As mentioned, Mivart (1874) promulgated this viewpoint, and it was emphatically endorsed even during recent times (Le Gros Clark, 1963). Support for the theory came from the discovery of the extinct subfossil indrioids, such as Hadropithecus and Archaeolemur, embraced by many researchers (e.g. Wood Jones, 1929; Le Gros Clark, 1963) as a sound example of how parallelism in primate evolutionary trends can produce anthropoid (or monkey-) grade taxa out of an ancestral stock of a lower grade. The interpretation of marmosets as anatomically primitive (e.g. Beattie, 1927) also seemed to suggest that platyrrhines had bridged the higher primate grade separately from the catarrhines, which were never suspected of having such primitive traits as digital claws and minimally convoluted brains.

Given this allegedly factual basis, it was logically inferred that anthropoids had independently evolved their similarities, in situ, in the New and Old Worlds. It was frequently surmised that much of Tertiary mammalian evolution was settled during the Eocene, when many modern families first appeared in northern continents. Coupled with the parallelism principle, and armed against the notion of imaginary land bridges across the oceans at middle latitudes, zoogeographers like Wallace (1876) were convinced that Eocene primates, rodents and others independently evolved into more advanced descendants after migrating into southern continents having similar tropical environments. Rather convincing palaeontological evidence for this view was supplied by Leidy (1873), Wortman (1903-4), and Gregory (1920) among others. They began to chart out the phylogenetic history of the major primate groups as an east-west hemispheric divide, which progressed over time via southward dispersal. In many of these studies, North American notharctines were promoted as platyrrhine ancestors, European michrochoerines were cited as catarrhine and tarsiid ancestors, and European adapines were possible strepsirhine ancestors. Following Wallace's explicit rationalization, the family-level segregation of these ancestral stocks was emphasized as proof that anthropoids evolved diphyletically in parallel, although Simpson (e.g. 1961) perpetuated the mixture of phylogeny and taxonomy by defining this case of parallelism as an example of monophyly.

In retrospect, one can appreciate how the parallelism hypothesis provided an elegant explanation of platyrrhine-catarrhine similarities before Darwin's monophyly concept became firmly established, and while the approach of character weighting, cladistic analysis and phylogeny reconstruction was not broadly understood. It would also have been solid proof of the theoretical evolutionary patterns championed by the 'New Synthesis'. Like other cases in the history of science, a rethinking of the diphyly theory was perhaps slow in coming because the data were presented as a paradox, to be answered by an unusual solution.

Since that time, revisions in systematic concepts and methods, improved knowledge of the affinities of all the primates, and a vastly improved fossil record have contributed to the rejection of the hypothesis that platyrrhines and catarrhines have evolved to anthropoid status independently. The genealogical relationships of the early Tertiary primates are better appreciated, but their detail does not resolve into anything resembling the schemes that were conducive to the diphyly theory elaborated by Gregory and others (see Szalay & Delson, 1979; Rosenberger et al., 1985). Geography, which was clearly a stronger barrier to ideas than to animal migration, now weighs less in phylogeny reconstruction. As extrinsic evidence, it is not amenable to tests of homology and polarity, and, at least initially, should be ignored.

The recovery of more early Oligocene catarrhines and platyrrhines has tended to blur their anatomical distinctions. Character analyses of shared platyrrhine-catarrhine traits (e.g. Szalay & Delson, 1979; Luckett, 1980; Delson & Rosenberger, 1980; Cartmill et al., 1981; Rosenberger et al., 1985) provide direct support for their monophyletic descent (see below), despite some nagging anatomical differences which require better evolutionary explanations. The 'arctopithecine' theory of marmoset evolution, which views callitrichines as primitive, has slowly eroded. More convincing analyses have supported the idea that they are a lineage of relatively apomorphic structure and behaviour (e.g. Rosenberger, 1983). The rampant parallelisms that so impressed earlier workers (e.g. toothcombs in lemurs and lorises, suspensory locomotion in gibbons and spider monkeys; incipient or complete postorbital closure in anthropoids, tarsiers and extinct indrioids; the manual dexterity of capuchin monkeys and cercopithecids) are now generally recognized as examples of incidental convergence, and not evidence of true affinity, or as true homology (e.g. in lemuriforms).

How, then, do recent advocates find support for the diphyly interpretation? Cachel (1981) questions studies dealing directly with the issues (see Ciochon & Chiarelli, 1980) and wrongly defines the platyrrhine-catarrhine riddle as "...the question of monophyly or diphyly of the anthropoid grade" (Cachel, 1981:168). It is the applicability of the 'grade' concept, in this particular case, or in general, which is in question; since Darwin's clear statement of the phylogeny concept, propinquity of descent has been the null hypothesis explaining similarities shared jointly by species. Darwin, who was a 'phylogeneticist' (as opposed to Huxley, who was a 'gradist'), wrote, in his Descent of Man:

> Every naturalist, who believes in the principle of evolution will grant that the two main divisions of the Simiadae, namely the Catarrhine and Platyrrhine monkeys, with their subgroups, have all proceeded from one extremely ancient progenitor, before they had diverged to any considerable extent from each other... The many characters which they possess in common can hardly have been independently acquired by so many distinct species..." (1871 p. 197-8).

To refute Darwin, one would have to successfully challenge the assumption that such similarities, especially if derived, are nonhomologous. With the single exception of the postorbital septum (see below), there are no potential anthropoid synapomorphies whose homology has been seriously questioned on the basis of anatomy. Nor does functional rationalization detract from the phyletic valence of potential synapomorphies simply because we can better envision why something evolved, which seems to be the premise of some arguments (Cachel, 1979). Rather, it makes pure similarity stronger evidence of affinity, because it implies the inextricability of phylogeny and adaptation.

There is another aspect that distinguishes these alternative views of platyrrhine-catarrhine relationships. Only one is subject to robust biological tests. The monophyly hypothesis is a relatively straightforward cladistic proposition. It may be wrong, but it lays out a set of facts that have predictive value. It can be corroborated by absorbing new data, explaining additional anthropoid synapomorphies. It is not weakened by zoogeographic uncertainties surrounding mechanisms that drive groups to disjunction. Nor is debate over the potential sister-groups of the anthropoids of relevance. Parenthetically, its credibility is increased now that the notharctine-platyrrhine/michrochoerine-catarrhine scenario has been thoroughly discredited, without any alternative candidates being proposed as the twin, separate ancestors to the platyrrhines and catarrhines.

On the other hand, the diphyly hypothesis is a more complex phylogenetic and adaptational argument. It rests entirely on the differences between platyrrhines and catarrhines, relegating their similarities to trivia. As an ancestral-descendant hypothesis with no clear statement of the identity of its dual antecedents, it is a phylogenetic hypothesis without roots in the world of experience. If framed as a cladistic hypothesis in which platyrrhines and catarrhines each have their own nonanthropoid sister-taxa, it would be amenable to test, but I know of no such proposition. If framed in purely gradistic terms, i.e. anthropoids are descendants of a single non-strepsirhine, non-tarsiiform species that had not yet evolved traits such as the fused mandibular symphysis, postorbital plate or cellular petrosal bulla, it would still not be testable so long as the classic tests of homology, analogy and polarity determination are deemed unacceptable (e.g. Cachel, 1981). The only recourse, given a disbelief in the validity of classic tests, would be to locate an actual common ancestral species and examine its morphology, a virtual impossibility.

Zoogeography of early anthropoids

Hoffstetter (1980) rekindled an old debate when he proposed that the Oligocene catarrhines from the Fayum supported the hypothesis that a transatlantic migration of Old World anthropoids gave rise to platyrrhines. Such crossings over land bridges stretching between widely separated continents were favourite images among 'philosophical' zoologists of the nineteenth century. It was rejected by more modernistic diphyly and monophyly advocates (e.g. Wallace, 1876; Matthew, 1915; Elliot Smith, 1924), all of whom preferred a dispersion of anthropoids via the northern continents. Hoffstetter, on the other hand, presented his case in a direct, comprehensive fashion, based upon the premises that: (1) anthropoid monophyly implied a common origin in the southern continents where they are basically endemic; (2) Egyptian parapithecids displayed morphology consistent with a hypothetical platyrrhine ancestor; (3) Africa and South

America were closer together during the middle and late Eocene, when platyrrhines and catarrhines probably emerged and (4) the same explanation is applicable to a second contentious group that may have migrated into South America simultaneously, the caviomoph rodents, thought by Hoffstetter and colleagues (Lavocat, 1980) to be the sister-group of the African phiomyid rodents.

Some of the weaknesses of this hypothesis have been outlined (e.g. Kay, 1980; Delson & Rosenberger, 1980) but let me cite several examples. Parapithecids are an unlikely ancestral stock for the platyrrhines because, as their anatomy becomes better known, so too grows the list of autapomorphous specializations (e.g. Szalay & Delson, 1979; Kay & Simons, 1983) that mark them as a divergent collateral catarrhine branch. Postulating a different catarrhine group as a possible platyrrhine sister-taxon, such as the pliopithecids (e.g. Fleagle & Bown, 1983), is questionable on similar grounds. Their general dental anatomy, which serves to unite catarrhines as monophyletic (e.g. Kay, 1977), is more derived than the platyrrhine pattern. The latter probably lacked such catarrhine traits as strongly differentiated talonid cusps, hypoconulids on first and second molars, auxiliary wear facets on the back of the trigonid, subequal trigonid-talonid elevation and advanced reduction of upper molar metaconules. Some of these features even suggest an extra-African origin for catarrhines, which negates that crux of the transatlantic argument, the African endemism of the catarrhines. The Burmese *Pondaungia* displays them, possibly because of a common ancestry shared with catarrhines after the platyrrhine-catarrhine split (Delson & Rosenberger, 1980).

Regarding the former positions of the continents, a factor that should be treated separately from the biological data, both transatlantic and transcarribean crossings seem to stretch the human imagination. Overwater distance during the Palaeogene would probably have been less of an impediment to primate dispersal than accessibility to island stepping-stones, now that sunken and/or resutured landmasses are thought to have been scattered between South America, and both North America and Africa (e.g. Sykes *et al.*, 1982; Tarling, 1980), in interrupted chains. Finally, the cladistic links between the African and South American rodents have been seriously challenged in a recent symposium on rodent phylogeny (Luckett & Hartenberger, 1985).

The timing of the arrival of primates in South America, and the allegedly equivalent emergence of related taxa and similar morphologies in Africa, is now also being reconsidered. A revised callibration of upper Fayum beds places them earlier in time, circa 32 million years ago, (Bown & Simons, 1984). New dates for the *Branisella* zone at LaSalla, Bolivia, are about 25 Mya (MacFadden, pers. comm.), roughly ten million years younger than previously thought (Marshall *et al.*, 1977). By comparison with the anthropoid morphotype, therefore, the earliest known catarrhines were highly modified dentally perhaps ten million years before the earliest known platyrrhines, which themselves are more primitive in some ways (Rosenberger 1981a, b) but more derived in others.

No resolution to the palaeozoogeography question is likely to come without the recovery of fossils from more African and South American localities, which presently represent nothing more than two oases in an otherwise desert of palaeontological ignorance. We might profit, however, by placing the question in a broader context. It appears that continental Africa and South America interacted with Eurasia and North America, respectively, throughout the Tertiary, giving passage to different mammals at various times. The Fayum contained a circum Tethyean fauna during the early Tertiary (e.g. Cooke, 1972; Savage & Russell, 1983), sharing many elements with southern Europe, the Indo-Pakistan region and central north Asia (Fig. 2). Identical genera, families and (probably) sister-taxa are present outside Africa and as far westward as North America, ranging in time from late Palaeocene to Oligocene (Table 1). This implies that the Fayum accumulated (and probably supplied) a rather cosmopolitan mammalian fauna, with the flux of the Tethys. The Fayum primates may have had an important

Fig. 2. A reconstruction of the world's continents during the late Eocene, made by Savage & Russell (1983). Several orders, families and genera of mammals (Table 1) were distributed across the northern continents and into Africa. Protoanthropoids could have been part of this fauna but were eventually divided, as when platyrrhines became isolated in South America. Eocene and Oligocene primate localities are emphasized.

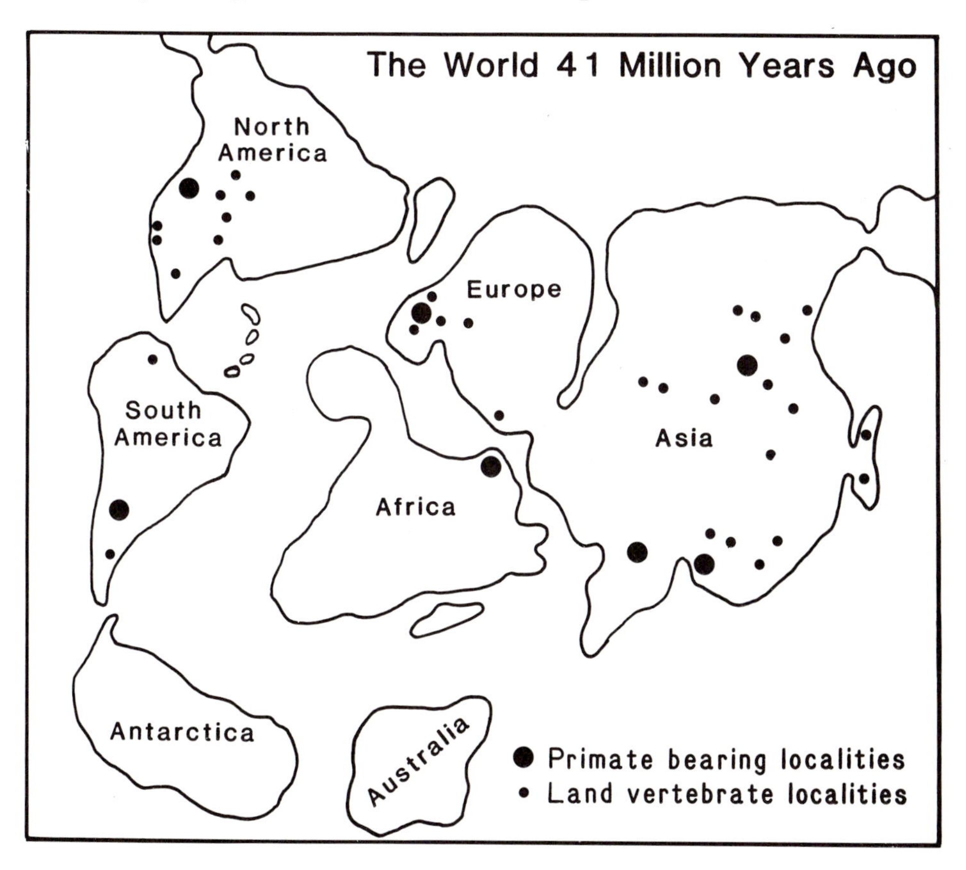

geographic linkage with Asian forms (not having close relatives in the Eocene-Oligocene Paris Basin) and may be represented by a diverse assemblage of taxa because they were affected en masse by palaeogeography. The parapithecids and eucatarrhines may have close ties with forms akin to Pondaungia, as already mentioned. Oligopithecus, which does not appear to have any catarrhine or anthropoid synapomorphies, (less even than Pondaungia possibly has) deserves more serious comparisons with Indian Indraloris (Szalay & Delson, 1979), with the Chinese Hoangonius and with Tarsius. There is also little doubt that the African ancestors of the Malagasy strepsirhines will some day turn up, as Africa probably supported them en route to Madagascar. These points tend to argue against an Africa to South America dispersal route in pleading the anthropoids as a special case, for they would be the only group suspected of bridging the Atlantic and surviving the trip, and they would be the only ones to head westward.

In the western hemisphere, a similar pattern prevailed, with important phyletic and geographic connections between North America and South America (see McKenna, 1980). The didelphid Alphadon was present in the late Cretaceous of both continents, when the water barrier between them was even greater than in the Eocene. The contemporaneous condylarth, Perutherium, resembles others from the Palaeocene and Eocene of North America, Europe and Asia. The exclusively North American soricomorph insectivores also seem to have contributed to the Neotropical realm during the Palaeogene, leaving Solenodon and Nesophontes as descendants that are now confined to the Greater Antilles (MacFadden, 1980). Thus, a smaller number of taxa are thought to have been involved in an interchange between the Americas than between Africa and Eurasia, but this may reflect a more difficult passage across the geophysically complex proto-Carribean Basin. Primates and rodents may simply represent one or two other cases of incidental dispersal.

Ancestors of the anthropoids

The three viable theories specifying the sister-group, or ancestral stock, from which anthropoids arose are respectively the adapid-anthropoid, tarsiid-anthropoid and omomyid-anthropoid hypotheses (Fig. 3). The adapid-anthropoid hypothesis is based on a variety of dental characteristics shared jointly by certain fossils and all anthropoids, and the case for it has been made most eloquently by Gingerich (e.g. 1975, 1977, 1980). Previous formulations of this position (e.g. Gregory, 1920; Le Gros Clark, 1963) were fallaciously influenced by the scala naturae doctrine, an (apparent or uncertain) acceptance of a diphyletic Anthropoidea, a misunderstanding of the affinities of Palaeogene primates, and the accidental nature of palaeontological discovery (Rosenberger et al., 1985). Some of the cranial evidence supporting the adapid-anthropoid hypothesis has been challenged recently (e.g. Rosenberger & Szalay, 1980; Delson & Rosenberger, 1980; Cartmill et al., 1981).

In the dentition, the essential phenetic resemblance linking adapids and anthropoids includes such features as a fused mandibular symphysis, spatulate incisors, canine sexual dimorphism, canine honing premolars and upper molar morphology (e.g. Gingerich, 1980). These have been reexamined critically (Rosenberger et al., 1985) and seriously challenged as a suite

Table 1. Comparison of the geographical distribution of Fayum mammals during the Eocene and Oligocene (compiled from various sources). The co-occurrence of genera in Europe, Asia, North America and Africa suggests the existence of a cosmopolitan Laurasian fauna, and significant interchange between Africa and Eurasia. The contrastingly sparse overlap between Fayum groups and South American eutherians, and their restriction to the ordinal level, suggests that Transatlantic crossings are inconsistent with the global zoogeographic pattern for nonvolant, terrestrial mammals. The presence of non-anthropoid primates in the Fayum, such as *Oligopithecus*, and the *possibility* that Eocene forms like *Pondaungia* of Burma are phyletically anthropoids - and more primitive than catarrhines - implies that catarrhines may not be endemic to Africa and that anthropoids arose on some other continent.

Fayum Mammalia	Europe	Asia	North America	South America
PROTOEUTHERIA	●	●	●	
INSECTIVORA	●	●	●	
MACROSCELIDEA				
CHIROPTERA	●		●	●
Phyllostomatidae				●
PRIMATES	●	●	●	●
Parapithecidae				
Quatrania				
Parapithecus				
Apidium				
Pliopithecidae				
Propliopithecus				
Aegyptopithecus				
Family indet.				
Oligopithecus				
RODENTIA	●	●	●	●
CREODONTA	●	●	●	
Hyaenodontidae	●	●	●	
Apterodon	●			
Pterodon	●	●		
Isohyaenodon	●	●		
PROBOSCIDEA		●		
Moeritheriidae		●		
SIRENIA				
EMBRITHOPODA		●		
HYRACOIDEA				
ARTIODACTYLA	●	●	●	
Cebochoeridae	●			
Mixotherium	●		●	
Anthracotheriidae	●	●	●	
Rhagatherium	●			
Brachyodus		●		
MARSUPIALIA				
Didelphidae	●		●	●

of potential synapomorphies. Rosenberger et al. (1985) attempted to show that the anterior dentitions of those adapids most similar to anthropoids manifest a non-homologous similarity, and are less comparable anatomically than our vague terminology allows. They interpreted the anterior dentitions of each group as reflecting divergent adaptive orientations. They claim, for example, that notharctines display a pattern laid over a bauplan that is strepsirhine and not anthropoid. The pattern exhibits a reduction in the importance of the anterior dentition in ancestral adapids, away from the primitive primate pattern where they play significant harvesting roles, towards a more lemuriform-like sniffing and grooming complex (Rosenberger & Strasser, 1985). This postulated preadaptation to a toothcombed anatomy rules out a phyletic adapid from anthropoid ancestry. Adapids are thus viewed as bona fide representatives of the autapomorphous strepsirhine clade. In contrast, anthropoids augment the plesiadapiform-like pattern (see below), where food harvesting predominates over the grooming or communicative faculties associated with the anterior dentition and snout.

The tarsiid-anthropoid hypothesis is based upon a number of cranial similarities thought to be exclusively shared by Tarsius and the anthropoids, to the exclusion of omomyids (Cartmill & Kay, 1978; Cartmill, 1980; Cartmill et al., 1981). These characters include details of the middle ear and the postorbital septum. Some suggested synapomorphies, such as the partitioning of an anterior bullar cavity and the relocation of the posterior carotid foramen, have been challenged as convergences (Rosenberger & Szalay, 1980; Packer & Sarmiento, 1984). The homologizing of an enlarged postorbital bar in Tarsius and a complete postorbital septum in anthropoids (see below) has also been disputed (Delson & Rosenberger, 1980). Added to these criticisms is the factor of phylogeny. Although the position

Fig. 3. Anthropoid monophyly, and the three current candidates for their ancestral stock. Omomyids (II) appear to be the most likely stem group.

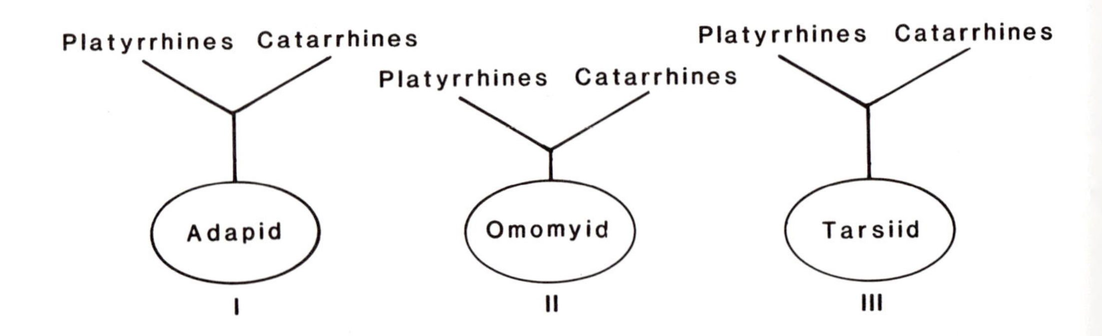

of tarsiers among the haplorhines is in dispute, advocates of the tarsiid-anthropoid hypothesis bear the burden of falsifying a series of possible synapomorphies in the skulls, dentitions and postcrania of Tarsius and the microchoerines (e.g., Simons, 1972; Gingerich, 1981; Rosenberger, in prep.) which would preclude them from sharing in a sister-group relationship with anthropoids.

The third option, the omomyid-anthropoid hypothesis, is based upon features demonstrating the monophyly of living haplorhines (Luckett & Szalay, 1978) and the presence in omomyids of apparently derived homologies shared with anthropoids (Rosenberger & Szalay, 1980) such as an apical interorbital septum, abbreviated face, enlarged brain and, possibly, fused frontal bones (Fleagle & Rosenberger, 1983). Those who object to this viewpoint, citing the presence of a fused tibiofibula and tarsal elongation (e.g. Gingerich, 1980), have been answered by the discovery of omomyid material showing neither of these derived, non-anthropoid conditions (Dagosto, 1985). This model is also supported by its ability to provide a preadaptive morphological substrate for the evolution of the anthropoid head.

The anthropoid transition

The list of shared derived features which characterize the Anthropoidea is drawn from diverse anatomical systems, ranging from the brain to the reproductive tract and the femur (e.g. Falk, 1980; Luckett, 1980; Ford, 1980). But as Cartmill (1982) pointed out, these still give us little insight into the lifestyle of early anthropoids, or the nature of the anthropoid transition. On the other hand, the cranial skeleton includes the highest concentration of anthropoid synapomorphies, which suggests that a study of the anthropoid head might shed more light on the subject. Several of these synapomorphies, such as the fused mandibular symphysis, the postorbital septum and the large, spatulate incisors have been discussed as significant contributions to a masticatory apparatus adapted to a frugivorous diet (e.g. Beecher, 1979; Cachel, 1979, Rosenberger et al. 1985). I am in essential agreement with this view, for reasons other than those given by Beecher, Cachel and others. Let me propose a model for the evolution of the anthropoid synapomorphies as adaptations to critical functions (see Rosenberger & Kinzey, 1976) for the harvesting of tough-coated fruits and, possibly, fruits with hard edible contents, such as seeds and nuts. The model is framed as a contrast of strepsirhine and anthropoid structure and function and uses forms like Lemur and Notharctus as representatives of the primitive euprimate anatomy (Fig. 4).

Anthropoid skulls are distinguished by features of the dentition, mandible, facial structure, craniofacial hafting and structure of the ossified petrosal bulla. I propose that the transition to the Anthropoidea involved the evolution of a masticatory apparatus designed to produce a powerful anterior bite employing the incisors and the anterior premolars effecting strong static stresses within the cranium. Further, the anatomical substrate for this complex was a haplorhine heritage; the particular mechanical solutions were conditioned by other architectural developments that emerged in the omomyid relatives of the anthropoids in response to different selective pressures.

Fig. 4. Comparison of (top) anthropoid (Cebus) and (bottom) euprimate (Lemur) skulls and dentitions to suggest some of the modifications involved in the anthropoid transition: (a) fused frontal bones, (b) recession of face, closure of orbit by enlargement and fusion of zygomatic bone to braincase. (c) enhanced grinding stroke of chewing cycle, (d) cancellous petrosal bones, (e) fused mandibular symphysis, (f) frontation and enlargement of incisors, blunting of premolars,

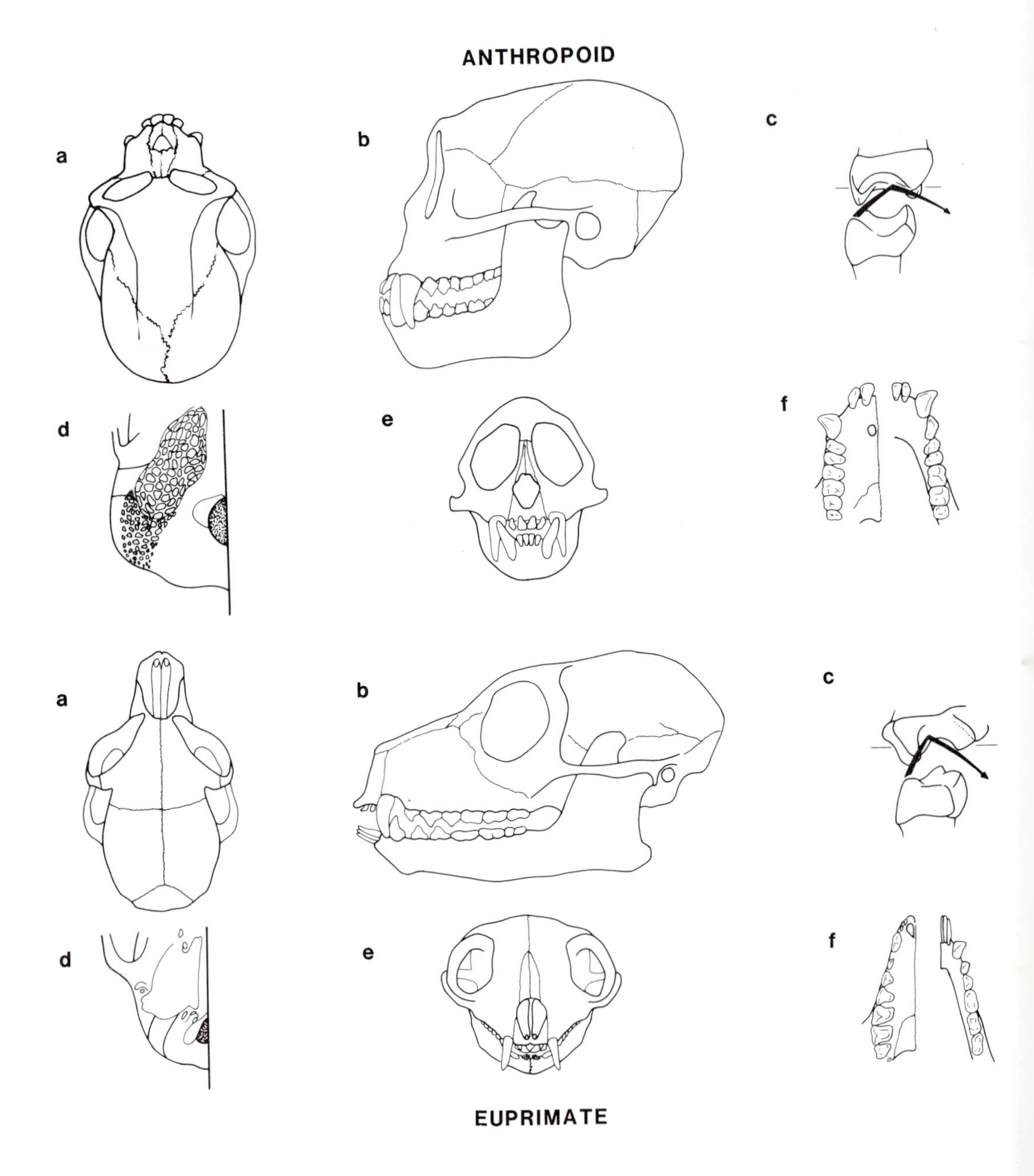

The most distinctive component of the anthropoid dentition is the morphology of the incisors. Anthropoid upper incisors are quite different from those of adapids (Rosenberger et al. 1985), with which they have been compared. They are relatively robust, high-crowned, bucco-lingually thickened teeth with strong roots, and are aligned mostly in the frontal plane (Fig. 4). They reciprocate with lowers that are solidly implanted across a fused mandibular symphysis. Anthropoid upper premolars tend to be more transversely extensive and anteroposteriorly compact than is the case among other primates. They also have subsequal protocones and paracones rather than a dominating buccal cusp, and there is a fairly large intervening occlusal basin. Molars tend to have larger occlusal basins and have crowns of lower relief than those of many omomyids, suggesting a transition at some point to a greater emphasis upon lingual phase processing during the chewing cycle (Kay & Hiiemae, 1974). Thus, in general, the molar teeth of anthropoids are designed for more crushing and grinding and less shearing, and the premolars for more crushing than puncturing during the preparatory cycle.

What is being proposed, in simple terms, is that the anthropoid head reflects a shift in design from a primitive euprimate pattern, in which the tooth-bearing facial skull is braced against the cerebral skull by an envelope of midline structures, to an architecture in which central and lateral trusses are more prominent (Fig. 5). Geometrically, this corresponds to a repositioning of the face from a precerebral to a more subcerebral location so that the face is hafted below the forebrain, rather than in front of it. The primitive euprimate condition, still exemplified by many primitive strepsirhines, has a cone-shaped face joined to the anterior cranial fossa at its base. Widely separated orbits are divided by an impressive inter-orbital plate that is continuous with the upper portion of the maxillary and nasal bones. This represents the outer, upper surface of the cone. The more important structures completing the cone laterally, inferiorly and internally are the medial walls of the orbits, the hard palate and connecting bones, e.g., the palatine and maxilla. During mastication, forces transmitted to the facial skull probably cause the face to bend and, to some extent, twist up against its moorings. Much of this load is probably distributed through the core of the cone. But, with the molar teeth and the temporomandibular joint and muscles of mastication positioned laterally, the postorbital bar will also probably be affected (see Endo, 1973; Roberts, 1979). The bar, being a T-shaped member connecting the frontal bone to the maxilla and temporal through the strut-like processes of the zygomatic, must also be loaded.

The contrasting anthropoid pattern is built around a greatly narrowed central complex, and a lower, recessed face (Figs. 4,5). The reduced nasal fossa and convergent orbits produce a relatively narrow interorbitum, eliminating the broad wedge between the eye sockets and reducing the capacity of this craniofacial junction to resist any twisting of the face upon the braincase. The medial orbital walls are more closely spaced and are less effective in bracing against lateral forces. The entire face tends to be tucked in below the frontal bone, making the toothrows more nearly perpendicular to the line of action of masseter and much of temporalis. The upper, lateral, aspect of the face is completely sutured to the sidewall of the cranium by the ossified postorbital septum. Thus, the anthropoid face is essentially hung from the neurocranium by a series of parallel pillars formed by the thin plates of the interorbitum and the postorbital septa.

Having a fully fused mandibular symphysis, anthropoids may transmit relatively large amounts of force to the mandible and, presumably, the rest of the masticatory periphery, in comparison with strepsirhines (see Hylander, 1979b for a contrast between Macaca and Galago). With enlarged, relatively vertical incisors, and premolars effecting relatively large amounts of resistance by virtue of their increased crushing-grinding surface area, the pattern of forces absorbed by the face of an anthropoid may be assumed to be different from that seen in strepsirhines. These distinctions are exaggerated because strong facial loadings occur antagonistically, and in unison. With a fused symphysis, the jaw can be powered by muscles on both sides of the head (Hylander, 1984) without dissipation of force through twisting of an open joint at the front of

Fig. 5. Schematic frontal section of hypothetical euprimate (left) and anthropoid (right) skulls at the craniofacial junction i.e., near optic foramen. The large nasal fossa acts as a central core of the face, bracing it against the neurocranium. The narrow inter-orbital septum, a consequence of olfactory reduction and orbital convergence in preadapted omomyids, is less able to resist twisting of the face about a central axis, as when masseter is active and the zygomatic is tensed against the resisting food and temporo-mandibular joint. The postorbital plate is a lateral pillar which compensates for loss of central stability.

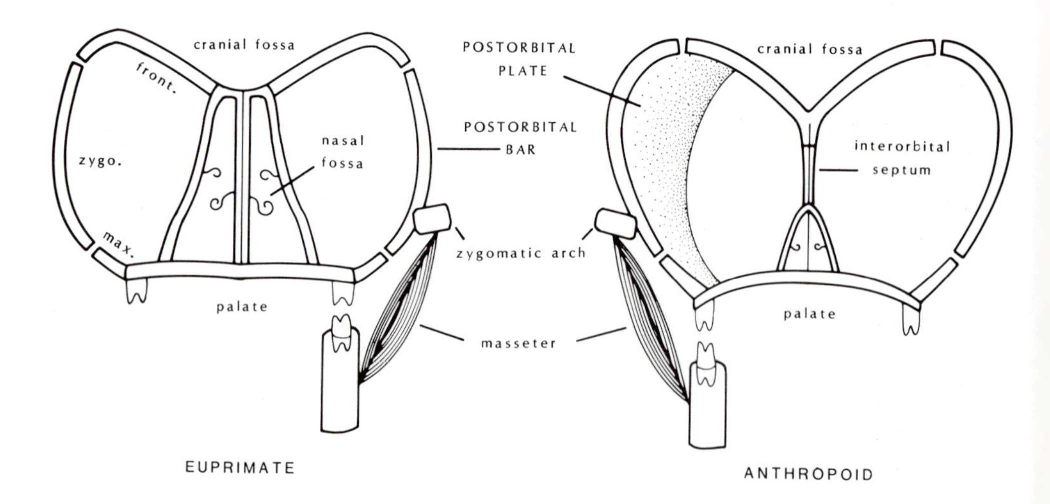

the lower jaw (e.g. Beecher, 1979) thus making parasagittal bite points more efficient. In the anthropoids, therefore, loads can be concentrated at the front of the face. With a fused symphysis, contralateral biting forces would be resisted by the parallel pillars at the craniofacial junction.

In this model the postorbital plate is viewed as a mainstay in the connection of cranial components, resisting the tendency to twist the face about the narrow central interorbital strut. Due to the fused anthropoid symphysis, contraction of the masseter (which arises along the lower border of the zygomatic arch) will produce a large tensile component in the postorbital bar, tending to separate it from the frontal at their suture. By increasing the length of the suture and, more importantly, adding a perpendicular extension that connects the postorbital bar to the sidewall of the skull, increasing the size of the zygomatic bone and giving it mechanical support, the tendency to pull or rotate the lateral pillar out of position is counteracted. The addition of a third, inferiorly placed, suture (i.e. the zygomaticomaxillary) adds mechanical integrity to the zygomatic plate. Thus, postorbital closure braces the facial skull against twisting produced by the system and reinforces the origin of the masseter muscle against enlarged forces.

The dentition is an important source of vibration. The zygomatic arch, under the bending influences of the masseter, and the articular surface of the temporomandibular joint, which is heavily loaded by the condyle (Hylander, 1979a) likewise contribute bone vibration. The transmission of such bone conducted sounds to the hearing mechanism via this heavily sutured and braced anthropoid skull must be insulated, possibly by the development of porous, spongy bone in the petrosal (cf. Fleischer, 1979).

Some comparative examples may be cited in support of the hypothesis that novel loading conditions influence a selectional response in the postorbital bar of strepsirhines, which by extension suggests that similar processes could have directed the evolution of full postorbital closure. For example, in *Loris* the orbits are extremely convergent and supporting central elements are correspondingly reduced. As compensation for the consequent reduction in static stability, the peripheral elements of the face are modified. The diameters of the lateral maxillary process, inferior and lateral aspect of the postorbital bar and zygomatic arch are all enlarged to increase their resistance against bending. In *Hadropithecus*, the fused mandibular symphysis increases the masticatory component of contralateral forces and adds to the amount of tension borne by the zygomatic arch via the masseter. The arch and lateral orbital pillar are consequently greatly strengthened. A similar condition occurs in *Adapis*, which also has a fused symphysis, although it retains the primitive elongate snout.

A number of explanations have been given for the evolution of the postorbital plate. Cartmill (1980) lists five: (1) support of the eye, (2) protection of the eye, (3) increased attachment for anterior temporalis, (4) bracing the eye and orbit against tension from masticatory muscles, and (5) insulation of the foveat eye from temporalis contractions (see Figure 6). These interpretations have been variously applied to the postorbital plate of anthropoids (2,3,5), the enlarged postorbital bar of tarsiers (1,5) and the unenlarged bar of ancestral euprimates (2,4). Perhaps the best developed arguments proposed in recent years are those given by Cachel (1979) and Cartmill (1980). Cachel explains the anthropoid condition as an adaptation to increase the surface of attachment for the anterior temporalis, thought to be especially useful in incisivation. Cartmill (1980) suggests that the explanation for the morphology in tarsiers and anthropoids is that posterior closure of the orbit is necessary to keep the fovea bearing eyeball from oscillating as temporalis contracts during chewing. The model proposed above is compatible with Cachel's hypothesis, though it emphasizes different factors. It is markedly different from Cartmill's, in part because our interpretations of tarsier affinities are mutually exclusive.

Fig. 6. Six theories for the evolution of the postorbital septum: (a) eyeball protection; (b) attachment surface for anterior temporalis; (c) eyeball support in tarsiers; (d) insulating the foveate eyeball from oscillating with temporalis activity; (e) resisting bending under muscular tension; (f) bracing the facial skull against twisting and securing the masseter against the non-rotating dentaries.

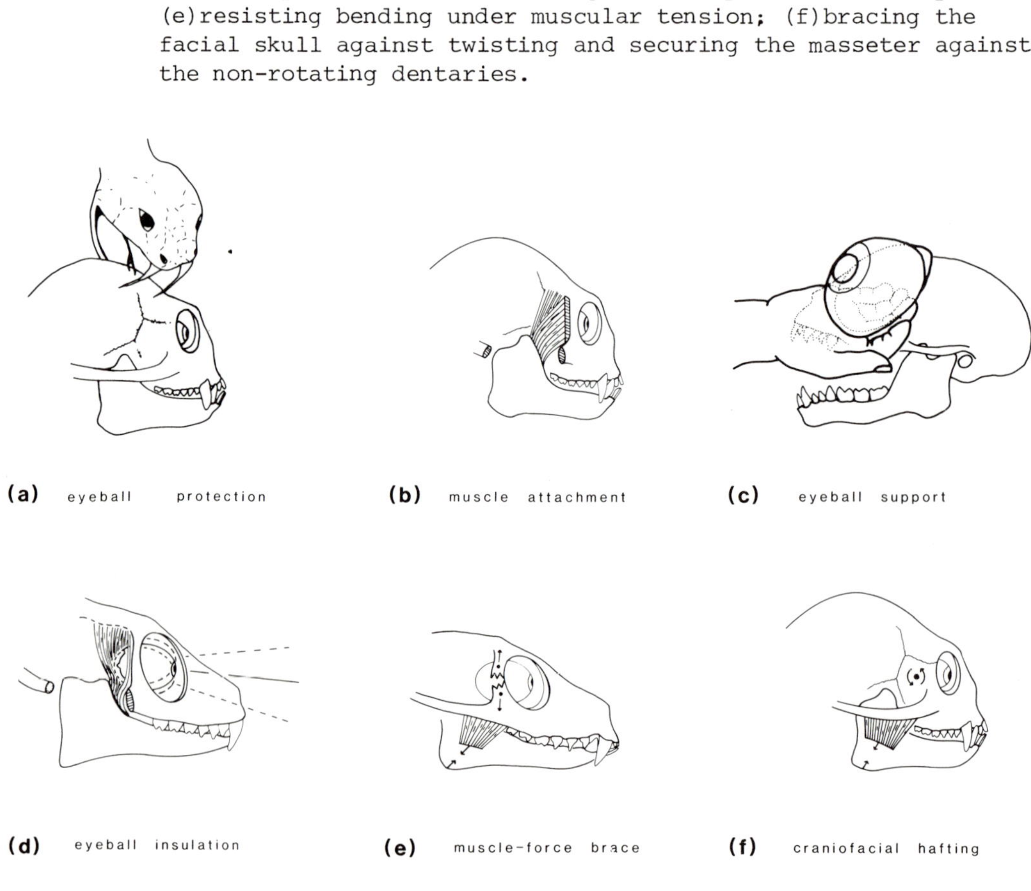

Summary and conclusions

Anthropoids are a monophyletic subgroup of the Haplorhini. The diphyly theory of anthropoid origins fails to address the contradictory implications of shared derived similarities found in platyrrhines and catarrhines. Instead, it focuses upon differences between the two groups, which have been considerably reduced by new information about the anatomy of the Oligocene African catarrhines. The diphyly theory is steeped in the gradistic tradition of primatology, which overemphasizes the possibilities of parallelism without falsifying the Darwinian null hypothesis that similarity in form and function is an indication of affinity.

The current geographical separation of platyrrhines and catarrhines into New and Old World realms postdates the emergence of the Anthropoidea. The morphological evidence indicates that catarrhines, of any sort, are too derived to be direct ancestors of the platyrrhines, and hints at the possibility that catarrhines have more primitive, extra-African relatives in the Indo-Pakistan region. Such a relationship favours Laurasia as the geographical source for protoanthropoids. Fayum primates are but one element of a changing Tethyean mammalian fauna. South America similarly absorbed various early Tertiary mammals that found their way across the tectonically-active nuclear Central America and proto-carribbean. Since the Tertiary history of mammals on both these southern continents mirrors one another in pattern, the invocation of a special circumstance, that is, a unique westward transatlantic dispersal, is not necessary to explain the disjunction of the anthropoids.

Anthropoids are probably the descendants of a haplorhine ancestral stock that would nominally be classified as omomyid. The latter were widespread in Laurasia during the Eocene and included cranial and dental morphs sufficiently primitive to be ancestral to the higher primates. Tarsiids are an unlikely sister taxon because they are highly autapomorphous and they are probably related to a different omomyid subgroup. Non-primitive similarities shared with anthropoids tend to be convergences. Adapids are probably the early members of the greater lemuriform clade, a modified group sharing no immediate ancestry with haplorhines after the differentiation of each from ancestral euprimates. A few adapids have apparently converged upon anthropoids, leading some to conclude that they are possibly anthropoid ancestors.

The anthropoid transition was adaptively predicated upon a haplorhine cranial morphology, typified by such features as an abbreviated, low face, a small nasal cavity and craniofacial hafting along a narrow interorbitum. Reinforcement of the craniofacial junction by the development of a postorbital plate enabled the anthropoid skull to absorb eccentric loads that tend to twist the face up against the neurocranium, to apply powerful biting force with the incisors and premolars, to secure the zygomatic bone against the tension of masseter, and to transfer forces across the fused mandibular symphysis to either side of the face and toothrows. The pneumatization of the petrosal bone may serve to insulate the hearing mechanism from vibrations transmitted through the more solidly fused anthropoid head. Thus, the adaptive shift of the masticatory apparatus was probably related to a critical reliance upon resistant fruits, such as legumes (with their hard coverings) and seeds, when less costly fruits were unavailable. Additional shared derived features of anthropoids help to delineate the other dimensions of their formative ecological niche and many of these are unrelated to feeding. For example, the coordination of a relatively large brain with acute vision made possible the coding of a huge amount of visual information stemming from the environment, which far exceeded the amount of 'smell' data cues that lemuriforms or primitive euprimates could extract. The sheer cognitive advantages of early anthropoids should not be ignored in models detailing their mode of origin.

By drawing together the approaches of phylogenetic reconstruction and functional analysis, rather than perpetuating the false dichotomy that has divided them in systematic endeavours, future work will add clarity to discussions on the major topics of anthropoid evolution. More pointed tests of homology, and more satisfactory interpretations of character polarity, should help generate powerful heuristic models of the adaptive transition that resulted in Anthropoidea, and the separate radiations of the New and Old World lineages. If the fossil record continues to grow as it has done recently in both hemispheres, the next decade of research on platyrrhines, catarrhines and the anthropoid transition will prove even more rewarding than the past century of excitement, discovery and controversy.

Acknowledgements

Jack Prost and Eric Delson contributed importantly to some of the ideas expressed in this paper, for which I am grateful. I acknowledge support from the National Science Foundation, BNS 810359, the Offices of Social Science Research and of Sponsored Research, UIC, for financial assistance. The illustrations were drawn by Lori Groves and Ray Brod. Stella Wrightsell, Kathleen Rizzo and Catherine Becker helped me complete the manuscript.

References

Key references

Cartmill, M. (1980). Morphology, function and evolution of the anthropoid postorbital septum. *In*: Evolutionary Biology of the New World and Continental Drift, eds., R.L. Ciochon and A.B. Chiarelli, pp. 243-74. New York, Plenum Press.

Gregory, W.K. (1920). On the structure and relations of Notharctus, an American Eocene primate. Mem. Am. Mus. Nat. Host., New series *3*, 49-243.

Hoffstetter, R. (1980). Origin and deployment of New World monkeys emphasizing the southern continents route. *In* Evolutionary Biology of the New World Monkeys and Continental Drifts, eds., R.L. Ciochon and A.B. Chiarelli, pp. 103-22. New York, Plenum Press.

Le Gros Clark, W.E. (1963). The Antecedants of Man. 2 ed. New York, Harper & Row.

Rosenberger, A.L. & Szalay, F.S. (1980). On the tarsiiform origins of Anthropoidea *In*: Evolutionary Biology of the New World Monkeys and Continental Drift., pp. 139-57. New York, Plenum Press.
Rosenberger, A.L., Strasser, E. & Delson, E. (1985). The anterior dentition of Notharctus and the adapid-anthropoid hypothesis. Folia primatol., *44*, 15-39.

Main References

Beattie, J. (1927). The Anatomy of the common marmoset (Hapale jacchus Kuhl). Proc. Zool. Soc. Lond., 593-718.
Beecher, R.M. (1979). Functional significance of the mandibular symphysis. J. Morph., *159*, 117-30.
Bown, T.M. & Simons, E.L. (1984). First record of marsupials (Metatheria: Polyprotodonta) from the Oligocene in Africa. Nature *308*, 447-9.
Cachel, S.M. (1979). A functional analysis of the primate masticatory system and the origin of anthropoid post-orbital septum. Am. J. Phys. Anthrop., *50*, 1-17.
Cachel, S. (1981). Plate tectonics and the problem of anthropoid origins. Yrbk. Phys. Anthrop., *24*, 139-72.
Cartmill, M. (1982). Basic primatology and prosimian evolution. *In* A History of American Physical Anthropology, ed. F. Spencer, pp. 147-86. New York: Academic Press.
Cartmill, M. & Kay, R.F. (1978). Craniodental morphology, tarsier affinities, and primate suborders. *In* Recent Advances in Primatology Evolution, (eds.) Chivers, D.J. & Joysey, K.A., pp. 205-14. London: Academic Press.
Cartmill, M., MacPhee, R.D.E. & Simons, E.L. (1981). Anatomy of the temporal bone in early anthropoids, with remarks on the problem of anthropoid origins. Am. J. Phys. Anthrop. *56*, 3-21.
Ciochon, R.L. & Chiarelli, B. (1980). Evolutionary Biology of New World Monkeys and Continental Drift. New York: Plenum Press.
Cooke, H.B.S. (1972). The fossil mammal fauna of Africa. *In* Evolution, Mammals, and Southern Continents, eds. A. Keast, F.C. Erk & B. Glass, pp. 89-139. Albany: State University of New York Press.
Dagosto, M. (1985). The distal tibia of primates with special reference to the omomyidae. Int. J. Primatol., *6* 45-75.
Darwin, C. (1871). The Descent of Man. London: Murray.
Delson, E.D. & Rosenberger, A.L. (1980). Phyletic perspectives on platyrrhine origins and anthropoid relationships. *In* Evolutionary Biology of New World Monkeys and Continental Drift, eds. R.L. Ciochan & Chiarelli, A.B., pp. 445-58.
Delson, E. & Rosenberger, A.L. (1984). Are there any anthropoid primate "living fossils"? *In* Living Fossils, eds. N. Eldredge & S. Stanley, pp. 50-61. New York: Springer-Verlag.
Elliot-Smith, G. (1924). The Evolution of Man, London: Oxford Univ. Press.
Endo, B. (1973). Stress analysis on the facial skeleton of gorilla by means of the wire strain gauge method. Primates, *14*, 37-45.

Falk, D. (1980). Comparative study of the endocranial casts of New and Old World monkeys. Evolutionary Biology of New World Monkeys and Continental Drift, eds. R.L. Ciochon & Chiarelli, A.B. pp. 275-92. New York: Plenum Press.

Fleagle, J.G. & Bown, T.M. (1983). New primate fossils from Late Oligocene (Colhuehaupian) localities of Chubut Province, Argentina. Folia primatologia, *41*, 240-266.

Fleagle, J.G. & Rosenberger, A.L. (1983). Cranial morphology of the earliest anthropoids. Morphologie, Evolution, Morphogenèse du Crane et Anthropogenèse, pp. 141-53. Paris: Centre National de la Recherche Scientifique.

Fleischer, G. (1978). Evolutionary principles of the mammalian middle ear. Adv. Anat. Embryol. Cell. Biol. *55*, 1-69.

Flower, W.H. (1866). An Introduction to the Osteology of the Mammalia. Amsterdam: A. Asher & Co.

Ford, S.M. (1980). Callitrichids as phyletic dwarfs, and the place of the Callitrichidae in Platyrrhini.Primates, *21*, 31-43.

Geoffroy, Saint-Hilare, E. (1812). Tableau des quadrumanes, I. Ord. Quadrumanes Ann. do Mus. d'Hist. Nat., Paris, *19*, 85-122.

Gingerich, P.D. (1975). A new genus of Adapidae (Mammalia,Primates) from the late Eocene of southern France and its significance for the origin of higher primates. Contrib. Mus. Paleontol. Univ. Michigan, *24*, 163-70.

Gingerich, P.D. (1977). Radiation of Eocene Adapidae in Europe. Geobios, Mem. sp. *1*, 165-82.

Gingerich, P.D. (1980). Eocene Adapidae, paleobiogeography, and the origin of South American Platyrrhini. *In* Evolutionary Biology of New World Monkeys and Continental Drift, eds. R.L. Ciochon & A.B. Chiarelli: pp. 123-38. New York: Plenum Press.

Gingerich, P.D. (1981). Early Cenozoic Omomyidae and the evolutionary history of tarsiiform primates. J. Hum. Evol., *10*, 345-74.

Gray, J.E. (1821). On the natural arrangement of vertebrose animals. London Med. Repost. *15*. 296-310.

Groves, C.P. (1972). Phylogeny and classification of primates. *In* Pathology of Simian Primates.Part I, ed. R. Fiennes, pp. 11-57. Basel, Karger.

Haeckel, E. (1899). The Last Link. London: Black.

Huxley, T.H. (1863). Evidence as to Man's Place in Nature. London: Williams & Norgate.

Hylander, W.L. (1979a). An experimental analysis of temporomandibibular joint reaction force in macaques. Am. J. Phys. Anthrop. *51*, 433-55.

Hylander, W.L. (1979b). Mandibular function in Galago crassicaudatus and Macaca fascicularis: an in vivo approach to stress analysis of the mandible. J. Morph., *159*, 253-96.

Hylander, W.L. (1984). Stress and strain in the mandibular symphysis of primates: A test of competing hypotheses. Am. J. Phys. Anthrop., *64* , 1-47.

Kay, R.F. (1977). The evolution of molar occlusion in the Cercopithecidae and early catarrhines. Am. J. Phys. Anthrop., *46*, 327-52.

Kay, R.F. (1980). Platyrrhine origins: a critical reappraisal of the dental evidence. *In* Evolutionary Biology of New World Monkeys and Continental Drift, eds. R.L. Ciochon & A.B. Chiarelli, pp. 159-88. New York: Plenum Press.
Kay, R.F. & Hiiemae, K.M. (1974). Jaw movement and tooth use in recent and fossil primates. Am. J. Phys. Anthrop. *40*, 227-56.
Kay, R.F. & Simons, E.L. (1983). Dental formulae and dental eruption patterns in Parapithecidae (Primates, Anthropoidea). Am. J. Phys. Anthrop. *62*, 363-76.
Keith, A. (1934). The Construction of Man's Family Tree. London: Watts & Co.
Lavocat, R. (1980). The implication of rodent paleontology and biogeography to the geographical sources and origin of platyrrhine primates. *In* Evolutionary Biology of New World Monkeys and Continental Drift, eds. R.L. Ciochon and A.B. Chiarelli, pp. 93-102. New York: Plenum Press.
Leidy, J. (1873). Contributions to the extinct vertebrate fauna of the Western Territories. Rep. U.S. Geol. Surv. Terr. (hayden) I.
Le Gros Clark, W.E. (1934). Early Forerunners of Man. Bailliere, London.
Luckett, W.P. (1980). Monophyletic or diphyletic origins of Anthropoidea and Hystricognathi: *In* Evolutionary Biology of New World Monkeys and Continental Drift, eds. R.L. Ciochon & A.B. Chiarelli, pp. 347-68. New York: Plenum.
Luckett, W.P. & Szalay, F.S. (1978) Clades versus grades in primate Evolutionary Relationships Among Rodents. New York: Plenum Press.
Luckett, W.P., Szalay, F.S. (1978). Clades versus grades in primate phylogeny. *In* Recent Advances in Primatology, Vol. 3 Evolution. eds. D.J. Chivers & K.A. Joysey, pp. 228-37, New York: Academic Press.
MacFadden, B.J. (1980). Rafting mammals or drifting islands?: biogeography of the greater antillean insectivores Nesophontes and Salendon. J. Biogeography, *7*, 11-22.
MacPhee, R.D.E., Cartmill, M. & Gingerich, P.D. (1983). Paleogene primate basicrania and the definition of the order Primates. Nature, London, *301*, 509-11.
Marshall, L.G., Pascual, R., Curtis, G.H. (1977). South American Geochronology: Radiometric Time Scale for Middle to Late Tertiary Mammal-bearing Horizons in Patagonia. Science, *195*, 1325-28.
Martin, R.D. (1973). Comparative anatomy and primate systematics. Symp. Zool. Soc. Lond. *33*, 301-37.
McKenna, M.D. (1980). Early history and biogeography of South America's extinct land mammals. *In* Evolutionary Biology of New World Monkeys and Continental Drift, eds. R.L. Ciochon & A.B. Chiarelli, pp. 43-77. New York: Plenum Press.
Matthew, W.D. (1915). Climate and evolution. Ann. New York, Acad. Sci: *24*, 171-318. Reprinted (1939) as Special Publ. New York Acad. Sci., 1.
Mayr, E. (1982). The Growth of Biological Thought. Cambridge, Mass. Belknap Press.
Mivart, St. George, (1874). Man and Apes. New York: D. Appleton & Co.

Packer, D. & Sarmiento, E.E. (1984). External and middle ear characteristics of primates, with reference to tarsier-anthropoid affinities. Am. Mus. Novitates, *2787*, 1-23.

Roberts, D. (1979). Mechanical structure and function of the craniofacial skeleton of the domestic dog. Acta anat., *103*, 422-33.

Rosenberger, A.L. (1981a). A mandible of Branisella boliviana (Platyrrhini, Primates) from the Oligocene of South America. Int. J. Primatol. *2*, 1-7.

Rosenberger, A.L. (1981b). Systematics: the higher taxa. *In* Ecology and Behaviour of Neotropical Primates, eds. A.F. Coimbra-Filho and R.A. Mittermeier, pp. 9-27. Rio de Janeiro: Academia Brasiliera de Ciencias.

Rosenberger, A.L. (1983). Aspects of the systematics and evolution of the marmosets. *In* A Primatologia No. Brazil, ed. M.T. de Mello, pp. 159-80. Belo Horizonte: UFMG.

Rosenberger, A.L. & Kinzey, W.G. (1976). Functional patterns of molar occlusion in platyrrhine primates. Am. J. Phys. Anthrop., *45*, 281-98.

Rosenberger, A.L. & Strasser, E. (1985). Toothcomb origins: Support for the grooming hypothesis. Primates.

Savage, D.E. & Russell, D.E., (1983). Mammalian Paleofaunas of the World: London: Addison-Wesley Publishing Co.

Schultz, A.H. (1969). The Life of Primates. New York: Universe Books.

Simons, E.L. (1972). Primate Evolution: An Introduction to Man's Place in Nature. New York: MacMillan.

Simpson, G.G. (1961). Principles of Animal Taxonomy. New York: Columbia Univ. Press.

Sykes, L.R., McCann, W.R. & Kafka, A.L. (1982). Motion of Caribbean plate during last 7 million years and implications for earlier Cenozoic movements. J. Geophys. Res. *87*, 10,656-76.

Szalay, F.S. & Delson, E. (1979). Evolutionary History of the Primates. New York: Academic Press.

Tarling, D.H. (1980). The geologic evolution of South America with special reference to the last 200 million years. *In* Evolutionary Biology of the New World Monkeys and Continental Drift, eds. R.L. Ciocho[n] and A.B. Chiarelli, pp. 1-41. New York, Plenum Press.

Wallace, A.R. (1876). The Geographical Distribution of Animals. 2 Vols. London: MacMillan.

Wood Jones, F. (1929). Man's Place Among the Mammals. London: Edward Arnold & Co.

Wortman, J.L. (1903-04). Studies of Eocene Mammalia in the Marsh Collectio[n] Peabody Museum. Part 2. Primates. Amer. J. Sc., Ser. 4, *15*, 163-76, 399-414, 419-36; *16*, 345-68; *17*, 23-33, 133-40, 203-14.

5 PROBLEMS OF DENTAL EVOLUTION IN THE HIGHER PRIMATES

P.M. Butler,
Department of Zoology, Royal Holloway College,
Egham, Surrey, TW20 OEX, England.

INTRODUCTION

Because they make up such a large part of the fossil evidence, teeth feature very prominently in discussions of primate evolution. They have been studied from many different aspects. Fossil teeth are used not only to distinguish species, but to estimate body size (e.g. Gingerich et al. 1982) and diet (e.g. Covert & Kay, 1981), and currently there is an increased interest in the microstructure of enamel (e.g. Gantt, 1982; Martin, 1983). Here I will discuss, from an evolutionary standpoint, the macroscopic morphology of the molar crown, its ontogenetic development and its bearing on occlusal relations. Molar evolution is an old topic. In the early part of this century it was mainly centred on problems of cusp homology, under the influence of the Tritubercular Theory (reviewed by Gregory, 1920, 1934). Today we are more interested in problems of ontogeny and function. We want to know how tooth development has become modified in the course of evolution, and what is the adaptive significance of the changes that have occurred.

The enamel-dentine junction

Looking at the crown of an unworn tooth one sees a folded surface, with peaks, ridges and valleys that together constitute the crown pattern. It is the surface of the enamel. Internally there is another folded surface, the surface of the dentine, which comes to view only where enamel is removed by wear. This dentine surface, or enamel-dentine junction, is preformed in the epithelial stage of tooth development as the basement membrane of the inner dental epithelium of the enamel organ. It represents the surface of contact between enamel organ and papilla, and its folding is the result of an interaction between epithelium and mesenchyme, such as takes place in the development of glands and hair follicles. Detailed resemblances between right and left teeth, between adjacent molars in the same jaw, and between the teeth of human identical twins, show that the folding process is under close genetic control, but the mechanism is far from clear. There is experimental evidence from the mouse that the information determining whether a tooth is to be an incisor or a molar comes from the mesenchyme (Kollar & Baird, 1969), but it seems unlikely that genes in the epithelial cells play no part in the process of pattern formation. However this may be, the folds are preserved by the deposition of dentine internally to the membrane and enamel externally to it.

The pattern develops sequentially, starting with the mesiobuccal cusp, the paracone in upper teeth, and the protoconid in lower teeth. The other cusps arise in a marginal zone around the primary cusp (Butler, 1956, 1982). Calcification commences at the tips of the cusps, first on the paracone and protoconid, and it spreads downwards, while the more basal parts of the crown, between and around the cusps, continue to grow. Thus the enamel-dentine junction does not correspond in form to the basement membrane of the dental epithelium at any one stage of development, but it represents an accretion over a period of time.

Basal growth continues to modify the crown pattern even during the spread of calcification. Folds and wrinkles develop in the basins, and the cingulum is formed at the margins. Basal growth can also change the relative positions of the cusps. On human molars the entoconid remains separate after the other cusps have become joined by calcification, and it moves away from them due to growth in the talonid basin (Butler, 1968). Similarly on dm^3, the protocone becomes more widely separated from the paracone. Growth in the basin can tilt cusps towards the margin of the crown. Thus in the human M^1 the distance between the apices of the cusps increases faster than the diameter across their base (Butler, 1967).

Very little is known about the development of the crown pattern in non-human hominoids, but it is reasonable to suppose that the same processes go on as in man. Compared with primitive forms like *Pliopithecus* and *Proconsul*, the molars of modern hominoids have more marginally-situated cusps and larger crown basins. This is true also of *Dryopithecus*, *Sivapithecus* and *Australopithecus*. It suggests that all these forms share the process of cusp-tilting found in man, but that this process was lacking in early Miocene forms (Butler & Mills, 1959).

The enamel layer

Developing human tooth germs have pointed cusps, sharp crests and rounded valleys, whereas on the enamel surface the cusps are blunt, the crests are rounded, and the valleys are represented by steep-sided grooves or fissures. The difference is due to the enamel layer, which is exceptionally thick in man. In the gorilla it is much thinner, and the difference in cusp form between man and gorilla is probably due to this. Not much is known about the shape of the dentine surface in non-human primates. Kraus & Oka (1967) illustrated M_1 of newborn gorilla and orang, with incomplete calcification, and Martin (1983) sectioned completed teeth of these species. In the orang, the dentine cusps are not as high as in man and the gorilla, and the same is true of *Sivapithecus sivalensis*. The human-like appearance of *Sivapithecus* molars is mainly due to their thick enamel, but this is formed on a different dentine base and the resemblance may be convergent. The low dentine cusps of *Sivapithecus* and the orang might be a shared derived character, not found in man, gorilla and gibbons whose high dentine cusps are probably primitive. It would be interesting to know whether *Dryopithecus* and *Gigantopithecus* also have low dentine cusps.

Korenhof (1960, 1978, 1982) studied teeth from a mediaeval cemetery in Java in which the dentine had rotted away but the enamel remained. He made casts of the internal surface of the enamel and so was able to compare the topography of the enamel-dentine junction with that of the outer enamel surface. His figures show that enamel not only modifies the form of the cusps but it affects the proportions of the crown as a whole, increasing the length more than the width. He also noted that the internal surface showed features which are visible on the outer surface in forms such as Pliopithecus, in which the enamel is thinner. The enamel-dentine junction is comparatively conservative, and differences in the appearance of hominoid molars are largely due to the enamel. In primitive forms the outer surface more closely reflects the form of the enamel-dentine junction. Sakai & Hanamura (1971, 1973), in Japan, removed the enamel from human teeth with acid and confirmed Korenhof's results.

The wrinkles and crenations that typically occur on the surface of unworn teeth are generally regarded as local thickenings of the enamel, but Korenhof (1960) found that they usually correspond to irregularities of the dentine surface; this had previously been noted by Weidenreich (1937). Minor elevations appear to be exaggerated by the formation of thick enamel, while in the intervening hollows the enamel is thin, resulting in grooves. Wrinkles take the form of ridges that branch off from the main crests and run down their slopes, especially into the basins. They vary in conspicuousness in modern man, but they were said to be highly developed in Australopithecus and in Homo erectus from Pekin (Weidenreich, 1937). They occur in Pliopithecus and the early Miocene apes, and wrinkling must be considered a primitive hominoid character. The modern great apes have closely wrinkled unworn molars, but Walkhoff (1925) found that in the orang the wrinkles were not represented in the enamel-dentine junction. It is possible that a distinction should be drawn between two sorts of wrinkling, and the matter needs further investigation. Faint traces of wrinkling can be seen in gibbons, suggesting that there it has been secondarily lost.

Probably related to wrinkling is the development of minor cusps on the edges of crests. Kraus (1963) found a number of these on the anterior marginal crest of human upper molars, and noted that some of them formed separate centres of calcification. They appear on unworn teeth as crenulations on the enamel. Subsidiary cusps also develop on the crests of the main cusps, for example one on the posterior crest of the paracone in man. Of particular interest is a cusp on the posterior edge of the anterior fovea which I called the foveal cusp (Butler, 1979). This can be seen on unworn molars of Limnopithecus and the specimen of Proconsul from Meswa Bridge (Andrews et al., 1981)(Figure 1). It is well developed in Parapithecus where it has been wrongly identified as the paraconule. The true paraconule is the cusp labelled 'cusp a' in Kay (1977).

The primitive hominoid molar

A common molar pattern, with minor variations, is shared by the early Miocene apes Dendropithecus, Limnopithecus and Proconsul. It survived into the middle Miocene with Pliopithecus, and had Oligocene forerunners in Propliopithecus and Aegyptopithecus. There can be little doubt that this molar pattern is morphotypic for the Hominoidea, and I will refer to its possessors as primitive hominoids. This usage is phenetic rather than cladistic; it will be shown later that the distinctive cercopithecoid molar pattern may have been derived from this primitive hominoid type.

It is a safe assumption that the primitive hominoid molar has evolved from one nearer to the tribosphenic pattern, such as is found in Eocene prosimians like Omomys. In Omomys the upper molars are triangular, the hypocone being a small cusp on the posterior cingulum. In addition to the three main cusps there are distinct V-shaped conules. On the lower molars the trigonid is clearly marked off from the talonid, and it has three cusps, including a paraconid. On the talonids of the first two molars the hypoconid and entoconid are well developed, but the hypoconulid is merely an indistinct elevation in the middle of the posterior crest; only on M_3 is the hypoconulid enlarged. If we compare Aegyptopithecus with Omomys (Figure 2), we note that the posterior cingulum of the upper molar has greatly expanded and it carries a well-developed hypocone. The conules are indistinct. The posterior arm of the metaconule has disappeared, and the anterior arm forms part of the oblique crest. On the lower molar the paraconid is rudimentary, the talonid is equal in height to the trigonid, the protoconid is joined to the hypoconid by a crest, the hypoconulid is enlarged on all the molars,

Fig. 1. Right upper teeth, illustrating the distinction between the foveal cusp (f) and the paraconule (p). A, Homo sapiens, dp^4; B, Proconsul sp. (Meswa Bridge), M^1; C, Oreopithecus bambolii, M^2; D, Apidium phiomense, M^1.

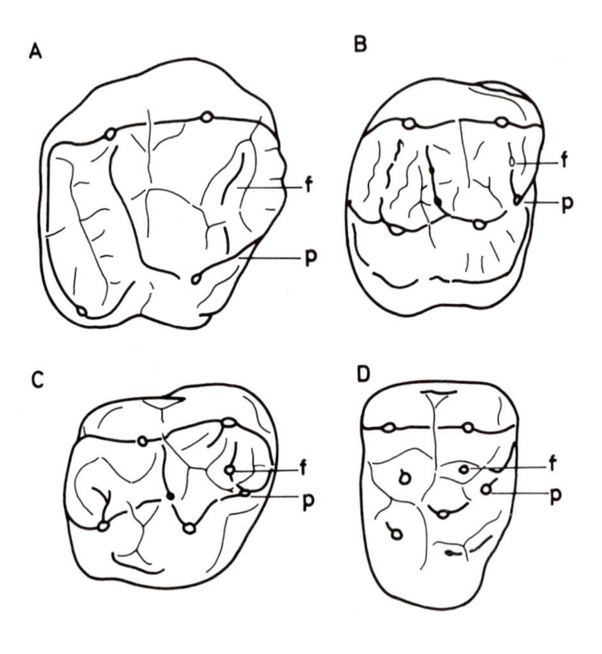

and there is a posterior fovea behind the entoconid.

A functional meaning can be given to these differences by a study of the wear facets, produced when the lower teeth are moved across the upper teeth in mastication. Wear facets of recent primates were first surveyed by Mills (1955, 1963, 1978), using museum material, and Kay & Hiiemae (1974), by making cineradiographic observations of living animals, have investigated the relationship between facets and jaw movements. The power stroke, i.e. the part of the chewing cycle when the opposing teeth are in contact, is divisible into two phases. In phase 1 (buccal phase of Mills) the lower teeth travel medially and upwards until they are maximally interlocked in centric occlusion. In phase 2 (lingual phase) they continue past the centric position and now move anteromedially and somewhat downwards. The direction of movement is shown by fine parallel scratches on the surface of the facets. Corresponding to the two phases there are two sets of facets. For example, the hypoconid in phase 1 passes between the paracone and the metacone into the trigon basin, producing facets 3 and 4 (of Kay's

Fig. 2. Right upper and left lower molars of *Omomys carteri* and *Aegyptopithecus zeuxis*. END, entoconid; fp, posterior fovea; HLD, hypoconulid; HY, hypocone; HYD, hypoconid; ME, metacone; MED, metaconid; ML, metaconule; PA, paracone; PAD, paraconid; PL, paraconule; PR, protocone; PRD, protoconid.

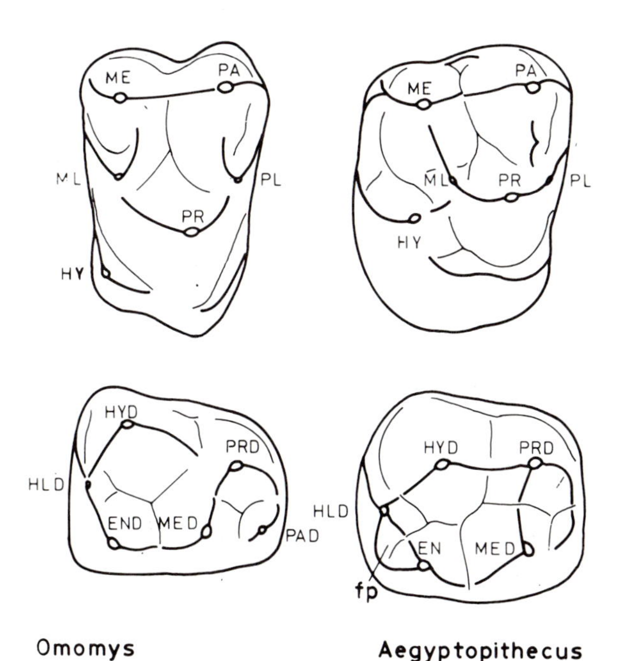

(1977) nomenclature), and then in phase 2 it travels anteromedially across the protocone, producing facet 9 (Figure 3). Thus wear facets on fossil teeth show, not only how upper and lower teeth fitted together, but how the jaw moved when the animal chewed.

The facets of Omomys and Aegyptopithecus were described and compared by Kay (1977); I described those of Omomys and other fossil prosimians

Fig. 3. Wear facets of Omomys and Aegyptopithecus, and relation of the teeth in centric occlusion. Arrows indicate the relative movement of the hypoconid across the upper molar, and that of the protocone across the lower molar. Facet numenclature based on Kay (1977).

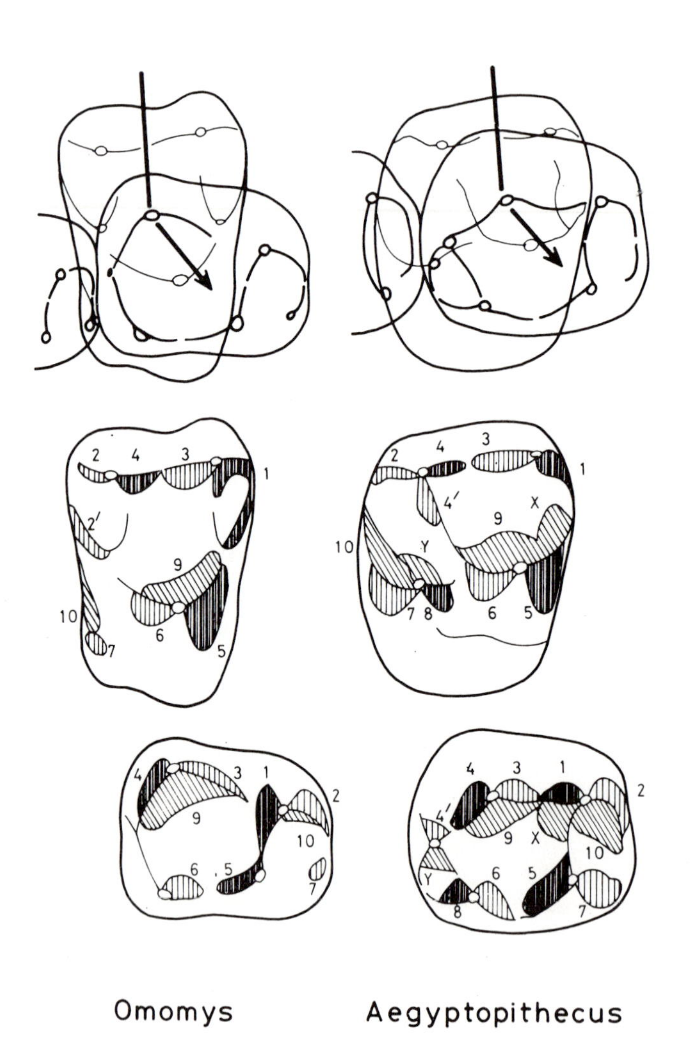

(Butler, 1973), and Maier & Schneck (1981) surveyed the facets of hominoids, including *Dendropithecus* and *Proconsul*. Most of the facets of *Omomys* can be identified in *Aegyptopithecus* and other primitive hominoids, but there are some differences. In *Omomys*, 1 to 5 are steeply inclined facets that form along cutting edges, that is the crests of the buccal cusps and the posterior trigonid crest. In *Aegyptopithecus* these crests are less sharp and the facets are less steeply inclined, implying a more horizontal jaw movement in phase 1. The posterior trigonid crest has almost lost its primitive cutting function, and it carries a new facet, produced by contact with the anterior crest of the protocone during phase 2; Kay calls this facet X. Also the posterior arm of the metaconule has been lost, and the widened posterior cingulum and enlarged hypocone are worn by the protoconid in phase 2 (facet 10). Whereas *Omomys* probably ate mainly insects, *Aegyptopithecus* seems better adapted for eating fruit.

Two further changes on the lower hominoid molar require explanation: the trigonid has shortened, with loss of the paraconid, and the posterior part of the crown has lengthened, with enlargement of the hypoconulid. The second change compensates for the first: the metacone has to pass between the hypoconid and the protoconid of the more posterior molar (Figure 4), and as the protoconid comes to stand more anteriorly the

Fig. 4. *Aegyptopithecus*, M^2 and $M_{2\ 3}$, illustrating the facets produced by the hypocone and the hypoconulid. Paths of the hypoconulid (hld), metacone (me) and hypocone (hy) are indicated.

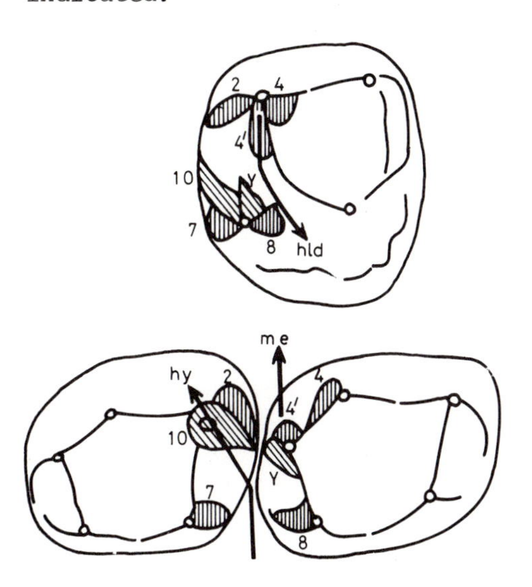

region behind the hypoconid is lengthened, preserving the distance between hypoconid and protoconid. The effect of this is to enlarge the hypoconulid, which is now placed more posteriorly in relation to the hypoconid. The metacone passes down its buccal surface, producing a facet in phase 1 that Maier & Schneck (1981) called 4'. On the lingual side, space is created behind the entoconid and the posterior fovea is formed. This provides room for the enlarged hypocone, which passes in phase 1 behind the entoconid (facet 8) and in front of the metaconid of the next tooth (facet 7). The hypocone-metaconid contact is an extension of the hypocone-paraconid contact that exists in *Omomys*; as the hypocone is enlarged, the paraconid is reduced, and the metaconid is shifted forwards so that the facet spreads onto the metaconid. Early in phase 2, as the hypocone passes out of the posterior fovea it contacts the hypoconulid, forming a facet that I will call Y (≡12 of Maier & Schneck, 1981). It will be referred to later in connection with the cercopithecoids.

Parapithecids

Of the derived characters of primitive hominoid molars, two are of particular phylogenetic interest: facet X and the hypoconulid. Facet X has been regarded by Kay (1977, 1980) as a synapomorphy that hominoids share with parapithecids and also with *Pondaungia*. It is absent in *Oligopithecus* and in most ceboids, but present in *Cebus* in which it must have developed independently, as Kay (1980) recognised. This example of parallelism reduces the value of facet X as an indicator of phyletic relationship; its occurrence in *Pondaungia* may well be another parallelism. Facet X may be regarded as a forward extension of facet 9 (hypoconid-protocone) which occurs when the trigonid is low. In such cases the chewing movement approaches the horizontal and the cutting action of the protoconid-metaconid crest has been lost. This is perhaps an adaptation to frugivory. The case for parapithecid-hominoid relationship is strengthened by a second shared-derived character, enlargement of the hypoconulid on the first two molars. In *Pondaungia*, *Oligopithecus* and the ceboids the hypoconulid is only a minor elevation of the posterior talonid crest, thus resembling omomyids. However, in parapithecids, as in hominoids, it approaches the other two talonid cusps in size, and there is an indication of a posterior fovea (Figure 5).

Parapithecid molars differ from those of primitive hominoids in having rounded cusps, with weak crests and lack of wrinkling. This suggests that the enamel is rather thick. There is also a tendency to form additional cusps, such as the mesoconid, and the foveal cusp is more distinct than in hominoids. Evidently parapithecids do not represent an ancestral stage in hominoid evolution, but they have specialised in a divergent direction.

In having three premolars parapithecids are more primitive than hominoids, which have lost P_2 as well as P_1. However, P_2 is somewhat larger than P_3 in *Apidium* and in *Parapithecus fraasi*; in the latter it has been mistaken for the canine. Enlargement of P_2 seems to be adaptive for sharpening, or honing, the posterior cutting edge of the upper canine. The honing facet on the anterolateral surface of P_2 is represented on all the cheek teeth; it is in series with facet 2 on the protoconid of the molar (Figure 5).

Honing is therefore a development of normal occlusion, and the evolution of the upper canine is reflected in the lower premolar with which it makes contact. Parapithecid premolars resemble in proportions those of primitive ceboids such as Aotus in which the upper canine is comparatively small. Within the ceboids, when the upper canine is enlarged P_2 is more differentiated from P_3, and sexual dimorphism in P_2 reflects that of the canine. Eventually we reach forms like Cebus in which the long upper canine is sharpened against an anteriorly extended P_2, in a manner that closely parallels the C-P_3 relations of cercopithecids and gibbons (Figure 6).

In the hominoid ancestor the canine must have occluded with P_2, and at some stage it transferred its contact to P_3. It may be supposed that P_2

Fig. 5. Apidium phiomense: M_2 showing hypoconulid (hld), posterior fovea (fp) and facet X; M_1-C_1 in lateral view showing facets. Oligopithecus savagei, M_2-P_3, showing "honing" facets (broken surfaces on molars cross-hatched).

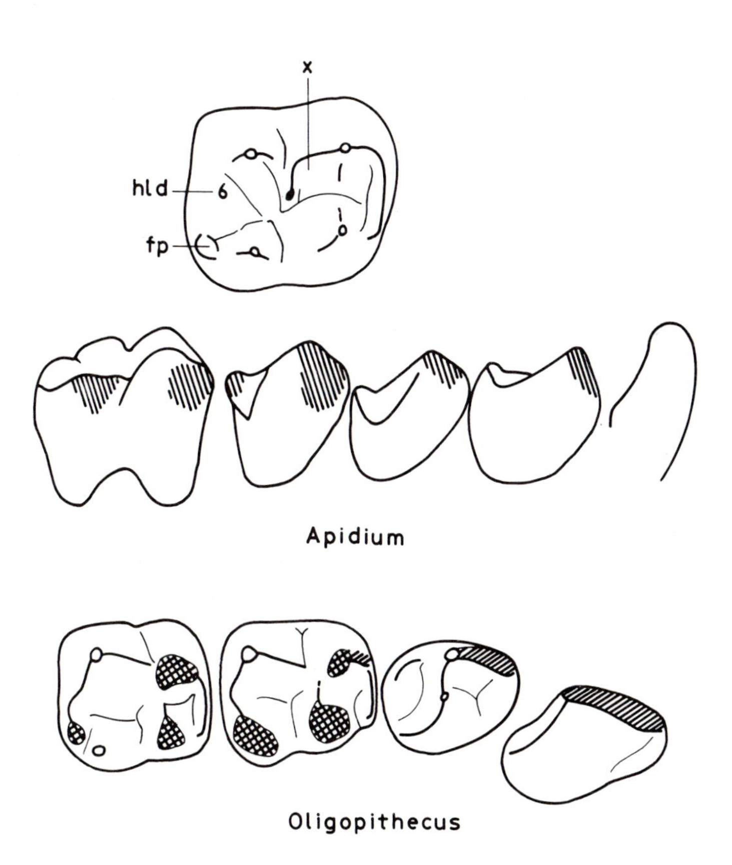

was reduced in size, and the canine facet extended beyond P_2, partly onto P_3. Then P_2 was lost, leaving the canine to occlude only with P_3. A transitional stage in which the canine occludes with two premolars (in this case P_1 and P_2) is exemplified by the adapid Leptadapis illustrated by Gingerich (1975). If this hypothesis is correct, hominoids would be derived from a form in which P_2 was smaller than P_3, which is the condition in Eocene primates generally. By enlarging P_2, parapithecids and ceboids departed from the primitive state in a direction away from that taken by the hominoids.

There is an interesting exception in Parapithecus grangeri, in which P_2 is smaller than P_3 (Kay & Simons, 1983). Has it retained a primitive character, or can the parapithecid trend be reversed? P. grangeri is specialised in having lost its lower incisors. Unfortunately P_2 is known only by its socket, and its occlusal relations cannot be determined.

Oreopithecus

Oreopithecus (Figure 7) has two premolars, and it also has facet X, an enlarged hypoconulid and a posterior fovea. It therefore appears to be a hominoid. It differs in many ways from the primitive hominoid morphotype, but this is not surprising in view of its late Miocene date. Thus the molars increase in size from M1 to M3 in both jaws. They are proportionately long and narrow, and this has an effect on the crown pattern. On the upper molars the metacone and hypocone are well separated from the paracone and protocone, the oblique crest does not run directly to the metacone but ends more anteriorly, the hypocone is enlarged, and the anterior fovea is restricted in its transverse development. On the

Fig. 6. Convergent adaptation for honing the upper canine in a ceboid (Cebus) and a cercopithecoid (Colobus).

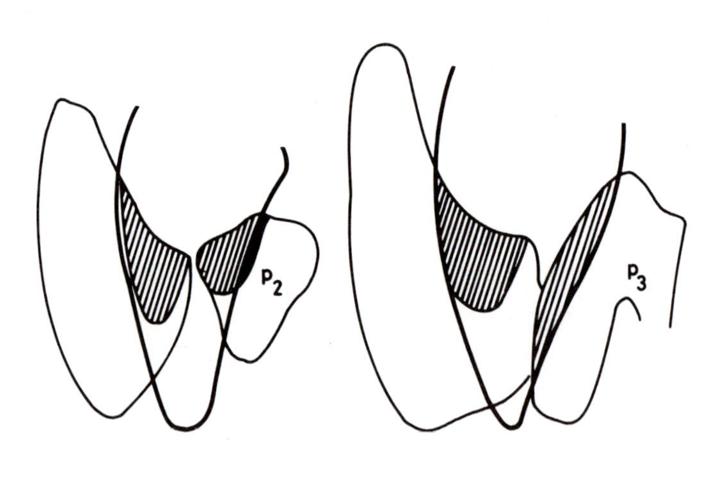

lower molars the talonid is extended, there is a central cusp ('mesoconid') which is joined to the protoconid and metaconid by crests, and the hypoconulid is placed well behind the hypoconid and entoconid.

These derived characters of Oreopithecus are foreshadowed in Rangwapithecus (Harrison, 1985), from the early Miocene of East Africa, described by Andrews (1978) as a subgenus of Proconsul. Again, the third molars are enlarged; the molars and last premolars are lengthened; the hypocone is large and the anterior fovea is restricted on the upper molars; on the lower molars there is an indication of a mesoconid (connected to the protoconid and metaconid by crests) and the posterior fovea is large. Rangwapithecus differs from Oreopithecus in having large canines and an elongated P_3, but the anterior teeth of Oreopithecus have probably been secondarily reduced. Whether or not Rangwapithecus is a direct ancestor, it helps to bridge the gap between Oreopithecus and more typical hominoids.

The cercopithecoid molar

Finally there is the problem of the origin of the cercopithecoids. They have two premolars like hominoids, but their molars are of a distinctive bilophodont pattern. They are elongated, with the cusps arranged in two pairs and with no oblique crest, buccal or lingual cingula,

Fig. 7. Oreopithecus bambolii, M_3-P_4 and M^2 (O), compared with Rangwapithecus gordoni (R).

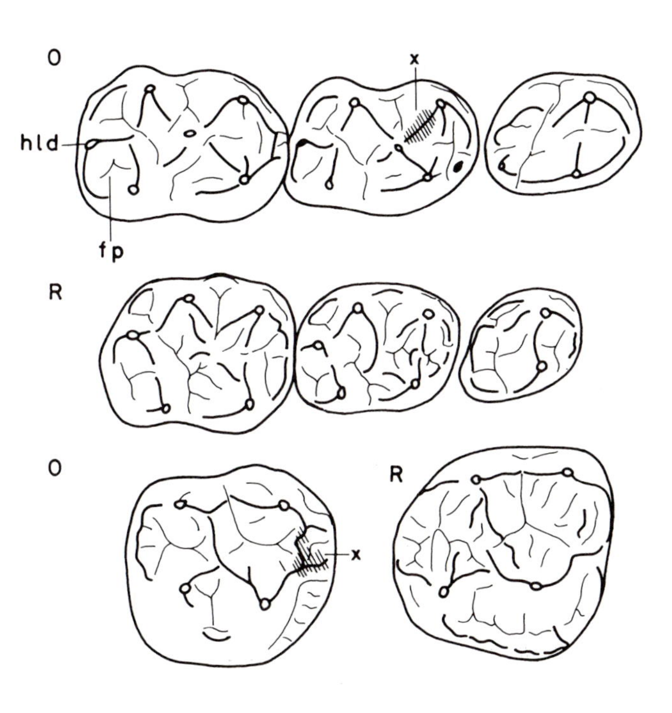

little wrinkling, and with at best a rudimentary hypoconulid on M_1 and M_2 (and in Cercopithecini even on M_3 (Swindler, 1983)). The dp^3 is more molariform than in hominoids; in colobines it frequently possesses remains of the oblique crest. Based on the variation of dP^3, Remane (1951) investigated the homologies of cercopithecoid crests with those of hominoids, and his results have been confirmed by studies of the wear facets by Mills (1955, 1963), Kay (1977) and Maier (1977).

Wear of the teeth shows that cercopithecoids and hominoids differ in the way in which the jaw moves when chewing. In hominoids the phase 2 movement is directed anteromedially, at an angle to the movement in phase 1, whereas in cercopithecoids phase 2 is a continuation of the transverse movement of phase 1. As a result, the hypoconid, instead of crossing the protocone, passes behind it, between the protocone and the hypocone. Facet 9 therefore is situated, not on the buccal surface of the protocone as in hominoids, but on its posterior surface (Figure 8). Similarly, facet 10, due to the protoconid, is on the posterior surface of the hypocone. This arrangement is found elsewhere among primates only in the lemuroid, Indri. As Mills (1963) pointed out, in cercopithecoids the mandibular condyles must move during chewing in a different way from those of hominoids. It seems that more is involved in the evolution of cercopithecoid molars than simply the joining of pairs of cusps by transverse crests.

Fig. 8. Comparison of Victoriapithecus with a hominoid (Dendropithecus) and a modern cercopithecoid (Colobus). Facets numbered as in Fig. 3; phase 2 facets shaded. Arrows indicate the path of the hypoconid across the upper molar, and that of the protocone across the lower molar.

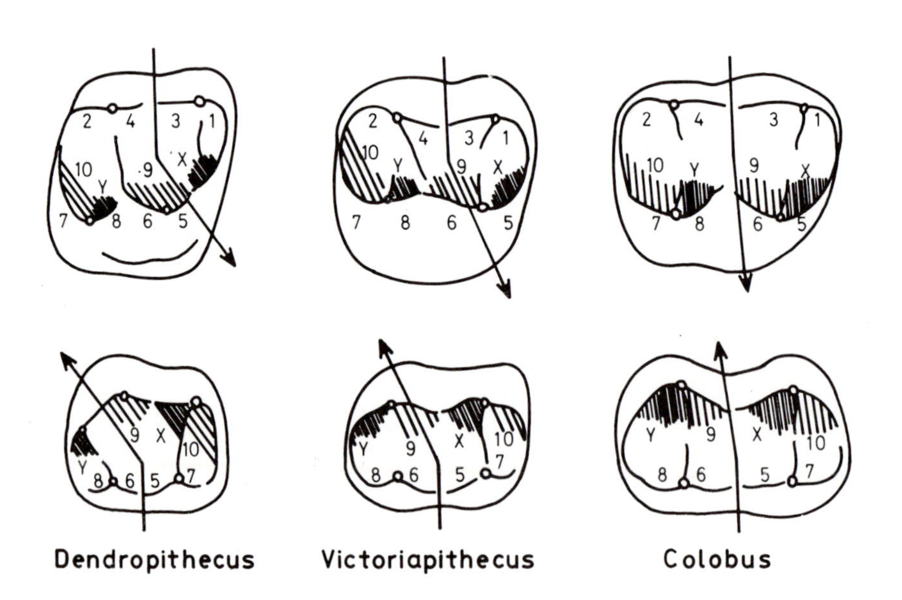

Two additional facets, on the anterior surfaces of the protocone and the hypocone, are of some interest. That on the protocone, called facet 11 by Maier (1977), is interpreted by Kay (1977) as an enlarged facet X. The anterior surface of the hypocone wears against the posterior surface of the hypoconid to produce Maier's facet 12. Maier & Schneck (1981) homologised it with a facet in hominoids due to contact of the hypocone with the hypoconulid; I have called this facet Y. To derive a cercopithecoid from a hominoid it would be necessary to absorb the hypoconulid into the posterior crest of the hypoconid.

The middle Miocene cercopithecoid Victoriapithecus is at an interesting transitional stage (Fig. 8). An oblique crest is retained in some specimens, though it is low and often incomplete. Facet 9 is tilted posteriorly, and the hypoconid seems to have passed behind the protocone, though its path was not as fully transverse as in modern forms. On the anterior crest of the protocone, in the area where the paraconule would be expected, is a facet 11, less developed than in the modern forms and thus more like facet X of hominoids. The anterior crest of the hypocone also shows a facet, smaller than the modern facet 12 and with more resemblance to facet Y of hominoids. On the lower tooth the corresponding facet is on the posterior crest of the hypoconid. In Prohylobates, known only by worn lower teeth, a hypoconulid seems to have been present in the position of this facet (Delson, 1979). Thus, the oldest known cercopithecoids provide some evidence that the bilophodont pattern was derived from one of hominoid type. As the phase 2 chewing movement became more transverse, the hypoconulid would be displaced buccally and secondarily lost, being absorbed into the posterior crest of the hypoconid. At the same time, on the upper molars the paraconule would be displaced lingually to merge with the protocone. These changes might be interpreted as an adaptive shift from a frugivorous to a leaf-eating diet, as preserved in modern colobines.

If what I have called the primitive hominoid molar pattern is ancestral also to the cercopithecoid pattern, primitive catarrhine might be a better term than primitive hominoid. However, the cercopithecoid molar has departed so widely from that of other catarrhines that phenetically a distinction between cercopithecoid and non-cercopithecoid (i.e. hominoid, in my sense) seems justified. Cladistically the situation is more complicated, as cercopithecoids are known only from the middle Miocene, and we do not know how much earlier they originated as a separate group. Were they just one of several products of the radiation that produced the variety of Miocene apes, including the oreopithecids? Or are they the sister group of all of them?

CONCLUSION

I have chosen to discuss the evolution of molar crown patterns because these seem to be phylogenetically the most informative aspect of dental anatomy at the higher taxonomic levels. The phylogenetic conclusions reached are summarised diagrammatically in Fig. 9. Evolutionary plasticity is probably greater in the anterior teeth, which are adapted mainly to the intake of food as distinct from its mastication, and inclusion of their characters would no doubt result in a more detailed picture. More information about the internal structure of the molars, particularly the enamel-dentine junction and the enamel, would help to give further information, and detailed studies of cusp development in non-human primates would also be of value.

It would appear that molar evolution in the Hominoidea is largely a matter of changes in the enamel, while the dentine is more conservative. The adaptive significance of enamel thickening is still disputed: is it to withstand abrasion, or to produce blunt cusps for breaking hard food, or to enable the animal to live longer? The function of wrinkles is also far from clear, especially as in many cases they wear off soon after the tooth erupts. The enamel-dentine junction has retained its essential features

Fig. 9. Phylogeny of the higher primates as suggested by molar crown patterns.

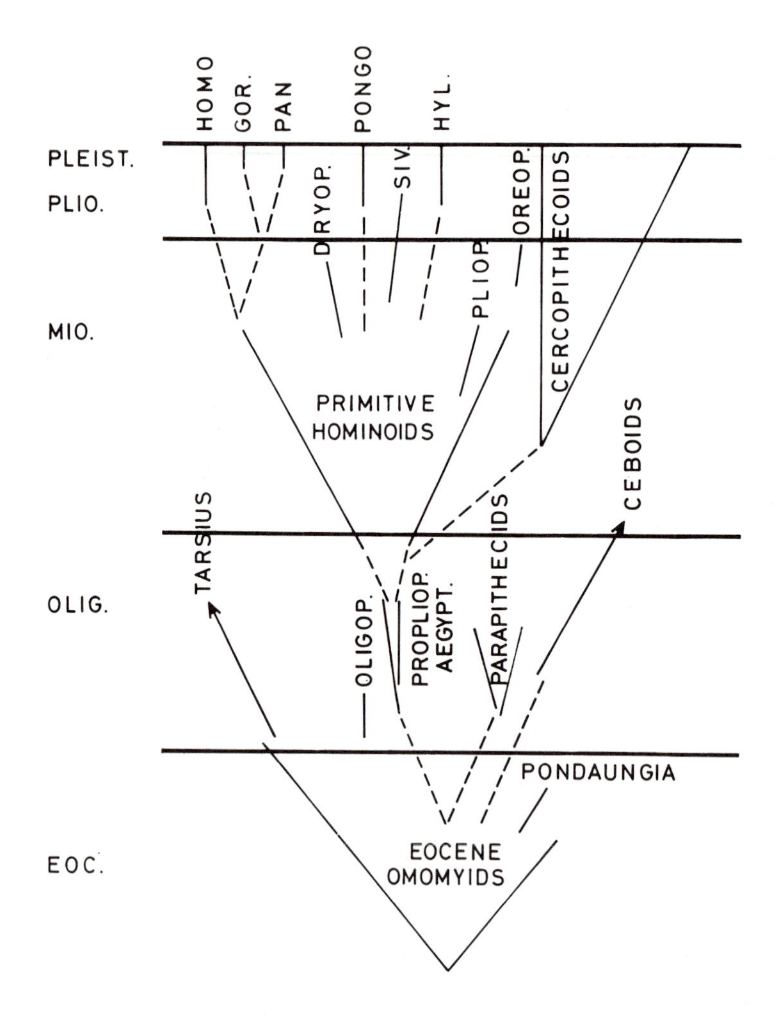

since the Oligocene, but some changes can be detected. Displacement of the cusps towards the margin, with enlargement of the basins, unites all the modern hominoids, from Dryopithecus onwards, and are features which contrast with the early Miocene forms. Does this indicate monophyly, or are the similarities due to parallelism? If the former, the separation of gibbons from great apes would be a comparatively late event. Another change seems to be a reduction in the height of the dentine cusps in the orang - Sivapithecus group, but that may be another parallelism.

The ancestral cercopithecoid probably possessed a molar pattern of hominoid type, but evidence from the teeth does not reveal the exact relationship between Cercopithecoidea and Hominoidea. Presumably Aegyptopithecus and Propliopithecus antedate the separation of the two superfamilies. The parapithecids seem to be the sister group of all other catarrhines, with which they share the derived character of an enlarged hypoconulid. They diverge particularly in premolar organisation, in which they parallel the ceboids. Further back we cannot go, but Oligopithecus may be a survivor of an earlier, perhaps prosimian stage. It has lost P_2, probably independently of later catarrhines.

Finally it is necessary to remember that any phylogeny based on a single anatomical system is very likely to contain errors. Dental evolution is notoriously subject to parallelism, convergence and reversals, and phylogenetic deductions from teeth should be checked wherever possible against evidence from skulls and skeletons.

Acknowledgments

I wish to thank the Primate Society of Great Britain for the honour of inviting me to give the Osman Hill Lecture. I am also grateful to Dr Peter Andrews and Dr Lawrence Martin for help with specimens and literature, and for stimulating discussions.

References

Key references

Butler, P.M. (1956). The ontogeny of molar pattern. Biol. Rev., *31*, 30-70.

Kay, R.F. (1977). The evolution of molar occlusion in the Cercopithecidae and early catarrhines. Am. J. phys. Anthrop., *46*, 327-52.

Korenhof, C.A.W. (1982). Evolutionary trends of the inner enamel anatomy of deciduous molars from Sangiran (Java, Indonesia). *In* Teeth: Form, Function and Evolution, ed. B. Kurtén, pp. 350-65. New York: Columbia University Press.

Maier, W. & Schneck, G. (1981). Konstruktions-morphologische Untersuchungen am Gebiss der hominoiden Primaten. Z. Morph. Anthrop., *72*, 127-69.

Mills, J.R.E. (1978). The relationship between tooth patterns and jaw movements in the Hominoidea. *In* Development, Function and Evolution of Teeth, ed. P.M. Butler & K.A. Joysey, pp. 341-54. London, New York, San Francisco: Academic Press.

Main references

Andrews, P. (1978). A revision of the Miocene Hominoidea of East Africa. Bull. Br. Mus. Nat. Hist. (Geol.), *30*, 85-224.

Andrews, P., Harrison, T., Martin, L. & Pickford, M. (1981). Hominoid primates from a new locality named Meswa Bridge, Kenya. J. Hum. Evol. *10*, 123-28.

Butler, P.M. (1956). The ontogeny of molar pattern. Biol. Rev., *31*, 30-70.

Butler, P.M. (1967). Relative growth within the human first upper permanent molar during the prenatal period. Arch. oral Biol. *12*, 983-92.

Butler, P.M. (1968). Growth of the human second lower deciduous molar. Arch. oral Biol., *13*, 671-82.

Butler, P.M. (1973). Molar wear facets of early Tertiary North American primates. Symp. 4th int. Cong. Primat., *3*, 1-27.

Butler, P.M. (1979). Some morphological observations on unerupted human deciduous molars. Ossa, *6*, 23-37.

Butler, P.M. (1982). Some problems of the ontogeny of tooth patterns. *In* Teeth: Form, Function and Evolution, ed. B. Kurtén, pp.44-51. New York: Columbia University Press.

Butler, P.M. & Mills, J.R.E. (1959). A contribution to the odontology of Oreopithecus. Bull. Br. Mus. Nat. Hist. (Geol.), *4*, 1-26.

Covert, H.H. & Kay, R.F. (1981). Dental microwear and diet: implications for determining the feeding behaviors of extinct primates, with a comment on the dietary pattern of Sivapithecus. Am. J. Phys. Anthrop., *55*, 331-36.

Delson, E. (1979). Prohylobates (Primates) from the early Miocene of Libya: a new species and its implications for cercopithecoid origins. Geobios, *12*, 725-733.

Gantt, D.G. (1982). Neogene hominoid evolution: a tooth's inside view. *In* Teeth: Form, Function and Evolution, ed. B. Kurtén, pp. 93-108. New York: Columbia University Press.

Gingerich, P.D. (1975). A new genus of Adapidae (Mammalia, Primates) from the late Eocene of southern France, and its significance for the origin of higher primates. Contr. Mus. Paleont. Univ. Mich., *24*, 163-70.

Gingerich, P.D., Holly, B.H. & Rosenberg, K. (1982). Allometric scaling in the dentition of primates and prediction of body weight from tooth size in fossils. Am. J. Phys. Anthrop., *58*, 81-100.

Gregory, W.K. (1920). The origin and evolution of the human dentition. J. Dent. Res., *2*, 89-233, 357-426, 607-717.

Gregory, W.K. (1934). A half-century of Trituberculy. The Cope-Osborn theory of dental evolution, with a revised summary of molar evolution from fish to man. Proc. Am. Phil. Soc., *73*, 169-317.

Harrison, T. (1985). African oreopithecids and the origin of the family. Am. J. Phys. Anthrop., *66*, 180.

Kay, R.F. (1977). The evolution of molar occlusion in the Cercopithecidae and early catarrhines. Am. J. Phys. Anthrop., *46*, 327-52.

Kay, R.F. (1980). Platyrrhine origins. A reappraisal of the dental evidence. *In* Evolutionary Biology of the New World Monkeys and Continental Drift., ed. R.L. Ciochan & A.B. Chiarelli, pp. 159-87. New York: Plenum Press.

Kay, R.F. & Hiiemae, K.M. (1974). Jaw movement and tooth use in recent and fossil primates. Am. J. phys. Anthrop., *40*, 227-56.

Kay, R.F. & Simons, E.L. (1983). Dental formulae and dental eruption patterns in Parapithecidae (Primates, Anthropoidea). Am. J. Phys. Anthrop., *62*, 363-75.

Kollar, E.J. & Baird, G.E. (1969). The influence of the dental papilla on the development of tooth shape in embryonic mouse tooth germs. J. Embryol. exp. Morph., *21*, 131-48.

Korenhof, C.A.W. (1960). Morphogenetical aspects of the human upper molar. Utrecht: Uitg. Neetlandia.

Korenhof, C.A.W. (1978). Remnants of the trigonid crests in medieval molars in man of Java. *In* Development, Function and Evolution of Teeth, ed. P.M. Butler & K.A. Joysey, pp. 157-70. London, New York, San Francisco: Academic Press.

Korenhof, C.A.W. (1982). Evolutionary trends of the inner enamel anatomy of deciduous molars from Sangiran (Java, Indonesia). *In* Teeth: Form, Function and Evolution, ed. B. Kurtén, pp. 350-65. New York: Columbia University Press.

Kraus, B.S. (1963). Morphogenesis of deciduous molar pattern in man. *In* Dental Anthropology, ed. D.R. Brothwell, pp. 87-104. Oxford, London, New York, Paris: Pergamon Press.

Kraus, B.S. & Oka, S.W. (1967). Wrinkling of molar crowns: new evidence. Science, *157*, 328-29.

Maier, W. (1977). Die Evolution der bilophodonten Molaren der Cercopithecoidea. Z. Morph. Anthrop., *68*, 26-56.

Maier, W. & Schneck, G. (1981). Konstruktions-morphologische Untersuchungen am Gebiss der hominoiden Primaten. Z. Morph. Anthrop. *72*, 127-69.

Martin, L.B. (1983). The relationships of the later Miocene Hominoidea. Ph.D. thesis, London University.

Mills, J.R.E. (1955). Ideal dental occlusion in the primates. Dent. Practit., *6*, 47-61.

Mills, J.R.E. (1963). Occlusion and malocclusion of the teeth of primates. *In* Dental Anthropology, ed. D.R. Brothwell, pp.29-51. Oxford, London, New York, Paris: Pergamon Press.

Mills, J.R.E. (1978). The relationship between tooth patterns and jaw movements in the Hominoidea. *In* Development, Function and Evolution of Teeth, ed. P.M. Butler & K.A. Joysey, pp. 341-54. London, New York, San Francisco: Academic Press.

Remane, A. (1951). Die Entstehung der Bilophodontie bei den Cercopithecidae. Anat. Anz., *98*, 161-65.

Sakai, T. & Hanamura, H. (1971). A morphological study of enamel-dentin border in the Japanese dentition. V. Maxillary molar. J. Anthrop. Soc. Nippon, *79*, 297-322.

Sakai, T. & Hanamura, H. (1973). A morphological study of enamel-dentin border in the Japanese dentition. VI. Mandibular molar. J. Anthrop. Soc. Nippon, *81*, 35-45.

Swindler, D.S. (1983). Variation and homology of the primate hypoconulid. Folia primatol., *41*, 112-123.
Walkhoff, O. (1925). Über das vermeintliche Leben im Zahnschmelz durchgebrochener Zähne. Z. Stomat., *23*, 93-110.
Weidenreich, F. (1937). The dentition of Sinanthropus pekinensis. A comparative odontography of the hominids. Palaeont. Sinica, *101*, 1-180.

6 MOLECULAR EVIDENCE FOR CATARRHINE EVOLUTION

Peter Andrews,
British Museum (Natural History)
Cromwell Road,
London SW7 5BD.

INTRODUCTION

The catarrhine primates include the living Old World monkeys and apes and a number of attributed fossils. Together with the New World Monkeys they comprise the Anthropoidea, and with the inclusion of the Tarsioidea, the Haplorhini. Many lines of evidence indicate that within these groups the Catarrhini is a monophyletic group, that is a group that is descended from a single common ancestor. This is inferred from the presence of many shared characters in the living catarrhines; characters that are shared as a result of descent from the common ancestor and which by further inference must have been present in the common ancestor. Recently I have listed the gross morphological characteristics of the catarrhine ancestral state (Andrews, 1985). These include both the primitive retentions from the anthropoid ancestor and the derived characters which are unique to the catarrhine clade and which therefore define it. In this paper I am going to consider characters at the protein or molecular (gene) level. Biochemical evidence adds a new dimension to the definition of groups of living organisms, and combined with morphology may lead to a more accurate understanding of their relationships and evolution (Andrews & Cronin, 1982).

Biochemical or molecular changes are reviewed here at several different levels. The information content varies depending on the methods used, and I will be grouping them by their results rather than by their methods. Measuring genetic distance by immunodiffusion and electrophoresis provides indices of change for whole proteins without necessarily knowing exactly what is changing. Similar investigations of DNA structure by hybridization techniques detect changes in the DNA molecule without knowing exactly which nucleotides have changed. All of these methods produce distance statistics as evidence of similarity or dissimilarity between taxa. At a second level, protein alteration is specified by amino acid sequencing, so that changes in protein can be defined in terms of known changes in amino acids at specific loci. In some cases, the DNA coding for the amino acids may also be known, and this is the third level of analysis, DNA sequencing. An approach to this is made by restriction enzyme mapping, which breaks the DNA chains into identifiable segments, but the most interesting results so far are from the actual sequencing data of DNA chains.

It is important to note that these data, both biochemical and morphological, provide the major source of information on catarrhine evolution. Fossils provide additional data, but they must always be secondary because they can only be interpreted in the context of extant biological groups. Fossil apes can be recognized as such only when we have understood and defined what an ape is, and this definition is based on the living species, not the fossils. Huxley (1863) recognized the kinship of the African apes and man without the aid of fossils, and fossils have contributed little to the identification of any anthropoid group. The need for caution is well illustrated by the history of study of Ramapithecus; its recognition as a hominid was long considered as critical in the emergence of the hominid lineage, but in fact it added so little to our understanding of human evolution that its removal from human ancestry has not altered our ideas on human evolution in any material sense. In making this point about fossils I am not trying to say that the information they provide is irrelevant. What I am trying to do is to recognize the limitations of fossil evidence so that we do not waste time asking questions of fossils that they cannot possibly answer. They can answer questions of time and place in evolution, and they can give indications of palaeoecology and the sequence of change in mosaic evolution, but they do not contribute significantly to questions of relationships and phylogeny. For this we must turn to modern comparative data, such as is reviewed here.

DISTANCE MEASURES

Immunology

Immunological techniques assess genetic divergence between taxa by measuring the degrees of antigenic distance separating them (Baba, et al., 1980). Procedures are described in Goodman & Moore (1971). The distance measure developed by Sarich & Wilson (1967) was based on an Index of Dissimilarity (ID) to distinguish between albumins. The Immunological Distance was calculated as 100 x logID, and they found that distances between pairs of species and along lineages were regular enough to approximate to a molecular clock for assessing the relative divergence times between lineages. For instance, the distance between man and a more distantly related non-catarrhine primate is similar to the distance between an ape or an Old World monkey and the same primate species. There are problems with this conclusion, as will be seen later for this and other aspects of Sarich & Wilson's work, but as a general principle, I believe it has stood the test of time and is a valid conclusion.

The results of immunological techniques, both with antisera directed against whole sera and with microcomplement fixation (Goodman, 1976; Sarich & Cronin, 1976) have been widely thought to have provided new insights into the relationships within the Catarrhini. They showed that the Catarrhini is a monophyletic group composed of two monophyletic superfamilies, the Hominoidea and the Cercopithecoidea. By this is meant that the taxa that compose these groups are all more similar to each other than they are to those belonging to other groups, but the translation of this straightforward similarity assertion into phylogenetic interpretation is far from straightforward. Within the Hominoidea the immunological data confirmed the similarity of the Hylobatidae with the rest of the Hominoidea, a point disputed by some of the chromosome evidence, and suggested the division of the remaining hominoids into two groups, one consisting of the orang utan and the other of the African apes and man (Goodman, 1963, 1976).

These should be grouped in a single family so as to be consistent with the Hylobatidae, and they should then be further sub-divided into two sub-families, the Ponginae for the orang utan and the Homininae for the African apes and man. The Cercopithecoidea were also split into two groups by the immunological data, the colobines and cercopithecines, but the former appears paraphyletic by all forms of analysis and there are a number of inconsistencies in the latter. Cercocebus is associated with Cercopithecus by Dene et al. (1976) and Goodman & Moore (1971); it is associated with Geladas by Goodman (1976); and it is shown to be paraphyletic by Sarich & Cronin (1976). It is not known which of these alternatives is correct.

The validity of the molecular clock for the albumins is based on the assumptions that minimizing immunological distances is an adequate basis for inferring phylogeny and that the constancy of rates of change along different phyletic lines had been established. On this basis, Sarich & Wilson suggested a divergence time of 5 million years (my) for the African apes and man. The relative timing of the clock always requires calibration from an independent source, and Sarich & Wilson used the monkey/ape divergence time of 30my from the fossil record. This argument could also be turned around, so that if fossil evidence could be provided to confirm the 5my date for the African apes and man it would also confirm the monkey/ape divergence time. More recently, Sarich & Cronin (1976) have reconfirmed their even younger date of 3.5 to 4my for the African ape/man divergence.

There are a number of problems with the immunological data which have to be mentioned briefly. The experimental technique has been questioned (Boyden, 1972; Friday, 1981), and peak level comparisons with constant antisera have been shown to be more discriminating than the ID value. They show the chimpanzee, for instance to be at a relatively greater distance from man than is indicated by the ID distance. Another problem is the assumption of an exponential straight-line relationship between logID and time, and Read & Lestrel (1970) and Read (1975) have shown that this assumption is not justified on either empirical or theoretical grounds. They showed that the power function gives a better fit line, and this greatly alters the projected divergence dates: 13my for the African ape and man divergence calibrated with the 30my divergence date of the monkeys and apes or 8-9my with a 20my monkey ape divergence. They achieved this result, however, by omitting the most divergent point (for the bull frog) from the data used to calculate the relation, and if that is included then the power function has a higher variance than the others (Uzzell & Pilbeam, 1971), and none of the relations, including the power function, give a straight line.

This is important point, for it is on these assumptions that the validity of the molecular clock rests. If the relationship is wrong, so also will be the inferred dates. Corruccini et al (1980) comment further on this problem, and they show that if the exponent in the power function deviates significantly from 1, a non-linear rate of change with time is indicated and that, regardless of time, if the different molecular models covary in a non-linear fashion, it is impossible for all of them to covary with time in linear fashion. For the primates as a whole, the immunological data have a differentially slow rate of change compared with other systems, whereas within just the Anthropoidea they covary in linear fashion with the other systems. Linear calibration of molecular time from outside the Anthropoidea would provide a wrong calibration, and this would be true also if the rates of protein change vary within the Anthropoidea.

What is most disturbing about these criticisms is that the regularity of the molecular clock should be so dependent on the taxa included, the divergence point used to calibrate the clock and on the statistical formulations used to relate ID change to time (Uzzell & Pilbeam, 1971).

Protein electrophoresis

There is little information on catarrhine relationships from this method. A recent paper examined 23 proteins for the six extant genera of the Hominoidea (Bruce & Ayala, 1979). They measured allelic frequencies for each locus for each species and calculated genetic distance as the number of electrophoretically detectable allelic substitutions per locus that have accumulated since pairs of species diverged. Distances given are all relative to man, and pairwise comparisons of all the species with man are the basis for calculating minimum distances for a similar parsimony analysis to that used for the immunodiffusion data (Sarich & Wilson, 1967; Sarich & Cronin, 1976).

The results of this method are not very illuminating. Gibbons appear to be part of the hominoid clade, but the great apes and man are almost equally divergent from each other, with no indication of an earlier divergence for the orang utan or closer relationship of man and chimpanzee (Bruce & Ayala, 1979). The discriminatory power of this method therefore appears to be less than that of immunology.

DNA hybridization

This method differs from the two preceding ones in that it is the DNA that is being compared rather than the proteins that are the product of the DNA coding. It might seem, therefore to be that much closer to the identification of 'real' genetic differences between taxa. In fact, the hybridization method produces distance statistics identical in form to those just described, so that the data produced must be considered in the same way. Fragments of disassociated single-stranded radioactively-tagged DNA from one species are reassociated with single strands from other species. The smaller the degree of correspondence between the DNA strands the less reassociation occurs and the less force is needed to separate them again. This force, which is usually measured by its heat products, provides a measure of genetic similarity (Sibley & Ahlquist, 1984), but in fact it is put forward as a measure of genetic relatedness, or genealogical distance; the problems of moving from 'similarity' to 'relatedness' are the same as those encountered for immunological methods and will be discussed below.

The results of this DNA similarity distance (delta $T_{50}H$) for the catarrhines are summarized in Table 1. Unfortunately there is no outgroup comparison for the hominoids, but there is clear evidence of similarity of the various taxa of the Hominoidea and of the great apes and African apes and man within the superfamily. The results are in agreement with the immunological results but differ from the more ambiguous results from protein electrophoresis. The DNA hybridization results also provide evidence for association between chimpanzee and man within the African ape and man group (Sibley & Ahlquist, 1984), but if the standard deviations of the distance are examined (Table 1) this result cannot be regarded as significant.

The delta $T_{50}H$ distances were translated into time separation by calibrating the orang utan divergence for two alternative dates: 13my and 16my (Sibley & Ahlquist, 1984). Unit distance of change corresponded to 3.5 and 4.3my respectively, and the divergence dates for the catarrhine lineages are shown in Table 2. The latter figure is very close to the rate of change found for three calibration points calculated for birds (Sibley & Ahlquist, 1984, and references therein), and this is taken to indicate the very great regularity of change in the DNA chains. The dates for the various branching points are generally older than those inferred from the immunological data except for the initial ape-monkey split, which at 27 to 33my spans exactly the date used for immunological calibration. This would suggest that the difference between the two methods is not the result of differences in calibration but is the result of a differently timed clock.

Sibley & Ahlquist (1984) state that their distance statistic is a measure of phylogenetic divergence. This is based on the assumption that the only reason for 'closeness' of fit is true homology of characters, so that

Table 1. DNA hybridization distances

Taxa comparisons	delta $T_{50}H$	standard deviation
man-chimpanzee	1.8 - 1.9	0.2 - 0.3
man-gorilla	2.4	0.2 - 0.3
chimpanzee-gorilla	2.1 - 2.3	0.2 - 0.3
man/chimp/gorilla-orang utan	3.6 - 3.8	0.2 - 0.4
man/great apes-gibbons	5.1 - 5.6	0.2 - 0.5
hominoids-cercopithecoids	7.4 - 8.0	0.2 - 0.6

Table 2. Catarrhine divergence dates: DNA hybridization

Taxa divergence

	millions of years	
delta $T_{50}H$ distance/time	3.5	4.3
Hominoid-cercopithecoid	27	33
gibbons-hominoid	18	22
orang utan-chimpanzee/gorilla/man	13*	16*
gorilla-chimpanzee/man	8	10
chimpanzee-man	6.3	7.7

*calibration dates from the fossil record

where the nucleotides are similar on two DNA chains, giving rise to thermal stability, the similarity is the result of homology. Even when some of the similarities may be the result of independent or convergent change the number of convergences is likely to be small compared with the number of homologies in a long DNA chain. Where this argument becomes suspect is when the differences between branching points is small, so that the occurrence of parallel change is more likely to be of greater significance relative to the small number of nucleotide differences.

The problem of convergence is illustrated in Figure 1. The set of

Figure 1. Theoretical example applied to DNA hybridization data. Three substitutions are supposed to be shared by man, chimpanzee and gorilla together and two each by each of the pairs of species. There is thus no information on the interrelationships among the three species. If two of the possible ancestral states are now considered, the first one (I) supposes all the substitutions shared both by man and gorilla and by man and chimpanzee to be primitive (that is, present in the ancestral genotype), whereas chimpanzee and gorilla share two substitutions uniquely (at positions 2 and 7). The second supposed ancestral state (II) shows man and chimpanzee as sharing the two unique substitutions, while man-gorilla and chimpanzee-gorilla share the primitive condition. If the sharing of unique derived characters is evidence of relationship, two quite different phylogenies result from these data, despite the fact that the characters of the three species remains the same.

DNA Hybridisation - theoretical example

	1	2	3	4	5	6	7	8	9
man	a	a	b	d	c	a	d	d	d
chimpanzee	a	b	b	d	d	c	a	d	d
gorilla	a	b	c	d	c	a	a	d	a

Possible Ancestral morphotypes	1	2	3	4	5	6	7	8	9
I	a	a	b	d	c	a	d	d	d
II	a	b	c	d	c	a	a	d	a

hypothetical data illustrated in the figure shows the gorilla, man and chimpanzee to be equally distant from each other in a manner akin to that of the DNA hybridization results. These data tell us nothing about the ancestral conditions from which these three species diverged, but if two possible sets of ancestral conditions are postulated, quite different phylogenetic relationships can be inferred from the similarity data. In the first (ancestral morphotype I) there are two synapomorphies linking gorilla and chimpanzee (at positions 2 and 7) and none linking man and chimpanzee or man and gorilla; while in the second there are two synapomorphies linking man and chimpanzee (at positions 3 and 9) but none linking gorilla and chimpanzee. Depending therefore on what is interpreted as the ancestral condition (or if that cannot be done, what is actually the ancestral condition) there is evidence from the same body of data either linking gorilla and chimpanzee or man and chimpanzee. In view of the fact that the differences between these options in the DNA hybridization distances are so small, these questions of phylogenetic interpretation cannot be ignored.

AMINO ACID SEQUENCING

By determining the sequences of amino acids for individual proteins it is possible to count the actual similarities and differences that make up the ID distance of immunology and electrophoresis. There is a useful analogy with the ways the characters of gross morphology are considered. Multivariate analysis of a complex structure such as the skull provides, like immunology at the protein level, a distance statistic such as D^2 which gives an overall distance between the species or specimens being tested but which of itself provides no information on how they differ. Individual assessment of characters provides this information, but more importantly it provides the basis for the phylogenetic interpretation of characters, so that it can be shown whether they are inherited from a more or less remote common ancestor or whether they are unique to a lineage. It is this latter information that is available at the protein level from amino acid sequencing.

The divisions within the Catarrhini indicated by amino acid sequencing are consistent with the results from DNA hybridization. The sequencing provides rather more information on processes of change, and this will be illustrated with two examples. Myoglobin sequence changes for the catarrhines are shown in Figure 2, with a number of amino acid residues characteristic of the catarrhine and hominoid divergence points. At position 23, serine is present in all anthropoids except man and the African apes whereas glycine is present in mammals generally (Romero-Herrera et al, 1976, 1978). It seems likely, therefore, that serine is a uniquely derived character (synapomorphy) of the Anthropoidea while the African apes and man are secondarily derived in returning to glycine at position 23. The alternative is that the ceboids, cercopithecoids, hylobatids and pongids (orang utan) all changed in parallel to serine.

Position 110 of the myoglobin chain is even more interesting (Romero-Herrera et al, 1976). Alanine is present in New World monkeys and most other mammals, serine in the Old World monkeys and the orang utan, and cysteine in gibbons, African apes and man. There must clearly be a parallelism or a reversal unless it be accepted that the gibbons are more closely related to the African apes and man than is the orang utan, but the key to this puzzle is the recognition that there is no point mutation by which alanine can change to cysteine. Two mutations in the nucleotides coding for the amino acids are required, so that for instance, a single codon mutation will change alanine to serine, and another will change serine to cysteine. The most parsimonious solution has alanine present in the ancestral anthropoid state and retained in the New World monkeys; one point mutation changed it to serine in the catarrhine ancestral state which is retained in the Old

Figure 2. Myoglobin sequence data. The catarrhines share four unique amino acids by which they differ from other anthropoids and primates. Four more amino acid changes characterize the Hominoidea, but there are none for the great ape and man group. The African apes and man share one change, while gorilla and chimpanzee have one each to themselves. Since man has no unique change, the additive differences between man and either of the African apes is one in both cases while the differences between the two apes is two, but since this is based on changes unique to single species it is not evidence of any particular set of relationships.

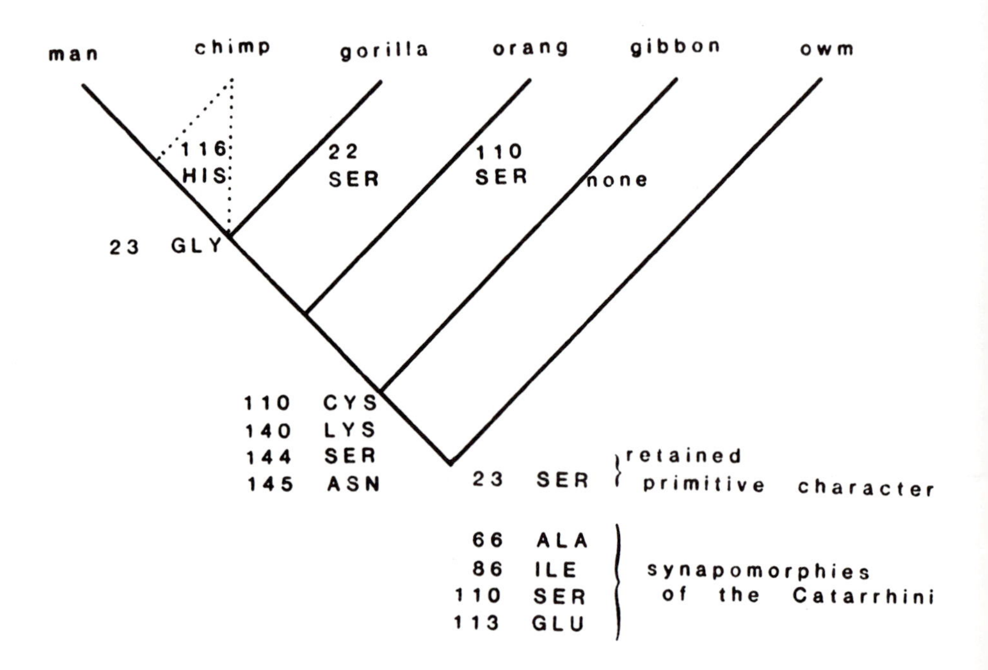

World monkeys (see Figure 3); and a second point mutation changed the serine to cysteine in the ancestral hominoid state, a condition retained in all hominoids except the orang utan in which a reversal occurred back to serine (Figure 2).

In addition to these shared residues at positions 23 and 110 in the myoglobin chain there are unique amino acid residue changes in a number of single lineages. For instance in Figure 2 it can be seen that the chimpanzee and gorilla both have single unique residue changes by which they differ from man. They thus differ from man by one residue each and from each other by two residues, which is presumably the basis on which Romero-Herrera _et al_ (1978) suggested that man and chimpanzee are more closely related than

Figure 3. Myoglobin amino acid sequence, position 110. Each main group of the Anthropoidea has a different amino acid at position 110. The New World monkeys (NWM) have alanine, the Old World monkeys (OWM) have serine, and the hominoids cysteine. Because two nucleotide substitutions are required for a change from alanine to cysteine, of which one could involve the initial change to serine, the most parsimonious model of change would require an intermediate stage in the ancestral catarrhine genotype not indicated by examination of the amino acids alone.

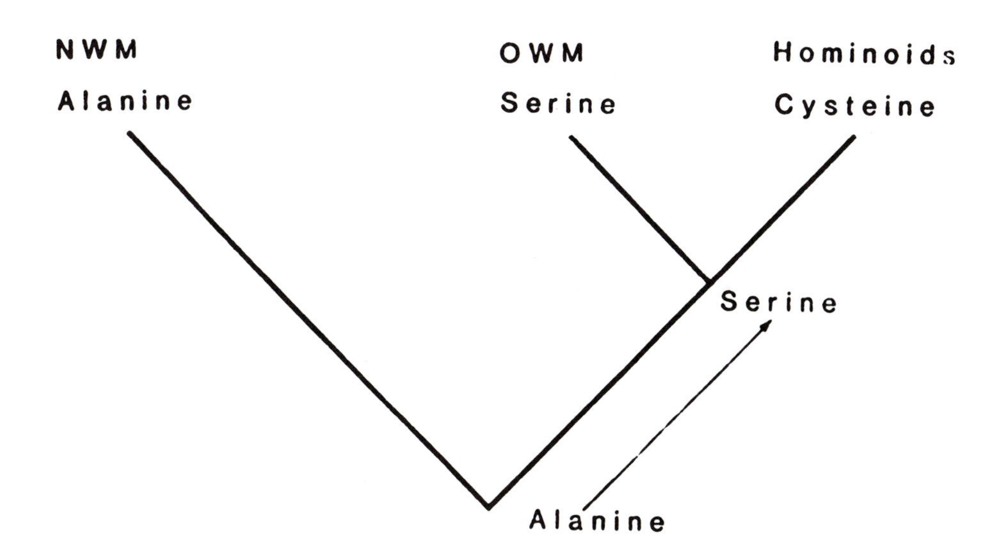

either is to the gorilla. These are unique characters, however, and tell us nothing about relationships of these three taxa because the gorilla is as close to man as is the chimpanzee (separated by a single mutation each), so that the trichotomy indicated by the dotted lines on Figure 2 is just as likely as the man-chimpanzee relationship.

Similar results are available for the alpha and beta haemoglobin chains (Goodman *et al*, 1983). Two amino acid residues are unique to the chimpanzee and man clade (Figure 4) and differ from the condition in anthropoids generally. These appear likely to be synapomorphies for this clade and constitute the strongest evidence available at present for the unique relationship of man with the chimpanzee.

The coding sequence for the fetal globin A_γ genes is identical for chimpanzee and man (Scott *et al*, 1984) whereas there are two substitutions at codons 73 and 104 in the gorilla. The gorilla condition at codon 104 is the same as is present in cercopithecoid monkeys, and Scott *et al* (1984) suggest that this condition is probably therefore a retained ancestral

Figure 4. Amino acid sequence data on haemoglobins. At least two amino acid substitutions are shared uniquely by chimpanzee and man, with all other catarrhines retaining the general primate condition, but it is not clear whether there are any unique catarrhine substitutions.

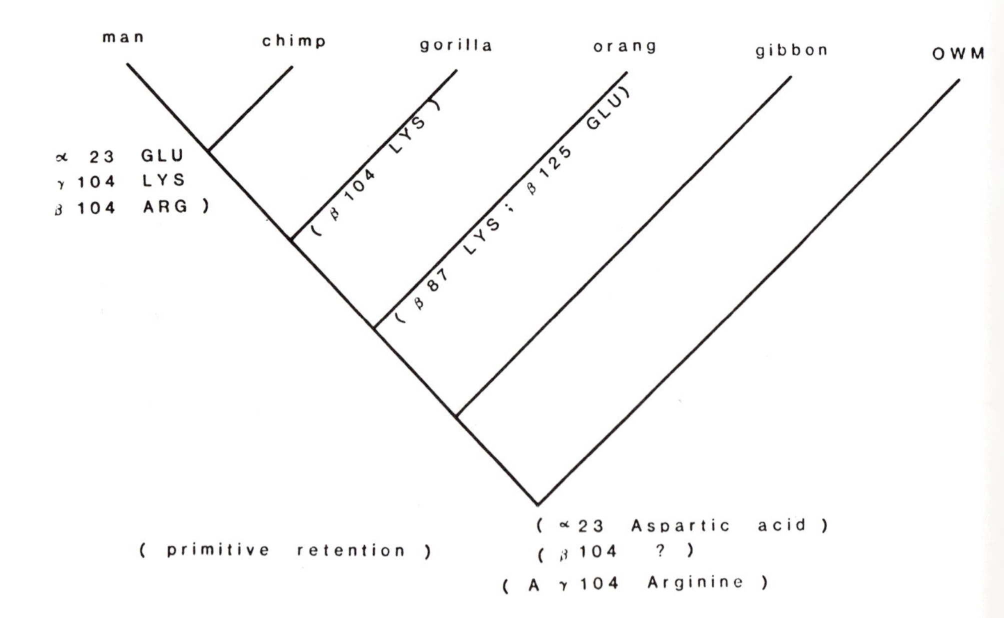

feature and the substitution in man and chimpanzee is a shared derived character.

Further evidence is available for the fibrinopeptides A&B amino acid sequences (Doolittle *et al*, 1971). The amino acid sequences are the same for the African apes and man; these differ from the orang utan by 2 residues, from gibbons by 3-4, from monkeys by 5-8 residues, and from New World monkeys by 9-10 residues (Goodman, 1973). The two differences from the orang utan are both uniquely shared characters of the African apes and man group, while other catarrhines retain what is interpreted as the primitive condition (Doolittle *et al*, 1971). Similarly, the great apes and man also share one amino acid uniquely, but the hylobatids have none. Within the hylobatids, the gibbons have several unique substitutions not present in siamangs or any other catarrhine, and this gives rise to the anomalous result that there is only one amino acid difference between siamangs and orang utans but two between orang utans and the African apes and man and at least two also between siamangs and gibbons. There is also an odd result in the suggested relationships of the Old World monkeys from the fibrinopeptide A&B sequences; baboons and macaques are separated, so that baboons are closer to *Cercopithecus* than to the macaques (Goodman *et al*, 1983a), whereas in all other protein analyses the macaques and baboons form a clade with mangabeys and geladas. This is not a serious problem, but it does show up the dangers of relying on a single protein (or any other single character, for that matter) in assessing similarity or relationship.

The carbonic anhydrase amino acid sequence has been reported by Tashian *et al* (1976). They show that man differs from the chimpanzee at only one locus and that they differ from the orang utan at five loci. Three of these, however, are uniquely derived for the orang utan lineage and two are primitive; the two latter can therefore be regarded as synapomorphies linking man and chimpanzee.

Changes in amino acid sequencing have not generally been used to establish a molecular clock, but Goodman (1981; Goodman *et al*, 1983a) has combined the sequence data from the proteins mentioned above, together with lens alpha-crystallin A and cytochrome c, to provide two sets of dates. One set, calibrated on a 90my divergence of Eutherian mammals gives divergence dates for the Anthropoidea of 20.5my, for the Catarrhini 13.4my, and for the African apes and man 1.3my. The other, calibrated on a 35my divergence of the Anthropoidea, gives dates for the Catarrhini of 22.9my and for the African apes and man of 2.2my. These dates conflict with the fossil evidence, and Goodman uses this as evidence for the slowing down of globin sequence evolution during later hominoid evolution.

There is also evidence for rate changes in protein evolution from myoglobin and haemoglobin chains, which are known from early on in vertebrate evolution. From the numbers of sequence changes observed for lower vertebrates, a nucleotide substitution rate (number of codon changes per 100 loci per 10^8 years) of 46NR% has been calculated by Goodman *et al* (1975). This compares with a rate of 15NR% for birds and mammals, so that the rate of change in vertebrate evolution has clearly slowed down with time. Within the catarrhines the rate slows down still further, and during the Plio-

Pleistocene the rate for African apes and man is only 6NR% compared, for instance, with rates for equids during the same period of 47NR%. This is direct evidence, therefore, that rates of protein change are not constant, as was required by the molecular clock model in its original simple form. It must be remembered, however, that it does not matter if the timing of the clock is not constant so long as its inconstancies are known: it could be set to a variable rate provided the variations are well enough known, although I do not know of any instance where this has been attempted.

mtDNA SEQUENCING

mtDNA sequencing is the most exciting development in recent years for increasing our understanding of hominoid evolution. It is well established that rates of change in DNA sequences are much more rapid in mtDNA than in nuclear DNA (Brown et al, 1982; Miyata et al, 1982; Ferris et al, 1981a&b; Hasegawa et al, 1984, in press). Rates of change are higher in silent or synonymous substitutions than in replacement substitutions (where any change may result in changed coding for an amino acid), but within these groups the rates are nearly uniform, in contrast to the variations in rate seen in different proteins. There are also differences between types of substitutions, with transitions being very much more abundant than transversions. DNA is a sequence of four alphabets, T(hymine) and C(ytosine) being pyrimidines and A(denine) and G(uanine) being purines, and any change within these two groups is termed a transition and any change between the groups is termed a transversion. Transitions accumulate so rapidly that within 20my they have 'saturated' all readily substituted positions and further changes simply obliterate earlier changes (Brown et al, 1982). Transitions either have to be discounted altogether, as is done by Hasegawa et al (1984, in press) who count only transversions; or else the periods of time covered have to be greatly curtailed (Brown et al, 1982).

Counting only transversions, Hasegawa et al (1984) obtained results consistent with the DNA hybridization results of Sibley & Ahlquist (1984). Their method was to calculate minimum distances between pairs of species based on adding together all substitutions, but their data show many uniquely shared substitutions and these are listed in Table 3. There is no information on whether any of these shared characters are primitive or derived, and the

Table 3. Numbers of codon substitutions of mtDNA shared uniquely by the branches of the Hominoidea

Group	Number of substitutions
all hominoids	46
great apes and man	19
African apes and man	16
chimpanzee and man	11
chimpanzee and gorilla	8
gorilla and man	2
chimpanzee, gorilla and orang utan	2
orang utan and all gibbons	2

lack of outgroup comparisons within the primates (they use cow and mouse) provides no help. Maximizing the uniquely shared substitutions makes clear how much convergence must have occurred in the mtDNA sequences - see Figure 5. This solution minimizes the degree of convergence, but it must be stressed again that there is no independent means of distinguishing between homoplasy and homology. When it is also considered whether the substitutions are transitions or transversions, the 11 shared substitutions between man and chimpanzee include 4 transversions while all 8 of the chimpanzee-gorilla similarities are transitions. These interpretations conform to Hasegawa's maximum parsimony pairwise comparisons, but Brown et al (1982), who used the same data and obtained similar results, surprisingly concluded

Figure 5. Cladogram maximizing numbers of uniquely shared mtDNA substitutions and minimizing homoplasy (data courtesy of M. Hasegawa). This solution shows man and chimpanzee sharing 11 substitutions as against chimpanzee and gorilla sharing 8, and it must be assumed that one or other of these two sets of substitutions are developed convergently or in parallel. The best fit solution can be taken as the one minimizing the necessary convergence.

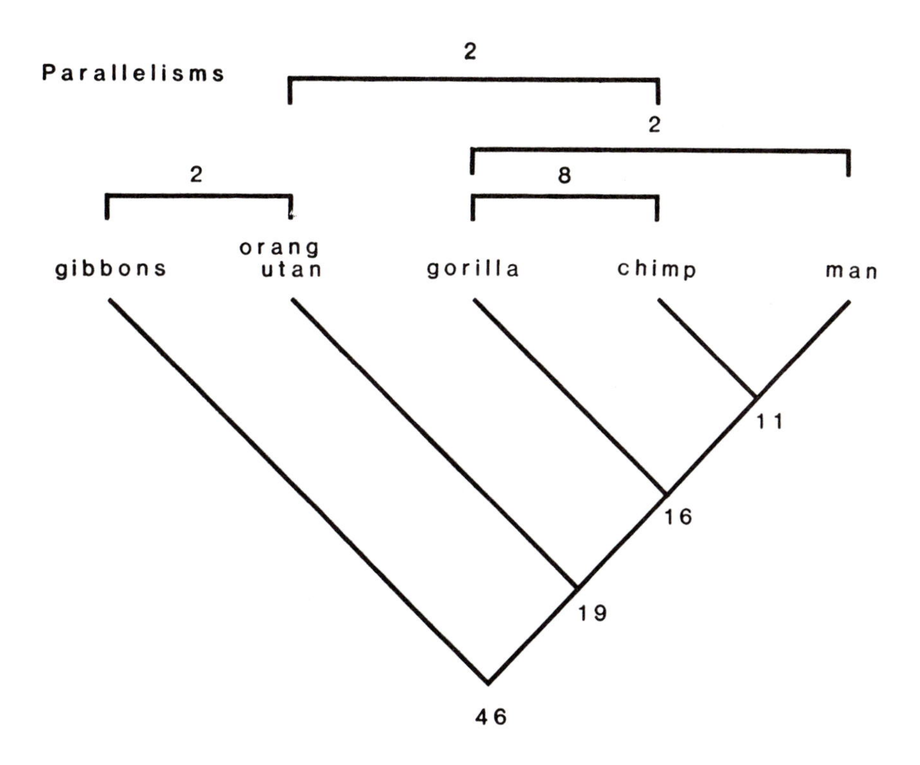

that chimpanzee and gorilla shared a common ancestor after the divergence of man.

Brown et al (1982) did not apply the clock model to their data, but Hasegawa et al (1984) have done so and have obtained the following dates for hominoid divergence events:

gibbons	19.1 ± 3.6 million years
orang utan	15.9 ± 2.9
gorilla	4.9 ± 1.0
chimpanzee	3.4 ± 0.7

These results are quite similar to those obtained from DNA hybridization. The mtDNA clock was calibrated using a bovine-primate divergence of 92my initially (Hasegawa, 1984), giving the above dates, and later recalibrated with 65my for the same calibration point, giving the younger dates of 13.8, 11.5, 3.6 and 2.5my for the same divergence events (Hasegawa, in press). Probably the calibration should lie somewhere in between these two, and this highlights the difficulties in having to calibrate any molecular clock from an independent, and in this case a most unreliable, source.

Additional work on restriction enzyme cleavage mapping of mtDNA provides an alternative approach and somewhat different results (Ferris et al, 1981a&b; Templeton, 1984). To try and get around the convergence problem, Templeton distinguishes between gain and loss at restriction sites, pointing out that while there are many ways of losing a specific restriction site there is only one way of gaining one. By this means he attempts to minimize convergence, and by including information on insertions, deletions, inversions and transpositions and keeping separate the results of the different enzymes used, Templeton concludes both that the molecular clock is not a valid model and that in the African ape and man divergence the chimpanzee and gorilla shared a common ancestor separate from man. The results of Ferris' work (Ferris et al, 1981a&b) show the same thing: during the evolution of the hominoids, a minimum of 147 mutations were found, with 67 shared by more than one species. The minimum number is achieved by the two species of chimpanzee forming the sister group to gorilla, with these the sister group to man, then orang utan, then gibbons. Despite the high rate of change in the mtDNA commented on earlier, it was concluded that there was still insufficient resolving power, since a man-chimpanzee link required only one more mutation. In fact, examination of the data in Ferris et al (1981a) shows that chimpanzee and gorilla share 5 mutations at 4 sites, chimpanzee and man share 6 mutations at 3 sites, and man and gorilla share 2 mutations at 1 site (all of these are exclusive, that is unique to that particular pairing, and these figures exclude similarities shared by additional species). It appears from this that the minimum pathway indicating a closer relationship between chimpanzee and gorilla may give an incorrect phylogenetic interpretation, and that these data support a closer link between man and the chimpanzee.

DISCUSSION

Molecular events in protein evolution consist, firstly, of mutation of one of the bases in a base pair and, secondly, its fixation in a population, either through genetic drift or natural selection (insertions, deletions, inversions and transpositions are usually not considered, although see Templeton, 1984). It is assumed that 'population differences reflect the length of time since the existence of a common ancestral population' (Thompson, 1975), but this is an invalid assumption where polymorphic variation is concerned (the subject of Thompson's book), because polymorphic genes may be inherited unaltered from an ancestral species. In this event, the polymorphism pre-dates the emergence of a new species. It may also be invalid at the species level, where, for instance, a species like the orang utan has considerable genetic variability between different populations of what is generally considered to be a single species (Bruce & Ayala, 1979; de Boer & Meera Khan, 1982; de Boer & Seuanez, 1982).

Molecular events can be recognized at first hand through DNA sequencing, at second hand through amino acid sequencing, or at third hand through the proteins built up by the amino acids (and also through the overall composition of the DNA chain where it has not been sequenced). These are the three categories of evidence that have been described in this paper. The accumulation of mutations seems to depend on the functional significance of the particular segment affected, and there are whole regions of the DNA chain that appear to have no function at all (Kohne, 1976). Mutations occurring in these non-functional regions are in effect neutral mutations and these can accumulate very rapidly whereas in other regions, for instance those regulating certain classes of RNA, any change would be likely to be deleterious and the mutations would therefore quickly be eliminated.

Implicit in most of the papers that have been reviewed here is the assumption that molecular events are somehow different from morphological events. For instance, Moore et al (1973) say that the use of the term primitive violates our intuitive understanding of mutation and fixation of characters at the DNA level since there is no reason to suppose that any nucleotide is more primitive than any other. This misunderstands the meaning of the term, which in phylogenetic context is always relative. It applies to the condition of a character in a specific ancestral state, and any taxa descended from that ancestral state and retaining that character are said to be primitive for that character. Thus a nucleotide substitution at a given branching point, such as the initial divergence of the Hominoidea, may be retained unchanged in most or all of the species of the group, so that while the substitution characterises the group as a whole it is primitive for any one of the taxa contained within it. There is no inherent value judgement or absolute connotation in the use of the term 'primitive' which simply stands for the unchanged retention of an ancestral state, whether molecular or morphological.

Many of the phylogenetic interpretations of molecular events are highly suspect because of this. Simply demonstrating homology is not enough: homologous characters include both those characters inherited unchanged from past ancestral states and characters inherited from the immediate (or unique) ancestral state. Similarity matrices and phenograms are produced by adding up the differences between all possible pairs of taxa and then minimizing the pairwise differences between them. There are problems with both the additive nature of phenetic change and the dependence on parsimony, and these have been recognized and discussed by many workers, but the assumptions underlying the interpretation of a phylogenetic cladogram from a similarity phenogram are still the subject of some controversy (Fitch, 1977; Felsenstein, 1981, 1983, 1984; Farris, 1981).

Parsimony is based on the intuitive idea that the projection of branching points from a known series of extant populations follows the minimum possible pathway (Cavalli-Sforza & Edwards, 1967; Friday, 1981). There is no theoretical or empirical justification for minimum evolution, and it is possible for it to be correct at one level of analysis but incorrect at another. For instance, the most economical nucleotide changes coding for the amino acids alanine, serine and cysteine involves two mutations (see Figure 3 and Romero-Herrera et al, 1976); these amino acids are present respectively in New World monkeys, Old World monkeys, and hominoids, but because two mutations are required to change from alanine to cysteine the minimum pathway for nucleotide change will not be the same as the possible minimum pathway for amino acids constructed in ignorance of DNA coding. This is a fairly trivial objection to the parsimony model, and it does not apply to data from DNA sequencing, but the existence of convergent change and back-mutations does constitute a more serious objection, and this will be considered in the context of all change being additive.

The additive change model depends on the assumptions that all change is divergent, independent, and that there are no parallel or convergent changes (Cavalli-Sforza & Edwards, 1967). The latter assumption could be extended to say that convergence could be accommodated if its distribution is random (Moore et al, 1973), or if its occurrence is minimal compared with a large number of homologies (Sibley & Ahlquist, 1984). Unfortunately, convergence reaches high levels in certain (if not all) proteins, and repetition of the same change in different lineages is not only common but is to be expected (Bishop, 1982). The cytochrome c cluster analysis performed by Fitch & Margoliash (1967) produced a phylogeny with 20% of the inferred changes being parallel, and extensive convergence has been shown for myoglobin (Romero-Herrera et al, 1978) and haemoglobins (Goodman et al, 1983). Back mutations occur in very high numbers in mtDNA (Brown et al, 1982; Hasegawa et al, 1984) so that a single observed substitution may be the product of many intermediate steps. Any or all of these may be randomly distributed, although there appears to be differential likelihood of substitution at different positions which may give rise to different convergence rates, but the error is likely to be small in groups with many differences (Uzzell & Pilbeam, 1971; Friday, 1981).

Maximum parsimony is applied to sequencing data to construct minimum distance matrices. Since the individual sequences are known, it is possible to construct minimum genealogical pathways and to reconstruct from this the hypothetical ancestral states required by these pathways: see Figures 1 and 2. Once this has been done it is then possible to recognize many similar sequence substitutions in separate lineages as being the result of convergence, but there is no way of distinguishing possible convergences that may have contributed to the maximum parsimony solution itself. In the case of the resolution of the African ape and man trichotomy, where there is often only a single substitution, the possibility that this may be a convergence built in to the best fit solution cannot be eliminated because the recognition of relationships and the recognition of convergence are interdependent and no independent source of information is known that can resolve this circularity: see Figure 5.

It is worth considering also at this point the problem of unequal rates of evolution in so far as it concerns additivity. Additive distances based on minimum distance will group two taxa with minimal change and separate a taxon with greater change, even if the change is unique to the taxon. For instance, the additive distance between the man and chimpanzee in Figure 2 is only one whereas the distance between chimpanzee and gorilla is two, but this is because both African apes have a unique substitution each where-whereas man has none. When added together both chimpanzee and gorilla therefore have one difference each from man and two differences from each other, but since the differences are unique to each species they do not signify relationship.

None of these assumptions and objections give rise to any problem as far as phenetic similarity is concerned. Minimum additive change produces sets of similarity-based clusters that are meaningful in their own right, but to extend this to claim that similarity denotes relationship when there is so much doubt about convergence and when characters are used indiscriminately without regard to their phylogenetic relevance seems to me to be highly dubious. It would be better, in fact, not to come to any phylogenetic interpretation at all but use the similarities as the basis for a phenetic or non-phylogenetic classification. This is the logic of the stamp collector, who may wish to classify stamps according to arbitrarily selected similarities based on country of origin, size, colour, etc. Questions about the origin and historical development of stamps, their spread from one place to another with advancing civilization, and the developing concepts in their production are peripheral to such a classification. These are interesting questions, however, and the ones most akin to biological questions about evolution, while the similarity-based phenetic classification, although useful in providing reasonably stable labels to biological entities, has no information content beyond the classification which is its end-product.

Statements about catarrhine evolution, with its corollary the molecular clock model, are entirely based on the phylogenetic interpretation of phenetic data, and so to the extent that this interpretation is doubtful so also must be the phylogenetic conclusions and the dates derived from the clock. The protein distance data has shown itself to have the least resolving power in defining catarrhine lineages, and it is also the most suspect in that it depends for its phylogenetic validity on the assumption that the observed protein distances are the result of genealogical distance (or, remoteness from the last common ancestor). The clock model also depends on this assumption, and it depends in addition on the recognition of variations in the rate of change of different proteins in single lineages and of the same proteins in different lineages. Both of these have been demonstrated many times (Goodman et al, 1983a), but they do not of themselves invalidate the protein clock provided the variations can be incorporated into the model. It is my belief from the evidence supplied in the papers reviewed here that far too little is known about these variations, with the result that the protein clock cannot at present be correctly timed. The DNA clock appears rather more regular, but the resolving power is no better than the protein distance data. Although it is claimed by Sibley & Ahlquist (1984) that there is evidence for a chimpanzee-human relationship in the Hominoidea, the discrimination is small compared with the errors involved and is certainly not statistically significant. All of the distance data agree, however, in suggesting recent common ancestry for the African apes and man (dates varying from 3.5my for the protein distance data and 8 to 10my for the DNA hybridization data). The evidence also agrees on the separate branching of the orang utan and gibbon lineages; the monophyly of the Hominoidea; the monophyly and subsequent division of the Cercopithecoidea into two branches, the colobines and cercopithecines; and, finally, the monophyly of the Catarrhini, for which divergence dates of 27 to 33my are given.

The sequence data provide testable sets of data, both on amino acid and DNA sequences, and some indication of polarity of change can be obtained. It is through work of this kind that some idea of the complexities involved in protein evolution can be gained, and it is worth pointing out at this stage that these complexities are not in any way restricted to analyses of sequence data: convergence and polarity uncertainties are present from DNA level all the way to gross morphology and behavioural studies, and it is the fact that they can be recognized at the DNA level but not at the others which leads to the emphasis on them here. It is no solution to reject the molecular data because of these difficulties simply because the difficulties can be identified, especially if the alternative chosen is morphological characters where we can do little more than guess about convergence. The sequence data, both amino acid sequences and mtDNA, provide clear indications of shared characters (amino acid residues or nucleotide substitutions) in man and the chimpanzee; these are characters which are shared uniquely and which differ from the character state present in other catarrhines, and as such they are taken to be evidence of relationship, but the circularity in the argument by which they are recognized as homologous must be remembered. All of the sequencing evidence supports the earlier divergence of the orang utan and gibbon clades, but there are only small differences in the maximum parsimony solution between linking the Asian hominoids or keeping them separate. The monophyly of the Hominoidea and the Cercopithecoidea is well supported. Dates for the branching points vary enormously: the protein clock based on the combined amino acid sequencing (Goodman et al, 1983) gives dates for the Catarrhini of 13 to 22my and of the African ape and human divergence of 1.3 to 2.2my, while the DNA clock gives dates of 4.9my for the latter and 19my for the Hominoidea (no date is calculated for the catarrhines). It is interesting that this range of values is similar to that given above for the protein and DNA distance data.

SUMMARY AND CONCLUSIONS

The distance measures described here all provide similar results. The Catarrhini appears to be a monophyletic group, and within it both Hominoidea and Cercopithecoidea also appear as monophyletic groups. The latter is divided into the two recognized subfamilies, but within both the Colobinae and the Cercopithecinae there is some ambiguity about the contained taxa: the colobines appear to be paraphyletic and the papionines do not always appear as a natural group. The Hominoidea separates into three groups, the Hylobatidae, the Pongidae (containing just the orang utan), and a third group that for consistency must be called the Hominidae, containing the African apes and man. Immunodiffusion methods cannot resolve the African ape and man clade any further than this. DNA hybridization is claimed to show closer relationship of man to chimpanzee, but the evidence is not convincing.

Distance methods rest on the assumptions that all similarities are homologous and additive. They take no account of possible homoplasy or of variations in rate of change in protein evolution from one lineage to another, both of which have been shown to occur. It can be concluded that while these methods provide convincing evidence of phenetic similarity, they are, at best, untestable approximations to phylogenetic relationship.

Amino acid sequencing results are generally similar to those of the distance methods, but they show rather greater resolution between taxa. The numbers of taxa that have been tested, however, is relatively small because of the difficulties and expense of the sequencing method, and very often there is simply not enough information to resolve interesting issues. Myoglobin sequencing confirms the results of the protein distance data in all respects, but it cannot resolve the chimpanzee/gorilla/man trichotomy. Both the adult haemoglobin and the fetal globin A sequences provide good evidence for a link between man and chimpanzee within the African ape and man clade, but other proteins provide no additional information because too few taxa have been tested.

The mtDNA sequencing also provides evidence for the man/chimpanzee link, although there is some conflict between the various workers in this field: Hasegawa et al (1984) provide good evidence at several different levels for the link; Brown et al (1982) using identical data suggest that chimpanzee and gorilla are more closely related; and Templeton (1984), using restriction site mapping of mtDNA combined with presence of inversions, deletions, transpositions and insertions, also showed that gorilla and chimpanzee are more closely related. Ferris et al (1981) again showed the same thing, but examination of their data suggests, on the contrary, that chimpanzee and man may be closer. I think that the balance of evidence is in favour of the chimpanzee/man link, but clearly the evidence leaves a lot to be desired.

Acknowledgements

I am very grateful to the following for commenting on various drafts of this paper: A. Friday, M. Goodman, S. Jones, L. Martin, C. Stringer, and D. Tills.

REFERENCES

Key References

Brown, W.M., Prager, E.M., Wang, A. & Wilson, A.C. (1982). Mitochondrial DNA sequences of primates: tempo and mode of evolution. J. Molec. Evol. *18*, 225-39.

Bruce, E.J. & Ayala, F.J. (1979). Phylogenetic relationships between man and the apes: electrophoretic evidence. Evolution *33*, 1040-45.

Goodman, M. (1976). Towards a genealogical description of the primates. *In* Molecular Anthropology, ed M. Goodman & R.E. Tashian, pp. 321-53, New York, Plenum.

Romero-Herrera, A.E., Lehman, H., Joysey, K.A. & Friday, A.E. (1978). On the evolution of myoglobin. Phil. Trans. Roy. Soc. *283*, 61-163.

Sibley, C.G. & Ahlquist, J.E. (1984). The phylogeny of the hominoid primates, as indicated by DNA-DNA hybridization. J. Molec. Evol. *20*, 2-15.

Main References

Andrews, P. (1985). Family group systematics and evolution among catarrhine primates. *In* Ancestors: the hard evidence, ed E. Delson, Alan Liss, New York.

Andrews, P. & Cronin, J. (1982). The relationships of Sivapithecus and Ramapithecus and the evolution of the orang-utan. Nature, Lond. *297*, 541-46.

Baba, M., Darga, L., & Goodman, M. (1980). Biochemical evidence on the phylogeny of the Anthropoidea. *In* Evolutionary Biology of the New World Monkeys and Continental Drift, ed. R. Ciochon and R. Corruccini, pp. 423-43, New York, Plenum.

Bishop, M.J. (1982). Criteria for the determination of the direction of character state changes. Zool. J. Linn. Soc. *74*, 197-206.

de Boer, L.E.M. & Meera Khan, P. (1982). Haemoglobin polymorphisms in Bornean and Sumatran oran utans. *In* The Orang utan, ed. L.E.M. de Boer, pp. 125-34, Junk, The Hague.

de Boer, L.E.M. & Seuanez, H.N. (1982). The chromosomes of the orang utan and their relevance to the conservation of the species. *In* The Orang utan, ed. L.E.M. de Boer, pp. 135-70, Junk, The Hague.

Boyden, A.A. (1972). A review of recent reports bearing on comparative immunochemistry and 'evolutionary' clocks. Bull Serol. Mus. *47*, 5-8; 48, 5-7.

Brown, W.M., Prager, E.M., Wang, A. & Wilson, A.C. (1982). Mitochondrial DNA sequences of primates: tempo and mode of evolution. J. Molec. Evol. *18*, 225-39.

Bruce, E.J. & Ayala, F.J. (1979). Phylogenetic relationships between man and the apes: electrophoretic evidence. Evolution *33*, 1040-45.

Cavalli-Sforza, L.L. & Edwards, A.W.F. (1967). Phylogenetic evidence: models and estimation procedures. Am. J. Hum. Genet. *19*, 233-57.

Corruccini, R., Baba, M., Goodman, M., Ciochon, R., & Cronin, J. (1980). Non-linear macromolecular evolution and the molecular clock. Evolution *34*, 1216-19.

Dene, H.T., Goodman, M. & Prychodko, W. (1976). Immunoduffusion evidence on the phylogeny of the primates. *In* Molecular Anthropology, ed M. Goodman & R.E. Tashian, pp. 171-95, New York, Plenum.

Doolittle, R.F., Wooding, G.L., Lin, Y. & Riley, M. (1971). Hominoid evolution as judged by fibrinopeptide structures. J. Molec. Evol. *1*, 74-83.

Farris, J.S. (1981). Distance data in phylogenetic analysis. *In* Advances in Cladistics, ed. V.A. Funk & D.R. Brook, pp 3-23, New York, New York Botanical Garden.

Felsenstein, J. (1981). A likelihood approach to character weighting and what it tells us about parsimony and compatability. Biol. J. Linn. Soc. *16*, 183-96.

Felsenstein, J. (1983). Statistical inference of phylogenies. J. Roy. Statist. Soc. *146*, 246-72.

Felsenstein, J. (1984). Distance methods for inferring phylogenies: a justification. Evolution, *38*, 16-24.

Ferris, S.D., Wilson, A.C. and Brown, W.M. (1981a). Evolutionary tree for apes and humans based on cleavage maps of mtDNA. Proc. Natl. Acad. Sci. *78*, 2432-36.

Ferris, S.D., Brown, W.M., Davidson, W.S. & Wilson, A.C. (1981b). Extensive polymorphism in the mtDNA of apes. Proc. Natl. Acad. Sci. *78*, 6319-23.

Fitch, W.M. (1977). The phyletic interpretation of macromolecular sequence information: simple methods. *In* Major Patterns of Vertebrate Evolution, ed M.K. Hecht, P.C. Goody & B.M. Hecht, pp 169-204, New York, Plenum.

Fitch, W.M. & Margoliash, E. (1967). The construction of phylogenetic trees. Science *155*, 279-84.

Friday, A.E. (1981). Hominoid evolution: the nature of the biochemical evidence. *In* Aspects of Human Evolution, ed C.B. Stringer, pp 1-23, London, Taylor & Francis.

Goodman, M. (1963). Man's place in the phylogeny of the primates as reflected in serum proteins. *In* Classification and Human Evolution, ed S.L. Washburn, pp. 204-34, Chicago, Aldine.

Goodman, M. (1973). The chronicle of primate phylogeny contained in proteins. Symp. Zool. Soc. Lond. *33*, 339-75.

Goodman, M. (1974). Biochemical evidence on hominid phylogeny. Ann. Rev. Anthrop. *3*, 203-226.

Goodman, M. (1976). Towards a genealogical description of the primates. *In* Molecular Anthropology, ed M. Goodman & R.E. Tashian, pp 321-53, New York, Plenum.

Goodman, M., Baba, M. & Darga, L. (1983a). The bearing of molecular data on the cladogenesis and times of divergence of hominoid lineages. *In* New Interpretations of ape and human evolution, ed. R. Ciochon & R. Corruccini, pp 67-86, New York, Plenum.

Goodman, M., Braunitzer, G., Stangl, A. & Schank, B. (1983b). Evidence on human origins from haemoglobins of African apes. Nature *303*, 546-48.

Goodman, M. & Moore, G.W. (1971). Immunodiffusion systematics of the primates: the Catarrhini. Syst. Zool. *20*, 19-62.

Goodman, M., Moore, G.W. & Matsuda, G. (1975). Darwinian evolution in the genealogy of haemoglobin. Nature *253*, 603-8.

Goodman, M. & Tashian, R.E. (1976). Molecular Anthropology, New York, Plenum.

Hasegawa, M., Yano, T. & Kishino, H. (1984). A new molecular clock of mtDNA and the evolution of hominoids. Proc. Japan Acad. Sci. *60*, 95-8.

Hasegawa, M., Kishino, H. & Yano, T. (in press). Dating of human-ape splitting by a molecular clock of mtDNA.

Hoyer, B.H., van de Velde, N.W., Goodman, M. & Roberts, R.B. (1972). Examination of hominoid evolution by DNA sequence homology. J. Hum. Evol. *1*, 645-49.

Huxley, T.H. (1863). Evidence as to Man's Place in Nature, London, Williams & Norgate.

Kohne, D. (1976). DNA evolution data and its relevance to mammalian phylogeny. *In* Phylogeny of the Primates, ed W.P. Luckett & F.S. Szalay, pp 249-261, New York, Plenum.

Miyata, T., Hayashida, H., Kikuno, R., Hasegawa, M., Kobayashi, M. & Koike, K. (1982). Molecular clock of silent substitution: at least six-fold preponderance of silent changes in mitochondrial genes over those in nuclear genes. J. Molec. Evol. *19*, 18-35.

Moore, G.W., Goodman, M. & Barnabas, J. (1973). An iterative approach from the standpoint of the additive hypothesis to the dendrogram problem posed by molecular data sets. J. Theor. Biol. *38*, 423-57.

Read, D.W. (1975). Primate phylogeny, neutral mutations and 'molecular clocks'. Syst. Zool. *24*, 209-221.

Read, D.W. & Lestrel, P.E. (1970). Hominid phylogeny and immunology: a critical appraisal. Science *168*, 578-80.

Romero-Herrera, A.E., Lehman, H., Castillo, O., Joysey, K.A. & Friday, A.E. (1976) Myoglobin of the orang utan as a phylogenetic enigma. Nature *261*, 162-4.

Romero-Herrera, A.E., Lehman, H., Joysey, K.A. & Friday, A.E. (1978). On the evolution of myoglobin. Phil. Trans. Roy. Soc. *283*, 61-163.

Sarich, V.M. & Cronin, J.E. (1976). Molecular systematics of the primates. *In* Molecular Anthropology, ed M. Goodman & R.E. Tashian, pp 141-70, New York, Plenum.

Sarich, V.M. & Wilson, A.C. (1967). Immunological time scale for hominid evolution. Science *158*, 1200-03.

Scott, A.F., Heath, P., Trusko, S., Boyer, S.H., Prass, W., Goodman, M., Czelusniak, J., Chang, L.-Y.E. & Slighton, J.L. (1984). The sequence of the gorilla fetal globin genes: evidence of multiple gene conversions in human evolution. Mol. Biol. Evol. *1*, 371-89.

Sibley, C.G. & Ahlquist, J.E. (1984). The phylogeny of the hominoid primates, as indicated by DNA-DNA hybridization. J. Molec. Evol. *20*, 2-15.

Tashian, R.E., Goodman, M., Ferrell, R.E. & Tanis, R.J. (1976) Evolution of carbonic anhydrase in primates and other mammals. *In* Molecular Anthropology, ed M. Goodman & R.E. Tashian, pp 301-19. New York, Plenum.

Templeton, A.R. (1984). Phylogenetic inference from the restriction endonuclease cleavage site maps with particular reference to the evolution of humans and the apes. Evolution *37*, 221-44.
Thompson, E.A. (1975). Human Evolutionary Trees. Cambridge, Cambridge University Press.
Uzzell, T. & Pilbeam, D.R. (1971). Phylogenetic divergence dates of hominoid primates: a comparison of fossil and molecular data. Evolution *25*, 615-35.

7 THE FOSSIL RECORD OF EARLY CATARRHINE EVOLUTION

John G. Fleagle,
Department of Anatomical Sciences,
School of Medicine,
State University of New York,
Stony Brook, New York 11974.

INTRODUCTION

This contribution reviews current interpretations of the early evolution of catarrhines, the Old World higher primates. Like many aspects of primate evolution, this area of research has undergone considerable revision, reformulation and reinterpretation in recent years, in response to both new palaeontological material and new, more critical methods of phylogenetic analysis (e.g. Ciochon, 1983). Since this contribution emphasizes the palaeontological evidence for early catarrhine evolution and complements the preceding paper on the neontological evidence regarding catarrhine phylogeny it seems appropriate to begin with a consideration of the unique nature of the information about primate evolution that the fossil record provides.

While it may not be the touchstone for understanding phylogeny that it has often been touted to be (e.g. Patterson, 1981), the fossil record provides several unique forms of information that are valuable for understanding the phyletic relationships among living forms, in addition to its role as our only record of primates from earlier epochs. As a source of information that can be used to understand the evolutionary relationships among extant taxa, the fossil record provides two types of data that can be useful for understanding the history (and polarity) of individual anatomical features. Like ontogeny the fossil record is ordered in a temporal sequence that is often useful in deciding which of several character states may be more primitive, and also may provide evidence of morphological transformations through time that demonstrate not only the sequence of evolutionary change, but a verification of homologies that may otherwise be obscure. In addition, the fossil record provides evidence of a diversity of primate morphologies and of otherwise unknown combinations of characters that are useful for determining the history of evolutionary novelties. Finally, the fossil record provides direct evidence for the minimum age of evolutionary divergences of extant taxa and some evidence of the tempo and mode of change during the evolution of extant groups (e.g. Delson & Rosenberger, 1984).

More significant, and, in my opinion, more fascinating than the information that the fossil record provides about the evolutionary history of extant taxa, is the information that it provides about the primates of the past, animals that by and large would be otherwise unknown. It frequently provides evidence of primates, such as *Australopithecus afarensis*, that are intermediate between extant taxa (e.g. Johanson & White, 1979; Susman *et al.*, 1984). It provides evidence of animals such as the subfossil Malagasy lemurs (e.g., Jungers, 1980) that extend the known morphological and adaptive diversity of living groups well beyond that indicated by extant members and cautions us that our extant sample of primate diversity is usually a depleted sample as a result of rather recent 'accidents'. Finally, it provides our only record of whole radiations of primates or primate-like mammals such as the plesiadapiformes (Gingerich, this volume) with no close living relatives. It is these features of the fossil record that are both its unique contribution to evolutionary biology and the aspects that frequently cause us trouble when we try to fit fossil taxa into systematic schemes based on our impoverished extant sample of primates.

This paper has several objectives. First, I will review the fossil record of early catarrhines, primarily those from the Oligocene and early Miocene of Africa, considering both the adaptations and the phyletic relationships of these extinct anthropoids with respect to extant catarrhines. I will then discuss current interpretations regarding some broader aspects of catarrhine evolution, emphasizing the information from the fossil record that is unavailable from other sources. Finally, I will briefly outline a few of the major unresolved issues in early catarrhine evolution that are likely to be the focus of research efforts in the coming years.

OLIGOCENE CATARRHINES

The best fossil record of early catarrhines comes from the early Oligocene Jebel Qatrani Formation in the Fayum Depression of Egypt now dated at older than 31± 1.0 million years (Simons, 1967, 1972; Fleagle *et al.*, 1985). As a result of recent palaeontological and geological studies of the Fayum depression, there is now considerable evidence to indicate that the Jebel Qatrani Formation was deposited in a tropical rainforest environment (Figure 1), and most of the primates come from point bar deposits formed by large meandering rivers (Bown *et al.*, 1982).

There are 11 species of primates currently known from the Egyptian Oligocene. Of these, two poorly-known species are likely to have only marginal relationships with later catarrhines. *Afrotarsius chatrathi* is a small tarsiiform primate known from a single jaw that seems to have closest phenetic affinities with the extant *Tarsius* and/or the European microchoerine *Pseudoloris* (Simons & Bown, 1985). *Oligopithecus savagei* has been considered by various authorities to be an early hominoid (Simons, 1972), an adapid prosimian (Gingerich, 1980), or a primitive catarrhine with uncertain affinities to other catarrhine taxa (Szalay & Delson, 1979). Until additional material of these two taxa is available, their relationships with later anthropoids, including catarrhines, must remain speculative. The remaining nine primate species are more clearly higher primates. They are normally placed in two separate families of higher primates, the Parapithecidae and the Pliopithecidae (Szalay & Delson, 1979).

There are five species of parapithecids from three distinct stratigraphic levels. *Qatrania wingi* (Simons & Kay, 1983), from the lower part of the Jebel Qatrani Formation is the oldest species; *Apidium moustafi* is from

Figure 1. A reconstruction of the tropical riverine environment of the Fayum Province, Egypt in the early Oligocene. On the left a group of Aegyptopithecus zeuxis; in the upper right a small group of Propliopithecus chirobates. In the lower part of the picture is a troop of Apidium phiomense.

an intermediate level; and there are (at least) three different species from the upper part of the formation - Apidium phiomense, Parapithecus fraasi, and Simonsius grangeri (Kay & Simons, 1983).

The parapithecids were small compared with most extant catarrhines, with estimated body weights of 300g for the smallest (Qatrania wingi) to 3000g for the largest (Simonsius grangeri). Their dentitions indicate that parapithecids were predominantly frugivorous. The one species for which numerous limb elements are known, Apidium phiomense, was an arboreal leaper. The few cranial remains of Apidium and Simonsius indicate that they were diurnal. Evidence of dental sexual dimorphism in several parapithecid species suggests some type of polygynous social organization (Fleagle & Kay, 1985; Kay & Simons, 1980).

There are four species of Oligocene Pliopithecidae, usually placed in two genera, Aegyptopithecus (1 species) and Propliopithecus (3 species). The stratigraphic relationships of pliopithecid species are less well-established than those of the parapithecids. Two species, Aegyptopithecus zeuxis and Propliopithecus chirobates are found synchronously in the uppermost levels of the Jebel Qatrani Formation; the other two species are known, with certainty, only from type specimens, collected earlier this century with no stratigraphic information (Simons, 1972; Kay et al., 1981).

The four pliopithecid species are more uniform in size than the parapithecids, ranging in estimated body size from roughly 4000g in the smallest (Propliopithecus haeckeli) to over 6000g in the largest (Aegyptopithecus zeuxis). Reconstructions of their behaviour from dental, cranial and skeletal remains indicate that all were diurnal, frugivorous, arboreal quadrupeds (Fleagle & Kay, 1985). Like the parapithecids, they show evidence of dental sexual dimorphism, suggesting a polygynous social organization (Fleagle et al., 1980).

In their adaptive diversity, both groups of Oligocene higher primates were notably different from extant catarrhines in lacking large species, folivorous species, and terrestrial species. In their adaptive diversity as in many aspects of their morphology they were more comparable to extant platyrrhines than to extant catarrhines (Figure 2). In addition to having overall adaptive differences from extant catarrhines, it has become increasingly evident that both groups of Egyptian anthropoids are phyletically rather distant from extant catarrhines. Both preserve primitive characters that have subsequently been lost or modified in the two groups of extant catarrhines, cercopithecoid monkeys and hominoid apes (e.g. Fleagle & Kay, 1983; Andrews, 1985).

The higher primate (or anthropoid) status of the parapithecids is well established by many aspects of the dental and cranial morphology including their postorbital closure and fused mandibular symphysis. However, in many aspects of their anatomy including their retention of three premolars, in the morphology of their auditory region (Cartmill et al., 1981), and in many aspects of their limb anatomy, parapithecids are more similar to extant platyrrhines and appear to preserve primitive anthropoid features (many of

Figure 2. Comparative bar graphs showing the adaptive features of living and fossil higher primates (modified from Fleagle & Kay, 1985).

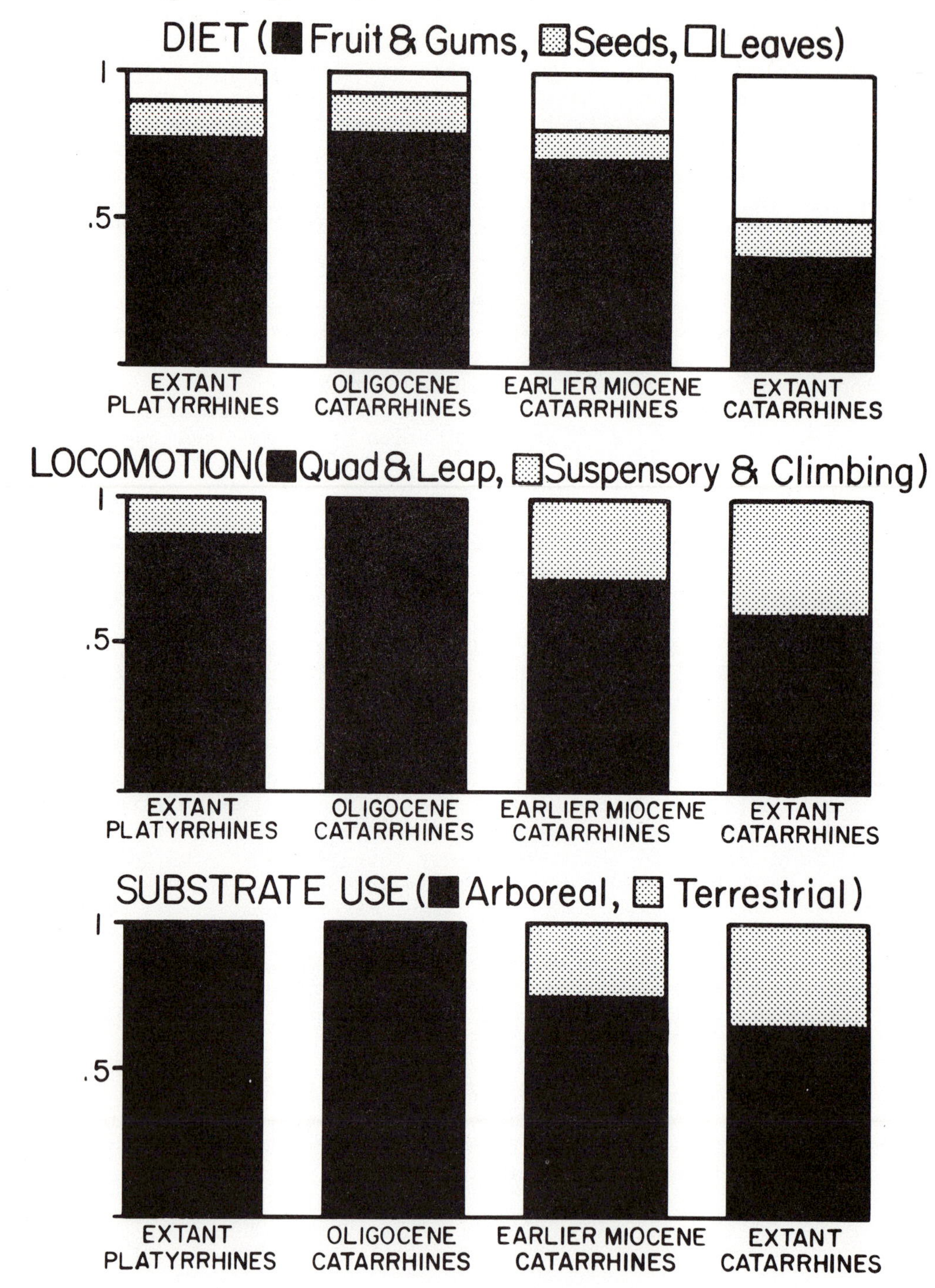

which are found in extant platyrrhines) rather than showing derived catarrhine features. The only features, aside from biogeography, that seem to link parapithecids uniquely with undoubted catarrhines among anthropoids are a few details of molar wear patterns (Kay, 1977a). In proportional features of the bony pelvis (Fleagle & Simons, 1979), and in the construction of the pterion region of the lateral wall of the skull (Fleagle & Rosenberger, 1983), parapithecids were probably more similar to platyrrhines than the catarrhines, as previously suggested. In fact, careful analysis of the dental, cranial and skeletal anatomy of parapithecids indicates that there is very little convincing evidence for features.

Thus present evidence suggests that parapithecids are only marginally, if at all, more closely related to catarrhines than to platyrrhines among known anthropoids (Figure 3). Earlier suggestions of a close phyletic relationship between parapithecids and cercopithecoid monkeys (e.g. Simons, 1972; Kay, 1977a), which would require parallel evolution of numerous catarrhine features (including premolar reduction, evolution of a tubular

Figure 3. A phyletic diagram for extant anthropoids showing the positions of the most primitive fossil catarrhines.

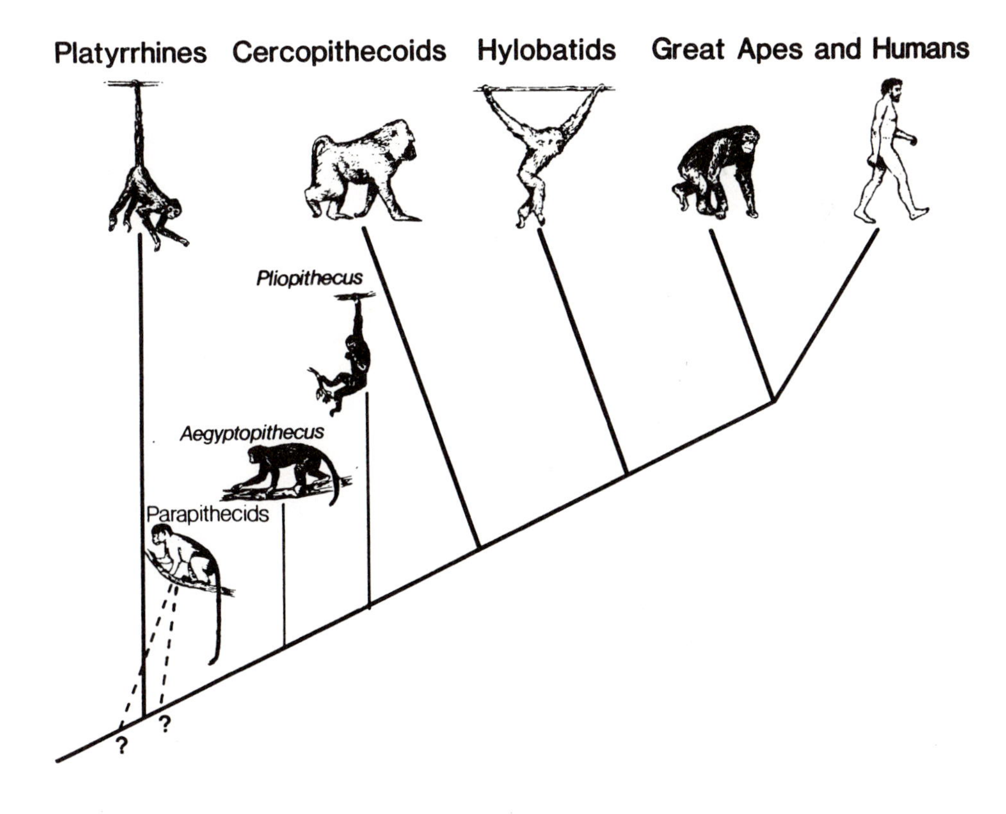

ectotympanic, deflation of the auditory bulla) independently in cercopithecoids and hominoids, have been abandoned by most authorities (Kay & Simons, 1983; Gingerich, 1984).

The pliopithecids Aegyptopithecus and Propliopithecus are more closely linked with later catarrhines by virtue of their similar dental formula (2.1.2.3) and numerous dental features they share with Miocene and later catarrhines. They also appear to have a catarrhine-like pterion region and a reduced postglenoid foramen. However, like the parapithecids, Aegyptopithecus and Propliopithecus retain numerous primitive anthropoid features in their cranial, auditory (Cartmill et al., 1981) and limb morphology (Fleagle, 1983) that distinguish them from extant catarrhines and link them phenetically with platyrrhines. Since these Oligocene anthropoids lack several catarrhine features shared by cercopithecoids and hominoids it seems most probable that the dental similarities they share with later hominoids are primitive for catarrhines (i.e. anthropoids with two premolars), and the Oligocene taxa are ancestral to both cercopithecoid monkeys and hominoid apes (e.g. Fleagle & Kay, 1983)(Fig. 3). The alternative view (Simons, 1972, 1985), that Aegyptopithecus and Propliopithecus are uniquely ancestral to 'apes' but not monkeys, requires parallel evolution of numerous cranial and skeletal features shared by extant catarrhines.

EARLY MIOCENE CATARRHINES

There is a temporal gap of approximately 15 million years between the early Oligocene anthropoids from Egypt and the next oldest catarrhines from the early Miocene of East (and North) Africa, which are approximately 19-17 million years old. The numerous early Miocene localities from Kenya and Uganda seem to sample a considerable diversity of environments from dense tropical rainforests at some sites to more open woodland and savannah environments in others (Bishop, 1967; Pickford, 1983). Each type of palaeoenvironment seems associated with a somewhat different fauna including the primate species (Figure 4), a factor that contributes greatly to the diversity of fossil catarrhines known from this time period. Two types of catarrhines are found in the numerous early Miocene localities of Kenya and Uganda, the more abundant and diverse (approximately 10 species and 6 genera) 'apelike' Proconsulidae and the less abundant and less diverse (3 species and 2 genera) early cercopithecoid monkeys (Harrison, 1982; Walker & Leakey, 1984; Leakey, Ms). The monkeys are also known from early Miocene sites in North Africa, and monkeys are more abundant, but no more diverse, in Middle Miocene sites of East Africa (Benefit, 1985).

The adaptive diversity of early Miocene catarrhines is much greater than that known for the Oligocene anthropoids and is comparable in gross measures to that found among extant catarrhines (e.g. Fleagle, 1978; Andrews, 1981; Fleagle & Kay, 1985)(Fig. 2). The proconsulid species ranged in estimated body size from about 4000g in the smallest (Micropithecus clarki) to more than 40,000g in the largest (Proconsul major). Their dentitions indicate that these early Miocene apes were mainly frugivorous, but a few species were folivorous (Kay, 1977b). They demonstrated a diversity of locomotor adaptations; at least one species was

Figure 4. A reconstruction of the early Miocene site of Rusinga Island, Kenya showing the diversity of proconsulids from this locality. In the upper left a small group of Proconsul africanus; in the upper right a family of Dendropithecus macinnessi, in the centre a small group of Limnopithecus legetet and on the ground, Proconsul nyanzae.

suspensory while others were arboreal and terrestrial quadrupeds (Andrews & Aiello, 1984; Fleagle & Kay, 1985).

The more poorly known and taxonomically less diverse early cercopithecoids Prohylobates and Victoriapithecus provide less evidence of adaptive diversity for monkeys during this time period (e.g. Andrews, 1981). The three described species ranged in estimated body weight from about 7,000g to about 20,000g. Dental morphology was remarkably similar in all species and indicates a folivorous/frugivorous diet (see Kay & Covert, 1984, for a discussion of the problems involved in reconstructing the dietary adaptations of early monkeys). The few limb bones of these early monkeys suggest quadrupedal abilities similar to extant species (e.g. Cercopithecus aethiops) with partly terrestrial and partly arboreal habits. The marked dental dimorphism suggests polygynous social organization.

Thus, in overall adaptive diversity, the early Miocene catarrhines appear to have been comparable to extant catarrhines (Figure 2). Early Miocene and Recent catarrhines overlap completely in size diversity and in diversity of presumed dietary abilities. Early Miocene proconsulids were distinctly different in skeletal morphology from extant catarrhines in that they lacked the skeletal specializations characteristic of either extant cercopithecoids or extant hominoids (e.g. Rose, 1983; Fleagle, 1983). However, in gross locomotor abilities and presumed substrate preferences they seem to have filled, in their own way, all of the major locomotor niches (e.g., suspensory, arboreal quadruped, terrestrial quadruped) utilized by extant catarrhines (Fleagle & Kay, 1985).

Although the proconsulids have been regarded traditionally as phyletically close to extant hominoids and the early monkeys are undoubtedly cercopithecoids (e.g. Szalay & Delson, 1979), there is increasing evidence that both groups are less closely allied with living monkeys and apes than has been previously believed (e.g., Harrison, 1982; Andrews, 1985; Leakey, Ms; Benefit, 1985). Some of the proconsulids possessed most of the features of the auditory region and limb skeleton that unite living catarrhines (but were lacking in Aegyptopithecus) although there are few derived features that link the proconsulids uniquely with living hominoids to the exclusion of cercopithecoids. Furthermore, there are numerous features uniting extant hominoids that the proconsulids lacked (Harrison, 1982; Andrews, 1985; Walker & Teaford, 1985). Thus, these early Miocene 'apes' were certainly catarrhines but less clearly hominoids in the modern sense (Figure 5). Feldesman (1982) has called them 'formative apes'.

In the case of Prohylobates and Victoriapithecus (from both the early Miocene and the middle Miocene), there is no doubt from the available dental evidence that they are catarrhines uniquely related to extant cercopithecoids. However, they do not readily fit into either of the extant subfamilies, Colobinae or Cercopithecinae. Rather, it seems that von Koenigswald's (1969) original allocation to a more primitive subfamily, the Victoriapithecinae, is a more accurate phyletic assessment of their relationship to living catarrhines since they retain several primitive features (such as trigonids on some upper molars, and retention of a dorsal epitrochlear fossa) that are absent in all known later cercopithecoids in both subfamilies (e.g. Leakey, 1985; Benefit, 1985).

DISCUSSION AND SUMMARY

The fossil record of early anthropoids in Africa provides us with considerable information about early catarrhine evolution that is unavailable (or at least was totally unsuspected by most workers) from our knowledge of extant primates. Above all, it shows that many of the Old World higher primates of previous epochs were notably different from the monkeys and apes of today. The ecological and morphological characteristics of extant catarrhines have appeared gradually over the last 35 million years.

Current fossil evidence indicates that a primitive anthropoid lineage leading to extant catarrhines was present in Africa by the early Oligocene. However, the earliest Old World higher primates, the parapithecids and pliopithecids from the early Oligocene of Egypt had very few features that would link them phenetically with extant catarrhines. Rather, a comparative study of catarrhine adaptations over the last 35 million years indicates that Old World anthropoids made a major adaptive breakthrough

Figure 5. A phyletic diagram for extant anthropoids showing the probable phyletic relationships of early and middle Miocene catarrhines.

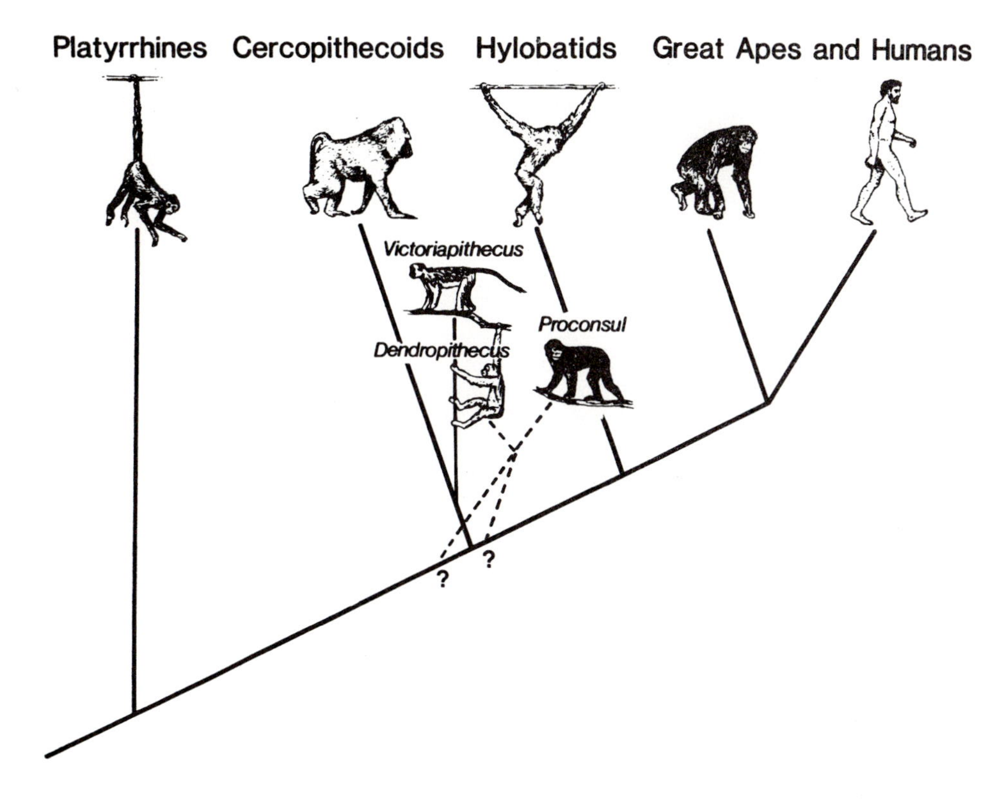

sometime in the late Oligocene (Fleagle & Kay, 1985). Gross adaptive features such as large size, folivory, and terrestrial locomotion are unknown among the Oligocene anthropoids, but are present among the early Miocene catarrhines (Figures 2 & 4). Since these adaptive features are also unknown or rare among platyrrhines as well, it seems quite likely they were only 'invented' by catarrhines sometime in the late Oligocene.

The initial divergence of the catarrhine lineage from a primitive anthropoid stock must have taken place by the early Oligocene, and the major ecological characteristics of living catarrhines seem to have been evolved by the early Miocene. However, the appearance of fossil catarrhines attributable to extant catarrhine higher taxa (e.g., Pongidae, Hominidae, Hylobatidae, Colobinae) appears to have been an even more recent phenomenon. No extant catarrhines seem to be living fossils. That is, no modern forms appear to have retained an Oligocene or early Miocene morphology with little modification for 20 or 30 million years (Delson & Rosenberger, 1984). The catarrhine fossil record provides us with a sequence of increasingly modern-looking taxa, or adaptive radiations, through time (Figure 6), not an early radiation of diverse types that have persisted separately over the course of catarrhine evolution.

Such a pattern of increasingly modern-looking taxa in catarrhine evolution unfortunately introduces considerable problems for standardizing the higher level taxonomy of the group. The phylogeny of early catarrhine evolution can only be reproduced accurately, if at all, by a systematic scheme that would entail a very unwieldy proliferation of terms (e.g. Gingerich, 1979). Alternatively, the use of taxonomic categories traditionally based on extant groups for fossil taxa that are much more primitive can imply an antiquity for extant taxa that is misleading. We can fairly easily mark the appearance of an individual morphological feature, but identifying the first appearance of a taxonomic group such as catarrhines or hominoids usually requires a more or less arbitrary decision about where within a lineage we wish to assign the traditional taxonomic labels.

For example, consider the possible alternatives in deciding which of the Oligocene and Miocene Old World anthropoids are catarrhines (Figure 7). If we wish to define catarrhines as those primates sharing the same suite of characters that unite extant cercopithecoids and hominoids, the victoriapithecines and the proconsulids may qualify, but certainly nothing more primitive would do so. If we allow those fossils that seem primitive enough in most characters to have given rise to all extant catarrhines, but to no other group of extant primates, *Pliopithecus*, *Aegyptopithecus*, and *Propliopithecus* would qualify. However, this group of catarrhines would be united by remarkably few characters, most obviously the possession of two premolars. Finally, if we wanted to include among the catarrhines all non-platyrrhine anthropoids, we could include the parapithecids, which at least seem to lack the derived features found among extant platyrrhines. Defining the Hominoidea is no less arbitrary a process (Fleagle & Kay, 1983; Ciochon, 1983).

In addition to the nomenclatural options created by this surfeit of intermediate forms, there are unresolved debates about the exact placement of many of the fossil taxa vis a vis others. There are two main reasons for these problems - lack of sufficient characters for analysis, and parallelisms. Lack of sufficient characters for a confident analysis of the phyletic position of fossils is generally due to their fragmentary nature. With additional material, the seemingly critical parts will hopefully

Figure 6. A general evolutionary tree showing the sequential nature of catarrhine evolution with increasingly modern-looking fossil taxa rather than many ancient lineages.

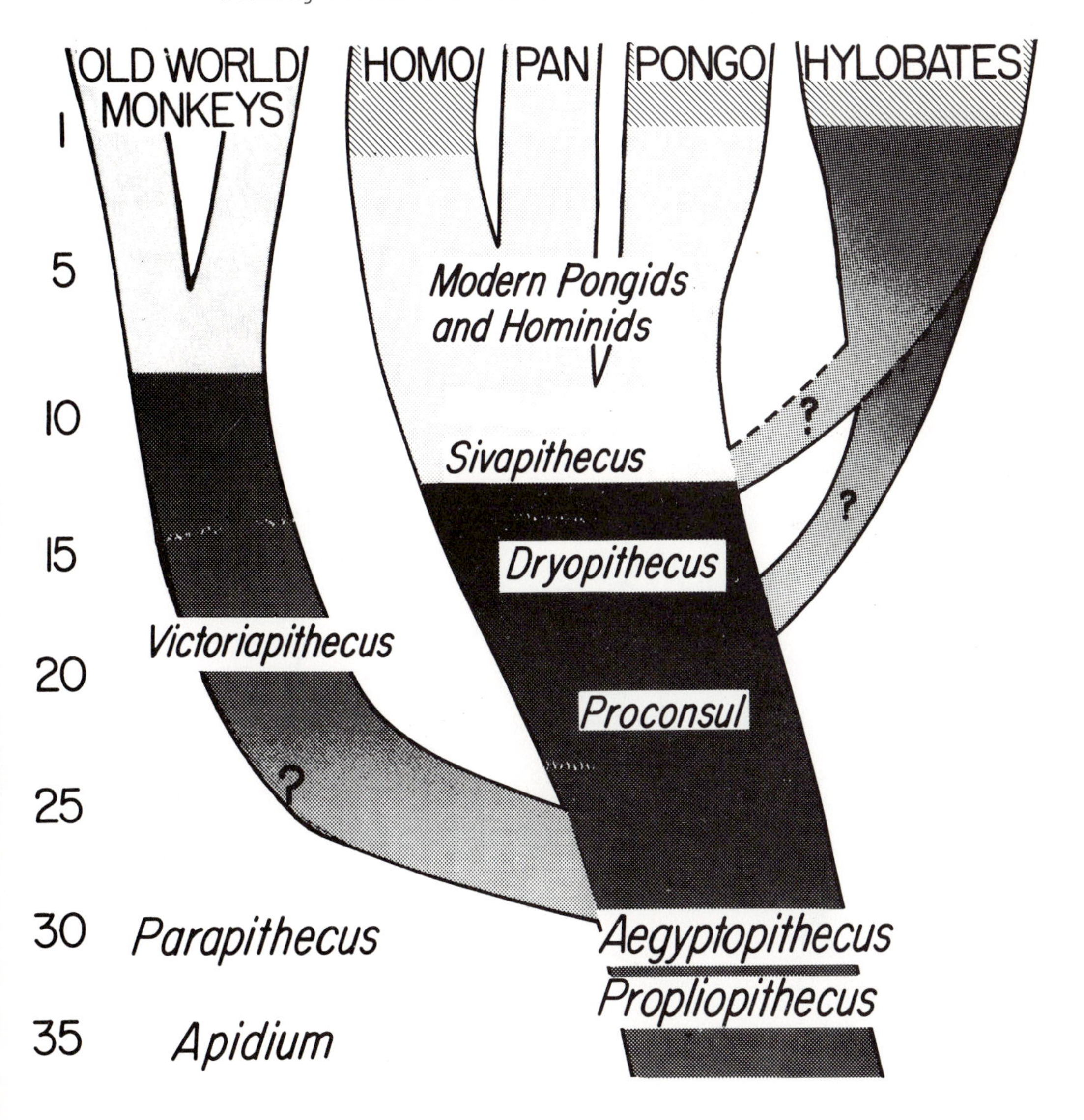

become available. For many extant taxa, lack of suitable characters for an appropriate analysis more frequently comes from another source - extensive unique specializations. Gibbons are the most notable example among extant catarrhines. Although there are thousands of complete skeletons of this ape available for scientists to study, so much of its skeletal anatomy is uniquely specialized that it has proved extremely difficult to place relative to more conservative hominoids (e.g. Delson & Andrews, 1975). Among prosimians, Daubentonia provides comparable difficulties (e.g. Oxnard, 1981; Tattersall, 1981).

Parallelism is the bane of phylogenetic analysis, and unfortunately, it is a serious problem that will not go away (Cartmill, 1982; Gosliner & Ghiselin, 1984). There is an unfortunate dullness to Occam's razor when the most parsimonious phylogeny has 25% reversals and the next 27%. Unfortunately, such comparisons are not unusual in phylogenetic analyses of extant taxa that consider numerous characters. In studies of fossil primates, the comparisons more usually involve so few characters that no one calculates the percentages, but parallelisms and reversals have unquestionably been rampant.

Figure 7. A phyletic diagram showing the position of early fossil catarrhines in relationship to extant anthropoids.

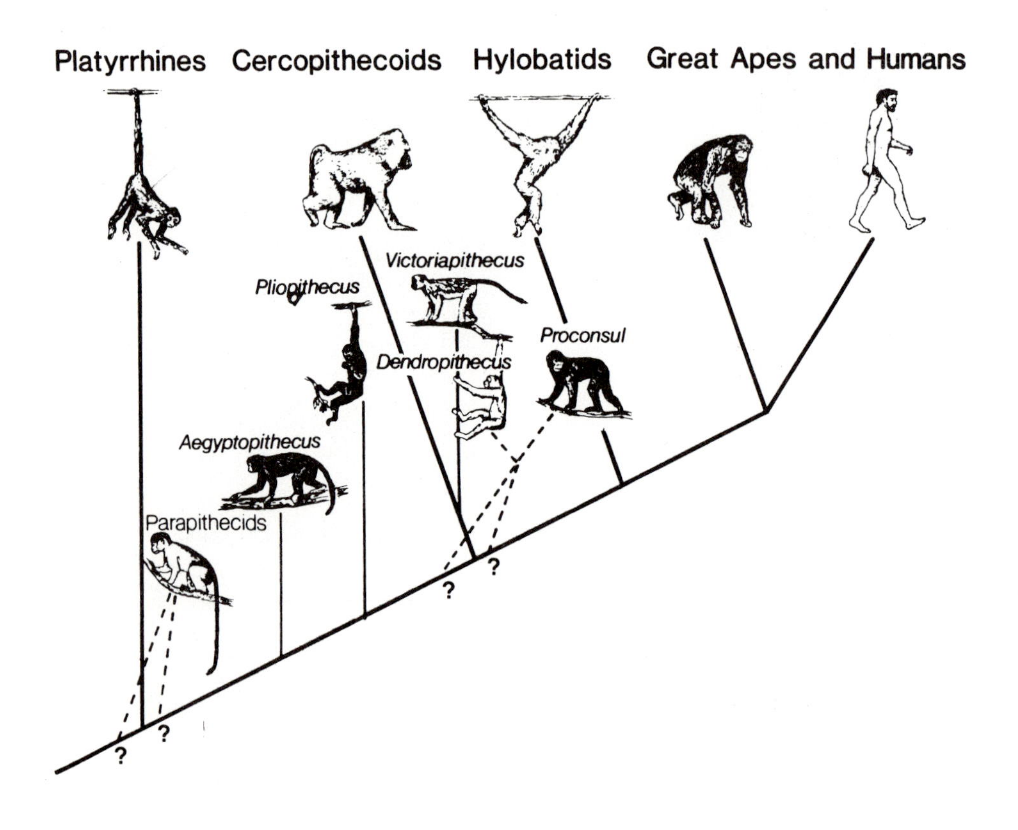

The apparently continuous nature of catarrhine evolution, in which modern-looking taxa evolved bit by bit over the millions of years provides systematists with some very difficult problems. However, it provides anatomists with a remarkable record of the way in which the bodies of extant catarrhines acquired their present-day form. In retrospect we can more or less date appearances of various aspects of our own anatomy over the last forty million years (Figure 8). The anthropoid features such as postorbital closure and fusion of the mandibular symphysis probably appeared sometime in the late Eocene or early Oligocene. Among the features that characterize extant catarrhines as a group, loss of the third premolar was one of the first to evolve, probably during the early Oligocene. Evolution of a tubular ectotympanic (and probably loss of an entepicondylar foramen) first occurred sometime in the late Oligocene. Evolution of the unique hominoid wrist joint (with a reduced ulnar styloid process) took place even later than that, and so forth.

Catarrhine evolution is full of intermediate forms or missing links. It is both ironic and unfortunate that in their attempts to identify the earliest gibbon, the earliest great ape, or the earliest hominid, palaeoanthropologists have unwittingly tried to portray these fossil primates as more like extant forms than they really were, and glossed over the most interesting aspect of the fossil record. Surely it is time we recognized this valuable record for what it is and used it to understand our evolutionary past.

PROGNOSIS

Despite the richness of the fossil record of early catarrhines compared with our record of other major radiations of primates (platyrrhines for example), our knowledge of this aspect of primate evolution is nevertheless based on a very limited sample in both time and geography. New fossils will undoubtedly bring new surprises and evidence of even more fossil catarrhines whose existence we could never predict. In addition, there are numerous aspects of the biology of known catarrhines, both living and fossil, that remain poorly understood, but are critical for understanding the evolution of the group. The following is a brief list of topics that, in my opinion, are poorly understood at present and deserve more intensive investigation:

1. The parapithecids are the earliest higher primates known in the fossil record and are the most primitive of all Old World anthropoids. However, their position relative to the common ancestor of all higher primates, to platyrrhines, to catarrhines, and even to tarsiiform primates is very poorly understood. Although they are known from relatively abundant dental and skeletal material, and less abundant cranial remains, most analyses have avoided their many primitive features and emphasised their supposed relationship to cercopithecoid monkeys. More careful study of this group in conjunction with fossil tarsiiform primates and primitive platyrrhines should help resolve their proper phyletic position and help clarify many of the relationships among early anthropoids and other haplorhines.

2. Although *Aegyptopithecus* and *Propliopithecus* are more clearly linked to later catarrhines than are the parapithecids, they have a remarkably large number of cranial and dental features that appear superficially to

Figure 8. A skeleton of Homo sapiens showing the relative antiquity of various features of the human skeleton in millions of years.

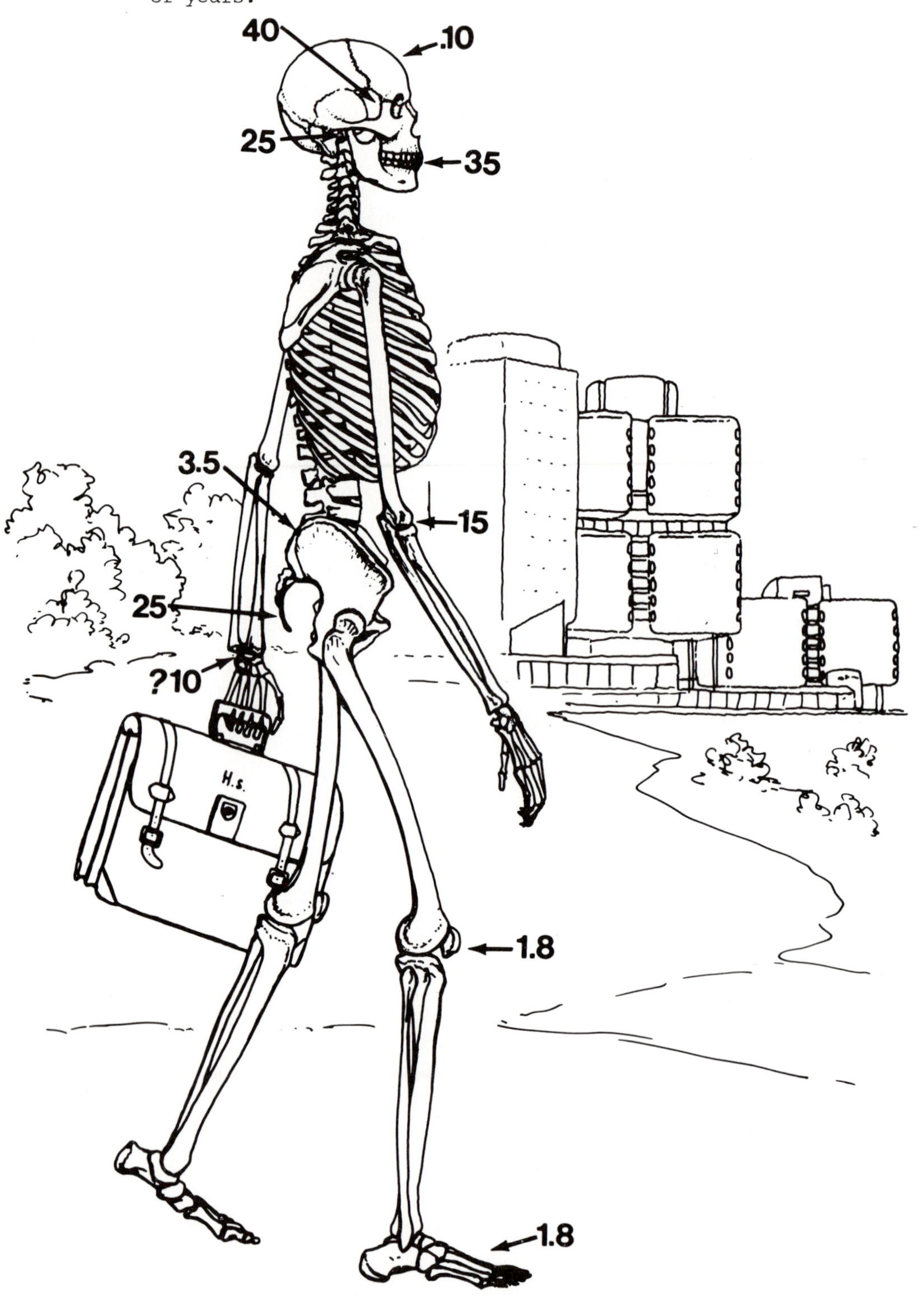

be more primitive than the conditions found among platyrrhines and consequently those expected in the last common ancestor of all extant anthropoids (e.g. Harrison, 1982; Fleagle & Simons, 1982). Both the morphological diversity among these Oligocene catarrhines (e.g. Simons, 1984) and the evolution of platyrrhines need more careful examination to resolve this paradox and its implications for early anthropoid evolution.

3. Most studies of early catarrhine evolution have focussed on the origin of hominoids (e.g. Simons, 1967) both because we are one and because the Fayum apes share dental features with living apes and humans. However, with the increasing realization that many ape-like features among early catarrhines are primitive for Old World anthropoids, the more critical question for catarrhine phylogeny is the origin of cercopithecoid monkeys (e.g. Delson, 1975). Perhaps more careful study of early cercopithecoids with an eye for primitive characters, in conjunction with detailed examination of the pliopithecids would yield more insight into cercopithecoid origins (e.g. Butler, this volume).

4. The early Miocene proconsulids were a very diverse group. Although it now seems clear that individual species are not ancestral to particular genera of extant hominoids, it is likely that some members of that radiation bear a closer relationship to later apes than do others. *Proconsul africanus* has been suggested to be more advanced in its skeletal morphology than many other species, if only because more of its anatomy is known. Pickford (1983) has argued that *Proconsul nyanzae* is a likely candidate for the ancestry of all later 'thick-enamelled apes', and other workers (e.g. Butler, this volume; Harrison, 1985) have argued for the origin of *Oreopithecus* in '*Rangwapithecus*'. The diversity of early Miocene proconsulids clearly deserves more careful investigation.

5. One of the most paradoxical of the early catarrhines is *Pliopithecus*. Recognized as an early gibbon for over a century, this genus is now generally regarded as a more primitive catarrhine, one that may have preceded the divergence of extant hominoids and cercopithecoids (e.g. Fleagle, 1984). However, the recent discovery of a *Pliopithecus*-like taxon from the late Miocene locality of Lufeng, China (Wu & Pan, 1984) again raises the issue of the relationship between *Pliopithecus* and gibbons.

6. Closely linked with the problem of *Pliopithecus* is the vagueness that surrounds our understanding of the evolutionary relationship of gibbons, and in particular their relationship to other catarrhines (Delson & Andrews, 1975). There is little doubt that gibbons are the most primitive of extant apes; however, we lack a detailed understanding of what morphological features probably characterized the last common ancestor of great apes and gibbons and how many of the similarities shared by gibbons and great apes are likely to be parallelisms. Most recent studies of gibbon anatomy have emphasized their functional uniqueness, rather than their phylogenetic position. A phyletically-oriented study of gibbon anatomy is desperately needed.

Acknowledgements

I am grateful to the Primate Society of Great Britain and the Anatomical Society of Great Britain and Ireland for inviting me to participate in the symposium that led to this book. Many of my colleagues have contributed their ideas and suggestions to the contents of this paper. In particular, I thank Richard Kay, Terry Harrison, and Alfred Rosenberger for many fruitful discussions about early catarrhine evolution, and William Jungers, David Krause, Susan Larson and John Watters for comments on this manuscript. Stephen Nash and Lucille Betti drew the figures. Much of the research described in this contribution was funded by a fellowship from the John Simon Guggenheim Memorial Foundation and research grants 8210949 and 8209937 from the National Science Foundation.

References

Key references

Andrews, P. (1985). Family group systematics and evolution among catarrhine primates. *In* Ancestors: The Hard Evidence, ed. E. Delson, pp. 14-22. New York: Alan R. Liss, Inc.

Fleagle, J.G. (1983). Locomotor adaptations of Oligocene and Miocene hominoids and their phyletic implications. *In* New Interpretations of Ape and Human Ancestry, eds. R.L. Ciochon & R.S. Corruccini, pp. 301-24. New York: Plenum Press.

Fleagle, J.G. & Kay, R.F. (1983). New interpretations of the phyletic position of Oligocene hominoids. *In* New Interpretations of Ape and Human Ancestry, eds. R.L. Ciochon & R.S. Corruccini, pp. 181-210. New York: Plenum Press.

Harrison, T. (1982). Small-bodied apes from the Miocene of East Africa. Unpublished Ph.D. Dissertation. University College, London.

Kay, R.F. & Covert, H.H. (1984). Anatomy and behavior of extinct primates. *In*, Food Acquisition and Processing in Primates, eds. D.J.Chivers B.A. Wood & A.Bilsborough, pp. 467-508. New York: Plenum Press.

Kay, R.F., Fleagle, J.G. & Simons, E.L. (1981). A revision of the Oligocene apes of the Fayum province, Egypt. Amer. J. Phys. Anthrop. *55*, 293-322.

Simons, E.L. (1967). The earliest apes. Sci. Amer., *217*, 28-35.

Main References

Andrews, P. (1981). Species diversity and diet in monkeys and apes during the Miocene. *In* Aspects of Human Evolution, ed. C.B. Stringer, pp. 25-61. London: Taylor & Francis.

Andrews, P. (1985). Family group systematics and evolution among catarrhine primates. *In* Ancestors: The Hard Evidence, eds. E. Delson, pp. 14-22. New York: Alan R. Liss, Inc.

Andrews, P. & Aiello, L. (1984). An evolutionary model for feeding and positional behavior. *In* Food Acquisition and Processing in Primates, eds. D.J. Chivers, B.A. Wood & A. Bilsborough, pp. 422-60. New York: Plenum Press.

Benefit, B. (1985). Dental remains of Victoriapithecus from the Maboko Formation. S.V.P. News Bulletin, *133*, 21.

Bishop, W.W. (1967). The later Tertiary in East Africa - volcanics, sediments, and faunal inventory. *In* Background to Evolution in Africa, eds. W.W. Bishop and J.D. Clark, pp. 31-56. Chicago: University of Chicago Press.

Bown, T.M., Kraus, M.J., Wing, S.L., Fleagle, J.G., Tiffany, B., Simons, E.L. & Vondra, C.F. (1982). The Fayum forest revisited. J. Human Evolution, *11*, 31-35.

Cartmill, M. (1982). Basic primatology and prosimian evolution. *In* Fifty Years of Physical Anthropology, ed. F. Spencer, pp. 147-86. New York: Academic Press.

Cartmill, M., MacPhee, R.D.E. & Simons, E.L. (1981). Anatomy of the temporal bone in early anthropoids with remarks on the problem of anthropoid origins. Am. J. Phys. Anthropol. *56*, 3-22.

Ciochon, R.L. (1983). Hominoid cladistics and the ancestry of modern apes and humans; a summary statement. *In* New Interpretations of Ape and Human Ancestry, eds. R.L. Ciochon & R.S. Corruccini, pp. 781-843. New York: Plenum Press.

Delson, E. (1975). Toward the origin of the Old World monkeys. Acts. C.N.R.S. Coll. Int., *218*, 839-50. Paris: CNRS.

Delson, E. & Andrews, P. (1975). Evolution and interrelationships of the catarrhine primates. *In* Phylogeny of the Primates: a Multidisciplinary Approach, eds. W.P. Luckett & F.S. Szalay, pp. 405-66. New York: Plenum Press.

Delson, E. & Rosenberger, A.L. (1984). Are there any anthropoid primate living fossils? *In* Living Fossils, eds. N. Eldridge & S.M. Stanley, pp. 50-61. New York: Springer Verlag.

Feldesman, M.R. (1982). Morphometric analysis of the distal humerus of some Cenozoic catarrhines; the late divergence hypothesis revisited. Amer. J. Phys. Anthropol., *59*, 173-95.

Fleagle, J.G. (1978). Size distributions of living and fossil primate faunas. Paleobiology, *4*, 67-76.

Fleagle, J.G. (1983). Locomotor adaptations of Oligocene and Miocene hominoids and their phyletic implications. *In* New Interpretations of Ape and Human Ancestry, eds. R.L. Ciochon & R.S. Corruccini, pp. 301-24. New York: Plenum Press.

Fleagle, J.G. (1984). Are there any fossil gibbons? *In* The Lesser Apes, eds. H. Preuschoft, D.J. Chivers, W.Y. Brockelman & N. Creel, pp. 431-47. Edinburgh: Edinburgh University Press.

Fleagle, J.G., Bown, T.M., Obradovich, J.D., & Simons, E.L. (1985). How old are the Fayum Primates? *In* Proc. Xth Cong. Int. Primat. Soc. eds. R.E.F. Leakey & J. Else. Cambridge: Cambridge University Press (in press).

Fleagle, J.G. & Kay, R.F. (1983). New interpretations of the phyletic position of Oligocene hominoids. *In* New Interpretations of Ape and Human Ancestry, eds. R.L Ciochon & R.S. Corruccini, pp. 181-210. New York: Plenum Press.

Fleagle, J.G. & Kay, R.F. (1985). The paleobiology of catarrhines. *In* Ancestors: The Hard Evidence, ed. E. Delson, pp. 23-36. New York: Alan R. Liss.

Fleagle, J.G., Kay, R.F. & Simons, E.L. (1980). Sexual Dimorphism in early anthropoids. Nature, *287*, 328-30.

Fleagle, J.G. & Rosenberger, A.L. (1983). Cranial morphology of the earliest anthropoids. *In* Morphologie Evolutive, Morphogenese Du

Crane et Origine de L Homme, ed. M. Sakka, pp. 141-53. Paris: CNRS.

Fleagle, J.G. & Simons, E.L. (1979). Anatomy of the bony pelvis in parapithecid primates. Folia primatol. *31*, 176-86.

Fleagle, J.G. & Simons, E.L. (1982). The Humerus of Aegyptopithecus zeuxis: A Primitive Anthropoid. Amer. J. Phys. Anthrop., *59*, 175-93.

Gingerich, P.D. (1979). Paleontology, Phylogeny and Classification: An example from the mammalian fossil record. Syst. Zool. *28*, 451-64.

Gingerich, P.D. (1980). Eocene adapidae, paleobiogeography, and the origin of South American Platyrrhini. *In* Evolutionary Biology of the New World Monkeys and Continental Drift, eds. R.L. Ciochon & A.B. Chiarelli, pp. 123-38. New York: Plenum Press.

Gingerich, P.D. (1984). Primate Evolution: Evidence from the fossil record, comparative morphology, and molecular biology. Yrbk. Phys. Anthrop. *27*, 57-72.

Gosliner, M.T. & Ghiselin, T.M. (1984). Parallel evolution in opisthobranch gastropods and its implication for phylogenetic methodology. Syst. Zool. *33*, 255-74.

Harrison, T. (1982). Small-bodied apes from the Miocene of East Africa. Unpublished Ph.D. Dissertation. University College, London.

Harrison, T. (1985). African oreopithecids and the origin of the family. Amer. J. Phys. Anthropol. *66*, 180.

Johanson, D.C. & White, T.D. (1979). A systematic assessment of early African hominids. Science *203*, 321-30.

Jungers, W.L. (1980). Adaptive diversity in subfossil Malagasy prosimians. Z. Morph. Anthrop., *71*, 177-86.

Kay, R.F. (1977a). The evolution of molar occlusion in the Cercopithecidae and early catarrhines. Amer. J. Phys. Anthropol., *46*, 327-52.

Kay, R.F. (1977b). Diets of early Miocene African hominoids. Nature, *268*, 628-30.

Kay, R.F. & Covert, H.H. (1984). Anatomy and behavior of extinct primates. *In* Food Acquisition and Processing in Primates, eds. D.J. Chivers B.A. Wood & A. Bilsborough, pp. 467-508. New York: Plenum Press.

Kay, R.F., Fleagle, J.G. & Simons, E.L. (1981). A revision of the Oligocene apes of the Fayum province, Egypt. Amer. J. Phys. Anthrop. *55*, 293-322.

Kay, R.F. & Simons, E.L. (1980). The ecology of Oligocene African anthropoidea. Int. J. Primatol., *1*, 21-37.

Kay, R.F. & Simons, E.L. (1983). Dental formulae and dental eruption patterns in Parapithecidae (Primates, Anthropoidea). Amer. J. Phys. Anthropol., *62*, 363-75.

Leakey, M. Early Miocene Cercopithecids from Nuluk, Northern Kenya. Folia primatol. *44*, 1-14.

Oxnard, C.E. (1981). The uniqueness of Daubentonia. Amer. J. Phys. Anthropol., *54*, 1-21.

Patterson, C. (1981). Significance of fossils in determining evolutionary relationships. Ann. Rev. Ecol. & Syst., *12*, 195-224.

Pickford, M. (1983). Sequence and environment of the lower and middle Miocene hominoids of Western Kenya. *In* New Interpretations of

Ape and Human Ancestry, eds. R.L. Ciochon & R.S. Corruccini, pp. 421-40. New York: Plenum Press.
Rose, M.D. (1983). Miocene hominoid postcranial morphology: monkey-like, ape-like, neither or both. *In* New Interpretations of Ape and Human Ancestry, eds. R.L. Ciochon & R.S. Corruccini, pp. 407-17. New York: Plenum Press.
Simons, E.L. (1967). The earliest apes. Sci. Amer., *217*, 28-35.
Simons, E.L. (1972). Primate Evolution. New York: Macmillan.
Simons, E.L. (1984). Dawn Ape of the Fayum. Nat. Hist., *93*, 18-21.
Simons, E.L. (1985). Origins and characteristics of the first hominoids. *In* Ancestors: The Hard Evidence, ed. E. Delson, pp. 37-41. New York: Alan R. Liss.
Simons, E.L. & Bown, T.M. (1985). Afrotarsius chatrathi, first tarsiiform primate (?Tarsiidae) from Africa. Nature *313*, 475-77.
Simons, E.L. & Kay, R.F. (1983). Qatrania, a new basal anthropoid from the Fayum, Oligocene of Egypt. Nature, *304*, 624-26.
Susman, R.L., Stern, J.T. Jr., & Jungers, W.L. (1984). Arboreality and bipedality in the Hadar Hominids, Folia Primatol., *43*, 113-56.
Szalay, F.S. & Delson, E. (1979). Evolutionary History of Primates. New York: Academic Press.
Tattersall, I. (1981). Two misconceptions of phylogeny and classification. Amer. J. Phys. Anthropol., *57*, 13.
von Koenigswald, G.H.R. (1969). Miocene Cercopithecoidea and Oreopithecidae from the Miocene of East Africa. Fossil Vertebrates of Africa, *1*, 39-52.
Walker, A. & Leakey, R.E.F. (1984). New fossil primates from the Lower Miocene site of Buluk, N. Kenya. Amer. J. Phys. Anthrop., *63*, 232.
Walker, A.C. & Teaford, M.F. (1985). New information concerning the R114 Proconsul site, Rusinga Island. *In* Proc. Xth Cong. Int. Primat. Soc., eds. R.E.F. Leakey & J. Else. Cambridge: Cambridge University Press, in press.
Wu Rukang & Pan Yuerong (1984). A late Miocene gibbon-like primate from Lufeng, Yunnan Province. Acta Anthropologica Sinica, *III*, 193-200.

8 MOLECULAR SEQUENCES AND HOMINOID PHYLOGENY

M.J. Bishop,
University of Cambridge,
Department of Zoology,
Downing Street, Cambridge CB2 3EJ, U.K.

A.E. Friday,
University of Cambridge,
Department of Zoology,
Downing Street, Cambridge CB2 3EJ, U.K.

INTRODUCTION

Attempts to reconstruct the evolutionary tree leading to extant hominoids have produced clear agreement neither on the topology of the tree nor on the dating of its branching points. A considerable amount of molecular sequence data is available from the extant hominoids, and progress continues to be made in collecting both protein and nucleic acid sequences. Recently, for example, Goodman *et al.* (1983) have reported fully determined alpha and beta haemoglobin sequences for *Pan troglodytes* and *Gorilla gorilla*, and previously undetermined sequences for *Pan paniscus*. Anderson *et al.* (1981) determined the entire sequence of the human mitochondrial genome (c16 kilo bases), and Brown *et al.* (1982) reported the sequences of a c900 base Hind III restriction fragment from the mitochondrial genome of man and other extant hominoids.

Comparative molecular data have provided some resolution of the hominoid evolutionary tree in the sense that earlier studies contributed to the rejection of schemes which grouped together chimpanzees, gorilla and orang utan but which excluded man from such a grouping. The immunochemical work of Sarich & Wilson (1967) and of Goodman & Moore (1971) had much to do with this change, and the subsequent analysis of comparative protein sequence data has tended to confirm their conclusions. A synthesis of molecular and morphological evidence supported the *Homo*-African ape grouping (Andrews & Cronin, 1982), but the separation of chimpanzees, gorilla and orang from the lineage leading to *Homo* has recently been given support by Kluge (1983).

When, however, one looks to molecular data for elucidation of other details of the hominoid evolutionary tree, then their contribution has been relatively poor. In the case of the globin proteins, for example, there are few differences between the species of extant hominoids. Accordingly, the information to support any particular pattern of branching over another may be insufficient or lacking. Rigorous application of the parsimony approach to myoglobin sequence data resulted in a tree which was judged unlikely on the available morphological evidence, and based on uncomfortably few reconstructed amino acid substitutions (Romero-Herrera *et al.* 1976).

MODELS OF EVOLUTIONARY CHANGE

We see most arguments about the estimation of evolutionary trees in terms of the underlying assumptions about the various models of change. Nearly all published studies of hominoid molecular sequence data have used the maximum parsimony approach for the reconstruction and evaluation of trees. This approach is based upon the assumption that all evolutionary change is divergent. In practice, of course, it is apparent that such an assumption is unrealistic, and it is relaxed so that the approach becomes a search for that tree which invokes the smallest number of convergent events. It is not clear that a formal model for the distribution of reconstructed events under the parsimony criterion can be formulated within a fully acceptable statistical framework (Felsenstein, 1983), or that, even if it can, whether such a model is 'biologically' realistic. More particularly, unless a model can be fully defined it is difficult to know how to choose between competing tree hypotheses by any accepted method of statistical inference.

It is important that progress be made in developing probabilistic models of the process of evolutionary change at the molecular level, both because such approaches are appropriate to modelling evolutionary events and because they force clear statements of the assumptions being made. As information becomes increasingly available it will be possible to refine existing probabilistic models.

Exponential failure model

We first consider a model of evolutionary change for DNA sequences. A nucleic acid sequence may be viewed as a family of independent, identically distributed, random variables (one for each site) which have the Markov property: conditional independence of the next transition and the past transition, given the present state.

Suppose that the disturbances which result in failure to copy DNA sequence correctly arise according to a Poisson process. In the simplest form of the model the failure rate is constant over all the sites. A given site, which may be in any of the states A,C,G or T, may remain in its previous state over a length of time t. When failure to copy correctly occurs, let the base be replaced by A,C,G or T with equal probability (1/4). For a Poisson process the probability $(1-e^{-ut})$ that failure will occur over time t (at rate u), implies the probability e^{-ut} that it will not. For a given site, the probability of finding base X at time zero and base Y at time t is therefore $(1-e^{-ut})/4$, and the probability of finding the same base at time t is $(1+3^{-ut})/4$. A model of this type was proposed by Felsenstein (1981), and the development of probabilistic models, their assumptions and current shortcomings have been dealt with by Bishop & Friday (in press).

Maximum likelihood estimation

Given some data and a probabilistic model of change, we may examine the likelihood of competing hypotheses of tree topology. Likelihoods of hypotheses may be compared for the same data. Felsenstein (1981) has dealt with the application of the method of maximum likelihood to the reconstruction of evolutionary trees using DNA sequence data, and we have further developed the procedure for investigating rooted trees (Bishop & Friday, in press). Our computer program undertakes joint estimation from DNA sequences by iteratively searching for those values of the times at each branching point which maximize the likelihood for a given tree pattern.

The problem of finding that tree pattern for which the likelihood is globally a maximum is intractable, and so some strategy is essential for systematically examining a subset of all possible patterns. In the example dealt with below we use data for five species. There are 263 rooted trees of different pattern when polychotomies are allowed, and 105 different patterns when branching is only dichotomous (Felsenstein, 1973). The computations for evaluating the likelihood and interactively moving the times are extensive, and our approach is initially to expand from the 'big-bang' tree (Thompson, 1975) to a fully bifurcating pattern. The 'big-bang' tree is the one in which all species originate at a single time in the past. The first row of trees in Figure 1 shows the results of such an approach. Species are elevated in turn from the root and paired in sequence with each other lineage proceeding immediately from the root. The pattern giving the highest likelihood is passed on to the next cycle of expansion. In this way dichotomies are generated progressively, and when a fully bifurcating topology is reached the search ends. The procedure cannot guarantee to find the global maximum likelihood solution, but our experience has been that it approaches quite closely any better solution found by further search. This further search we carry out partly by evaluating tree patterns suggested by comparative morphological and palaeontological evidence.

We have examined the DNA sequence data collected by Brown *et al.* (1982) from the mitochondrial genomes of hominoids. These data were used as blocks of 896 bases, rather than cut into their constituent genes. The alignments were as published by Brown *et al.* (1982). The species represented are *Homo sapiens* (Ho), *Pan troglodytes* plus *Pan paniscus* (Pa), *Gorilla gorilla* (Go), *Pongo pygmaeus* (Po), and *Hylobates lar* (Hy).

Some of the trees evaluated by the method of maximum likelihood under the exponential failure model of genetic change are shown in Figure 1. Each tree is drawn to a common relative time scale with the relative time written against each branching point. Under each tree is given the log maximum likelihood for that pattern. A less negative value indicates a tree of higher likelihood than a more negative value. The tree of highest likelihood is therefore tree (j), and that of lowest likelihood is tree (a), that is the tree corresponding to the 'big-bang' pattern.

There is as yet no formally developed method for assessing the significance of the difference in the likelihoods of different trees. Felsenstein (1983) has identified this area of the estimation procedure as one needing attention. Thompson (1975) has suggested that the likelihood of the 'big-bang' tree be compared with the likelihoods of tree patterns containing more information about branching points. The tree patterns may also be compared with one another by their likelihoods. A tentative approach is that two units of difference between log likelihoods suggests a significant difference as it is conventionally understood. On this basis, tree (j) of Figure 1, the best found, is not clearly superior as an estimate of evolutionary events to tree (d); the difference in pattern involves the location of the branching point leading to the orang utan (Po).

It is of interest, however, that tree (j) has been regarded as the 'best' pattern by Sibley & Ahlquist (1984) on the basis of parsimony analysis of DNA hybridization data. Two other recent accounts, those of Goodman et al. (1983) on haemoglobins and Yunis & Prakash (1982) on karyotype, whilst not including gibbon in their data, support this grouping of man and the chimpanzees. However, many other recent studies calling on a variety of types of evidence (for example, Kortlandt, 1972, Read, 1975, Delson et al., 1977, Andrews, 1982) including that of Brown et al. (1982) themselves, on the mitochondrial sequences, support the African ape grouping of chimpanzees and gorilla (as shown in tree (i) of Figure 1), or regard the man-chimpanzee-gorilla trichotomy as unresolved (Andrews & Cronin, 1982). Tree (e) of Figure 1 embodies the uncertainties over this trichotomy and over the grouping of orang and gibbon, but it has a substantially lower likelihood than trees (d) and (h), (i), (j).

Figure 1. Maximum likelihood trees for different hypotheses of evolutionary relationship between extant homonoids. (Homo, Ho; Pan, Pa; Gorilla, Go; Pongo, Po; Hylobates, Hy;). See text for details

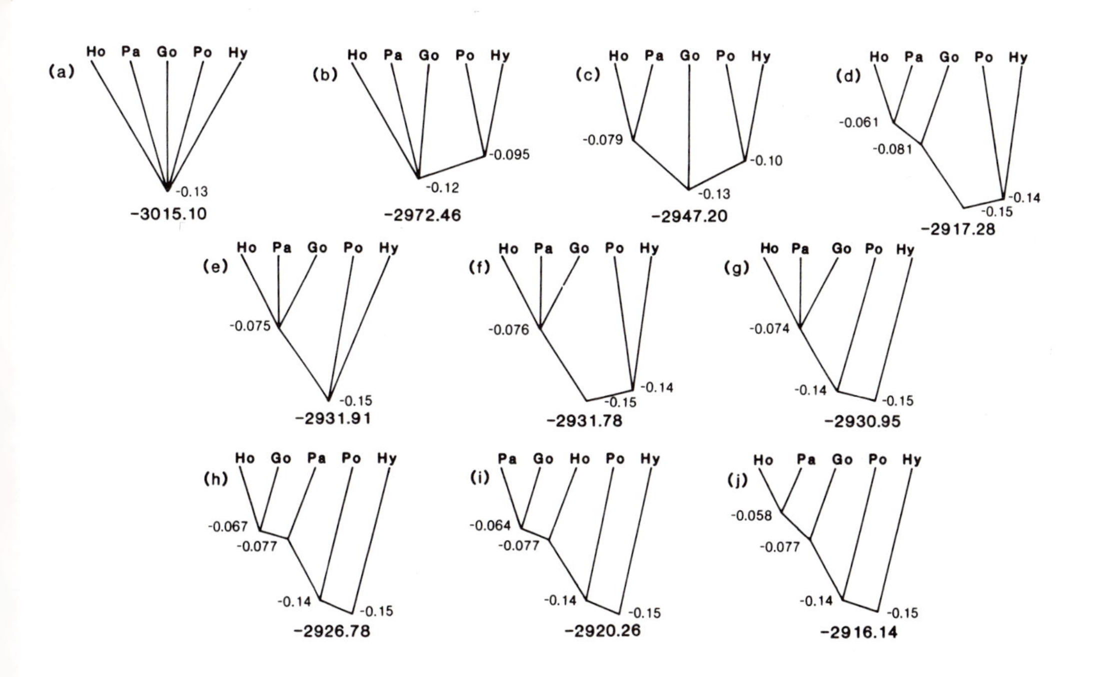

Suggestions that the grouping of chimpanzee-gorilla-orang be reinstated (Kluge, 1983) or that man and orang share more recent common ancestry than either does with any other extant hominoid (Schwartz, 1984) do not receive support from our examination of the mitochondrial DNA sequences. Trees reflecting these suggestions are unstable in the sense that they result in collision of the branching times on iteration.

Finally, the relative times of branching implied by tree (j) can all be converted to absolute time if the absolute time ago of one branching point is known with sufficient confidence.

SUMMARY AND CONCLUSIONS

Most attempts to use the data of molecular sequences to estimate the evolutionary relationships of hominoid primates have employed 'maximum parsimony' methods.

Different models of the process of evolutionary change may lead to different best estimates of phylogenetic trees for the same data. The model of evolutionary change implied by the parsimony approach involves unrealistic assumptions about the nature of molecular evolution. Stochastic models of change can be defended as more realistic, and estimations can be carried out under these models using the method of maximum likelihood. As further molecular biological information accumulates, the probabilistic models can be adjusted to take account of this. We have re-examined mitochondrial DNA sequences from hominoids, using a Poisson-based model for mutation, and have evaluated the relative likelihoods of a number of tree patterns proposed on the basis of comparative anatomy and the fossil record. The trees with the greatest likelihoods ((j) and (d) in Fig. 1) both group Homo sapiens and Pan together, and are uncertain about the location of Pongo. The tree (j) happens to correspond with the 'best fit' tree of Sibley & Ahlquist (1984) which was generated from a parsimony analysis of DNA hybridization data.

Acknowledgement

We thank Dr L.E. Friday for the preparation of Figure 1.

REFERENCES

Key references

Bishop, M.J. & Friday, A.E. (in press). Evolutionary trees from nucleic acid and protein sequences. Proc. R. Soc. B.

Brown, W.M., Prager, E.M., Wang, A. & Wilson, A.C. (1982). Mitochondrial DNA sequences of primates: tempo and mode of evolution. J. Molec. Evol. *18*, 225-39.

Felsenstein, J. (1983). Statistical inference of phylogenies. J. R. Statist. Soc. A, *146*, 246-72.

Goodman, M. & Moore, G.W. (1971). Immunodiffusion systematics of the primates I. The Catarrhini. Syst. Zool. *20*, 19-62.

Romero-Herrera, A.E., Lehmann, H.L., Castillo, O., Joysey, K.A. & Friday, A.E. (1976). Myoglobin of the orangutan as a phylogenetic enigma. Nature, Lond. *261*, 162-64.

Sibley, C.G. & Ahlquist, J.E. (1984). The phylogeny of the hominoid primates as indicated by DNA-DNA hybridization. J. Molec. Evol. *20*, 2-15.

Main references

Anderson, S., Bankier, A.T., Barell, B.G., de Bruijn, M.H.L., Coulson, A.R., Drouin, J., Eperon, I.C., Nierlich, D.P., Roe, B.A., Sanger, F., Schreier, P.H., Smith, A.J.H., Staden, R. & Young, I.G. (1981). Sequence and organisation of the human mitochondrial genome. Nature, Lond. *290*, 457-65.

Andrews, P. (1982). Ecological polarity in primate evolution. Zool J. Linn. Soc. *74*, 233-44.

Andrews, P. & Cronin, J.E. (1982). The relationships of Sivapithecus and Ramapithecus and the evolution of the orang-utan. Nature, Lond. *297*, 541-46.

Bishop, M.J. & Friday, A.E. (in press). Evolutionary trees from nucleic acid and protein sequences. Phil. Trans. R. Soc. B.

Brown, W.M., Prager, E.M., Wang, A. & Wilson, A.C. (1982). Mitochondrial DNA sequences of primates: tempo and mode of evolution. J. Molec. Evol. *18*, 225-39.

Delson, E., Eldredge, N. & Tattersall, I. (1977). Reconstruction of hominid phylogeny: a testable framework based on cladistic analysis. J. Hum. Evol. *6*, 263-78.

Felsenstein, J. (1973). The numbers of evolutionary trees. Syst. Zool. *27*, 27-33.

Felsenstein, J. (1981). Evolutionary trees from DNA sequences: a maximum likelihood approach. J. Molec. Evol. *17*, 368-76.

Felsenstein, J. (1983). Statistical inference of phylogenies. J. R. Statist. Soc. A, *146*, 246-72.

Goodman, M., Braunitzer, G., Stangl, A. & Shrank, B. (1983). Evidence on human origins from haemoglobins of African apes. Nature, Lond. *303*, 546-48.

Goodman, M. & Moore, G.W. (1971). Immunodiffusion systematics of the primates I. The Catarrhini. Syst. Zool. *20*, 19-62.

Kluge, A.G. (1983). Cladistics and the classification of the great apes. *In* New Interpretations of Ape and Human Ancestry, eds. R.L. Ciochon & R.S. Corruccini, pp. 151-77. New York: Plenum Press.

Kortlandt, A. (1972). New perspectives on ape and human evolution. Amsterdam: Stichting voor Psychobiologie, University of Amsterdam.

Read, D.W. (1975). Primate phylogeny, neutral mutations, and 'molecular clocks'. Syst. Zool. *24*, 209-21.

Romero-Herrera, A.E., Lehmann, H.L., Castillo, O., Joysey, K.A. & Friday, A.E. (1976). Myoglobin of the orangutan as a phylogenetic enigma. Nature, Lond. *261*, 162-64.

Sarich, V.M. & Wilson, A.C. (1967). Immunological time scale for hominid evolution. Science *158*, 1200-03.

Schwartz, J.H. (1984). The evolutionary relationships of man and orangutans. Nature, Lond. *308*, 501-05.

Sibley, C.G. & Ahlquist, J.E. (1984). The phylogeny of the hominoid primates as indicated by DNA-DNA hybridization. J. Molec. Evol. *20*, 2-15.

Thompson, E.A. (1975). Human evolutionary trees. Cambridge: University Press.

Yunis, J.J. & Prakash, O. (1982). The origin of man: a chromosomal pictorial legacy. Science *215*, 1525-30.

9 HOMINOID EVOLUTION: MOLECULAR AND PALAEONTOLOGICAL PATTERNS

Maryellen Ruvolo,
Department of Biological Chemistry,
Harvard Medical School, Boston, Mass. 02115, USA.

David Pilbeam
Peabody Museum,
Harvard University, Cambridge, Mass. 02138, USA.

Understanding phylogenetic patterns requires both comparative and palaeological-geological data, particularly genetic comparisons and fossil data. Branching patterns (branching sequence and branch lengths) are best determined using mainly genetic data, while their calibration (making ages actual rather than relative) depends on the fossil record.

A high degree of consensus has now been achieved on higher primate patterns based on morphological, genetic distance, and genetic character analyses (Goodman & Tashian, 1976). The principal modification of earlier morphological studies lies in the linkage of humans with African apes and the separation from them of *Pongo*.

Work on the comparative immunology of the single protein, albumin, using microcomplement fixation, is the best known of the attempts to use comparative genetics to date cladistic events. Table 1 lists divergence estimates from albumins (Sarich & Cronin, 1976), based on the assumption that 100 immunological distance (ID) units correspond to 60 my and that change is linear and uniform (with stochastic variation). The dates also reflect clustering techniques essentially similar to the unweighted pair group method (UPGMA).

Some of these assumptions have been questioned. ID is slightly but significantly nonlinear relative to time (Gingerich, n.d) and to DNA-DNA hybridization distances (Ruvolo, n.d), and other clustering techniques (distance Wagner; linear algebra) show rate variation by factors of three or more, although without changing branching patterns. These latter differences show how important it is to use a variety of clustering techniques (Farris, 1981).

More recent work focuses on DNA. Sequence analyses of almost 900 base pairs of hominoid mitochondrial DNA and restriction endonuclease mapping of longer segments show the same basic branching patterns (Brown *et al.*, 1982; Ferris *et al.*, 1981). Disagreements arise over the precise branching sequence of *Homo*, *Pan* and *Gorilla*. The most precisely framed analyses (Friday & Bishop, this volume; Hasegawa *et al.*, 1984) use explicit probabilistic models and a likelihood analytic method and show *Homo* and *Pan* as closest relatives. Friday & Bishop's study is very important in that it reminds us that the explicit statement of hypotheses of biological change are a critical and frequently neglected analytical step. Nei *et al.*

(1985), using UPGMA of mtDNA data, also show Pan and Homo as nearest relatives. That mtDNA data should be treated with some caution is shown by recent work suggesting that mtDNA sequences can be transferred across species boundaries (Lewin, 1983; 1984; Powell, 1983; Ferris et al., 1983; Spolsky & Uzzell, 1984).

Two recent DNA hybridization studies (Sibley & Ahlquist, 1984 and Benveniste, 1985) have produced high quality data (judged by consistency and reciprocity). These data, judged by the rate test, imply more uniform rates of change than any other set. It is also argued theoretically that analyses of all single copy DNA (more than 10^9 bp) should 'smooth out' rate variations in various genes and in non-expressed parts of the genome to produce an average rate of DNA change that is very uniform.

These analyses confirm earlier ones in showing that Pongo is twice as different from Homo and African apes as Homo is from Pan or Gorilla. They differ from albumin analyses in the relative lengths of branches to Hylobates and cercopithecoids.

Sibley & Ahlquist (1984) show that Pan-Homo distances are less than Pan-Gorilla and Homo-Gorilla. A smaller number of comparisons by Benveniste (1985) show no difference between Homo-Pan and Homo-Gorilla. Sibley & Ahlquist have produced a dated phylogeny using rate estimates based mainly on their previous bird comparisons, also shown in Table 1. Again, these estimates assume that DNA distance changed linearly with time, an assumption that is theoretically more plausible than for single expressed genes.

The fossil record, and its calibration, have also improved over the past two decades, and Table 1 includes a set of minimum divergence dates estimated from the fossil record.

Anthropologists interested in hominid origins can use these data to infer that earliest hominids are likely to have been African hominoids living in the late Miocene. Should we ever find these hominids, let us hope we can recognize them.

Table 1. Divergence estimates from genetic and palaeontological data

	Albumin[1]	DNA-DNA[2]	Fossil[3] (all minima)
Homo-Pan-Gorilla	4	8-10	4.5
HPG-Pongo	8	16	12
Large-Small hominoids	9	22	16
Hominoid-Cercopithecoid	20	33	22
Cercopithecoid-Ceboid	35	-	35

[1]Sarich & Cronin, 1976.

[2]Sibley & Ahlquist, 1984.

[3]Pilbeam.

References

Key references

Brown, W.M., Prager, E.M., Wang, A. & Wilson, A.C. (1982). Mitochondrial DNA Sequences of Primates: Tempo and Mode of Evolution. J. Mol. Evol. *18*:225-39.

Farris, J.S. (1981). Distance Data in Phylogenetic Analysis, pp.3-23. *In*: Advances in Cladistics (V.A. Funk & D.R. Brooks, eds.), The New York Botanical Garden, Bronx, New York.

Goodman, M. & Tashian, R.E. (1976). Molecular Anthropology. Plenum Press, New York.

Nei, M., Stephens, J.C. & Saitou, N. (1985). Methods for Computing the Standard Errors of Branching Points in an Evolutionary Tree and their Application to Molecular Data from Humans and Apes. Molecular Biology and Evolution *2*:66-85.

Sibley, C.G. & Ahlquist, J.E. (1984). The Phylogeny of the Hominoid Primates, as Indicated by DNA-DNA Hybridization. J. Mol. Evol. *20*:2-15.

Main references

Benveniste, R.E. (1985). The Contributions of Retroviruses to the Study of Mammalian Evolution. *In*: Molecular Evolutionary Genetics (Ross J. MacIntyre, ed.), Plenum Pub. Co., New York.

Brown, W.M., Prager, E.M., Wang, A. & Wilson, A.C. (1982). Mitochondrial DNA Sequences of Primates: Tempo and Mode of Evolution. J. Mol. Evol. *18*:225-39.

Farris, J.S. (1981). Distance Data in Phylogenetic Analysis, pp.3-23. *In*: Advances in Cladistics (V.A. Funk & D.R. Brooks, eds.), The New York Botanical Garden, Bronx, New York.

Ferris, S.D., Sage, R.D., Huang, C.-M., Nielsen, J.T., Ritte, V. & Wilson, A.C. (1983). Flow of Mitochondrial DNA across a Species Boundary. Proc. Natl. Acad. Sci. U.S.A. *80*:2290-94.

Ferris, S.D., Wilson, A.C. & Brown, W.M. (1981). Evolutionary Tree for Apes and Humans Based on Cleavage Maps of Mitochondrial DNA. Proc. Natl. Acad. Sci. U.S.A. *78*:2432-36.

Goodman, M. & Tashian, R.E. (1976). Molecular Anthropology. Plenum Press, New York.

Hasegawa, M., Yano, T. & Kishino, H. (1984). Phylogeny and Classification of Hominoidea as Inferred from DNA Sequence Data. Proc. Japan Acad. *60*:389-92.

Lewin, R. (1983). Invasion by Alien Genes. Science *220*:811.

Lewin, R. (1984). Frog Genes Jump Species. Science *226*:955.

Nei, M., Stephens, J.C. & Saitou, N. (1985). Methods for Computing the Standard Errors of Branching Points in an Evolutionary Tree and their Application to Molecular Data from Humans and Apes. Molecular Biology and Evolution *2*:66-85.

Powell, J. (1983). Interspecific Cytoplasmic Gene Flow in the Absence of Nuclear Gene Flow: Evidence from Drosophila. Proc. Natl. Acad. Sci. U.S.A. *80*:492-95.

Sarich, V.M. & Cronin, J.E. (1976). Molecular Systematics of Primates. *In*: Molecular Anthropology (Morris Goodman & Richard E. Tashian, eds.), Plenum Press, New York, pp. 141-70.

Sibley, C.G. & Ahlquist, J.E. (1984). The Phylogeny of the Hominoid Primates, as Indicated by DNA-DNA Hybridization. J. Mol. Evol. *20*:2-15.

Spolsky, C. & Uzzell, T. (1984). Natural Interspecies Transfer of Mitochondrial DNA in Amphibians. Proc. Natl. Acad. Sci. U.S.A. *81*:5802.

10 RELATIONSHIPS AMONG EXTANT AND EXTINCT GREAT APES AND HUMANS

Lawrence Martin,
Department of Anatomy and Embryology,
University College London,
Gower Street, London WC1E 6BT, England

Present address: Department of Anthropology,
State University of New York,
Stony Brook, NY 11794, USA.

INTRODUCTION

In recent years, the extant great apes (Pan troglodytes (the chimpanzee), Pan paniscus (the Bonobo), Gorilla gorilla, and Pongo pygmaeus (the orang-utan)) have been classified together into one family, the Pongidae. This view was advocated by Pilgrim (1927) and became prevalent by the 1940s, but has recently been questioned on several grounds, particularly as a result of the biomolecular studies initiated by Morris Goodman (Goodman, 1963). A number of alternative views regarding the phylogenetic relationships of hominoids are current in the literature (see Figure 1). The aim of this paper is to re-evaluate the evidence concerning relationships within the great ape and human clade, to determine cladistic relationships among the included genera and to establish whether there is any phylogenetic reason to maintain a family grouping for the extant great apes. This analysis includes both morphological characters and biochemical data, although particular emphasis is placed on those characters applicable to fossil hominoids, i.e. dental and skeletal morphology. The ancestral condition for the great ape and human clade has been determined as far as possible (Tables 1 and 2), and contained clades have been defined on the basis of shared derived characters with respect to the hypothetical common ancestor of the great ape and human clade (Table 3). Pertinent fossil species, i.e. those which possess at least some of the derived characters for the great apes and human clade, have been assessed on the basis of this analysis and their relationships determined.

Past and present views of great ape and human relationships

Early accounts of the gorilla regarded it as a second species of the genus to which the chimpanzee belonged (Savage & Wyman 1847a,b; Owen 1848, 1851), but in 1851 Saint-Hillaire proposed a separate genus for the gorilla and this position has been widely adopted since, with the exception of a period during the 1970s when it became popular to place these two into a single genus. I have used separate genera for the chimpanzee and the gorilla throughout. Many early scholars regarded the two African apes as being more closely related to man than was the orang utan (Huxley 1863; Darwin 1871; Keith 1910; and Gregory 1927), but others such as Haekel (1866) linked humans with orang utans. Pilgrim (1927) believed the great apes to be monophyletic and this view became prevalent by the 1940s and persisted to the 1960s (Tuttle, 1974) (see Figure 1e).

Even very early biochemical studies suggested that humans were most closely related to African apes (Nuttall, 1904), and this view was corroborated by Goodman (1963). This interpretation was supported by Sarich & Wilson (1967) who found a clear pattern of biochemical similarities linking man and African apes, with Pongo and Hylobates more distant (Figure 1b).

All recent molecular studies tell essentially the same story (see reviews by Andrews and Ruvolo & Pilbeam, this volume, for details and references): among the hominoids, the African apes and man form a monophyletic group with the orang utan next closest and gibbons and siamangs most distant (Figure 1b). There is some disagreement as to whether the branching of African apes and man is a trichotomy (Figure 1b) (e.g. Sarich & Wilson, 1967; Sarich & Cronin, 1976; Benveniste & Todaro, 1976; has Pan and Gorilla closest (Figure 1d) (Brown et al., 1982; Templeton, 1984) or shows humans closest to Pan (Figure 1c) (Goodman et al. 1983; Hasegawa et al. 1984; Sibley & Ahlquist, 1984). A case that the great apes should be seen as a monophyletic group (Figure 1e) has been proposed by Kluge (1983). Most recently, Schwartz (1984) has also questioned the basis for the recognition of an African ape and human clade. He has suggested that the morphological evidence supports, and the molecular evidence does not contradict, the recognition of Pongo as the sister genus to Homo, with the African apes the sister group of the man/orang utan clade (Figure 1f). It is therefore the case that all possible groupings of genera of great apes and humans are currently argued by some individuals, with the exception of any groupings of one African ape genus with the orang-utan (Figure 1).

Figure 1. Current views concerning relationships among extant great apes and humans. Ho = Homo; Pan = Pan (the chimpanzee and bonobo); Go = Gorilla; Po = Pongo (the orang utan); Hylo = Hylobates (the gibbons); and Cerc = Cercopithecidae (the Old World monkeys)
A: The concensus view for many years has been that the great apes and man form a clade (labelled G.A.) and that this is most closely related to the gibbon clade. Together, the apes and humans form the Hominoidea which is most closely related to the Cercopithecoidea. This set of relationships was adopted for the present work, and the gibbons and the Old World monkeys provided the major outgroups for character polarity assessment.
B: The undoubted concensus among scientists working with a molecular data base has been that the African apes and humans form a clade as shown here. Andrews (this volume) found convincing molecular evidence for these relationships.
C: A limited number of molecular studies have found that Pan and Homo form a clade, with the gorilla more distant, but still with an African ape/human clade separate from the orang utan. Andrews (this volume) did not find the evidence for this interpretation to be robust.

D: A majority of scientists working with a morphological data base have adopted the position shown here. They have accepted the molecular evidence for an African ape/human clade, but have generally considered the two African ape genera to be sister genera within that clade.
E: The view shown in this figure has been the traditional one for about 40 years since 1927. This was generally a grade, rather than a clade, view which saw the extant great apes as more similar to one another than any were to humans. A case that the great apes in fact represent a clade, the view necessary for this author to maintain a familial group for them, has recently been made by Kluge (1983).
F: This figure shows the orang utan as being the sister genus to humans and has recently been proposed by Schwartz (1984).

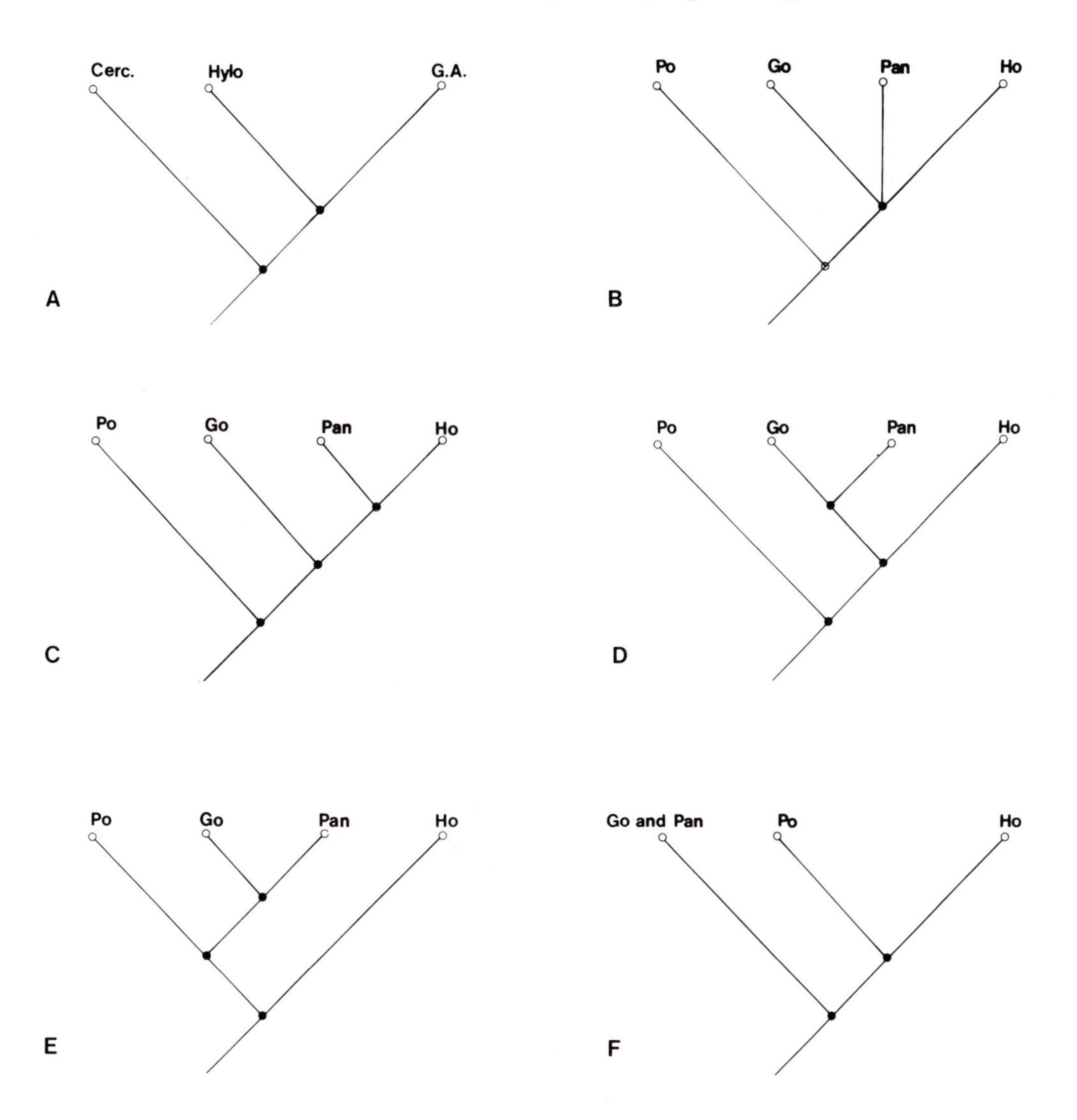

The morphological evidence relating to catarrhine phylogeny has recently been comprehensively reviewed by Andrews (1985). He found that the great apes and man formed a clade and that some Miocene fossil taxa could be shown to belong to that clade, but he did not address in any detail the relationships within the great ape and human clade. As the currently available biochemical data are equivocal concerning relationships among African apes and humans (although definitive concerning the closer affinity of man with the African apes than the orang utan) it seems that the morphological variation within this group, which is very well known, should form the basis for the assessment of their relationships. This is especially the case for those of us who are interested in using the fossil evidence to flesh out the picture of hominoid phylogeny which may be derived from studying the extant hominoids alone. This chapter therefore provides a consideration of the morphological evidence for relationships among the living and fossil members of the great ape and human clade.

PHYLOGENY RECONSTRUCTION

It is now widely accepted that phylogenetic relationships among taxa are best determined on the basis of the shared possession of derived characters, the cladistic method (Hennig, 1966; Delson & Andrews, 1975; Delson *et al.* 1977). There is some disagreement as to whether this is an attainable goal, mainly because of the problems of character polarity determination (i.e. the direction of change between character states). This problem is particularly acute for the biomolecular evidence (see Andrews, this volume). A major aim of the present work has been to re-examine the polarity of characters which vary among the great ape and human clade. Relationships have been determined among genera, the sister pairing of the chimpanzee and bonobo being accepted. Relationships between the great ape and human group and the gibbons have not been the focus of this work and have been assumed for this study. One reason for this omission, other than space, is that it would require the precise determination of the ancestral morphotype for the *Hylobates* clade, which should be the basis for comparison with the great apes rather than selected members of the clade which are likely to be distinct from the common ancestor of *Hylobates*.

Phylogenetic relationships are best determined on the basis of the analysis of living species wherever possible. This is for several reasons: firstly that the extant forms are much more fully known in all aspects of anatomy rather than skeleto-dental alone; secondly, the polarity of character change may be more easily established and parallelisms and convergences recognised when characters can be analysed in the light of both developmental and functional evidence; thirdly, that when fossils are excluded from the initial analysis it protects the scientist from falling into the trap of being influenced by the age of specimens in their interpretation of phylogeny. It is more productive to analyse the relationships of living forms, and to reconstruct their hypothetical common ancestral conditions, and then to assess the fossils on that basis. Inclusion of fossil material may require the modification of interpretations as fossils add to the known diversity, as well as providing unique evidence as to the where, when and how of evolutionary change, and about the sequence of change in mosaic evolution.

It may be thought that the question as to the naming of a family group is a purely semantic argument, but this is only the case if one accepts that classification should be completely divorced from phylogenetic relationships. It has been argued that the great apes represent a similar grade of organisation, and should be classified as a family on that basis regardless of probable phylogenetic relationships. This position might be justified if there was universal agreement about the definition of such a grade, but this is not the case. The weakness of grade based classifications is that they are subjectively based. As Darwin (1871) pointed out, the only objective basis for classification which we can hope to employ is one based on phylogenetic relationships, and the goal of taxonomists should be to establish a phylogenetic basis for classification.

Although I argue in favour of a phylogenetically based classification, I do not apply strict cladistic principles to the nomenclatorial side of taxonomy. It has been argued that a classification system should reflect each branching point at a higher taxonomic rank, but this system is inherently unstable when new forms are continuing to be recovered: the discovery of a new clade requires the modification of the nomenclature of all clades which branched off before. Instead, I am in favour of the strict use of cladistics for phylogeny reconstruction, but a more liberal approach to classification. Specifically, I shall only use taxonomic terms to describe monophyletic groups but would be prepared to have, for example, three or more families contained within a single superfamily. What this means in the present work is that the Pongidae will only be recognized as a valid family name for the great apes if they can be shown to be more closely related to one another than any are to *Homo*. The family name Pongidae is also available for use for the orang-utan alone, and this possibility will also be considered.

HOMINOID PHYLOGENY

It is almost universally accepted on the basis of a wide range of biomolecular and morphological studies that the Catarrhini is a monophyletic group (comprising the Old World monkeys, apes and man), and similarly that the Hominoidea is monophyletic (containing the gibbons, the great apes and man). That the great apes and man represent a clade is also widely accepted. The starting point adopted here was that shown in Figures 1a and 2, which accept all of these monophylies. The Catarrhini may be established on the basis of sharing a derived loss of P2, and the loss of the tympanic bulla in addition to many other morphological and molecular features (see in particular, Andrews, 1985, this volume; also other sources listed in Table 1). The Hominoidea is equally well defined on the basis of the loss of a tail and many further shared derived similarities (see Table 1 for details). The possible relationships among the great ape and human group are shown in Figures 1b-1f. Only Figure 1e conforms with the definition required for the Pongidae to be maintained as a family group for all of the great apes.

The molecular data have been extensively reviewed by Andrews (this volume). These undoubtedly support a great ape and human clade (Figure 2) and within that clade an African ape/human clade (Figure 1b) compared to a *Pongo* clade. Some molecular workers have argued for a chimp/human clade within the African ape/human clade (Figure 1c) but Andrews (this volume) found little reason to uphold this view. I shall follow Andrews' conclusion that the molecular evidence supports the interpretation of the great apes and man as a clade and that the weight of evidence supports an African ape/human clade within that. The resolution of the molecular data within the African ape and human clade is presently insufficient to address relationships within that clade. The greater known morphological diversity within that clade must therefore be the basis for determining relationships among Pan, Gorilla and Homo. In the light of Schwartz's (1984) work no initial assumptions have been made concerning the relationships of the orang-utan to the other great apes and man for the analysis of morphological data.

Character analysis

The morphological characters used in this study were compiled from a literature survey of morphological characters which have been suggested to be important in determining hominoid relationships. All of the cranio-dental characters have been re-evaluated and the polarities of postcranial and soft tissue characters have also been reassessed on the basis of published descriptions of variability in extant forms (see Table 1 for sources). The hypothetical common ancestral condition for the great ape and human clade has been reconstructed as listed in Table 1 (see Figure 2). This has been achieved by examining developmental evidence, and functional evidence as well as outgroup comparison with two accepted clades; firstly, the Hylobatidae (the gibbons) and secondly the Cercopithecidae (the Old World monkeys). Where appropriate reference was made to a third outgroup, the New World monkeys. The characters listed in Table 1 which are derived with respect to the common ancestor of the Hominoidea define the great ape and human clade. These form a basis for the assessment of fossil hominoids which may be more closely related to great apes and humans than are the gibbons. The other characters which are listed are those for which the living great apes and humans exhibit varying character states and which are therefore useful for determining relationships within the great ape and human clade. This is not a definitive list but provides 122 morphological characters pertinent to the determination of hominoid relationships most of which can be studied, at least potentially, in fossils. Characters which vary among the great ape and human clade have been compared to this hypothetical common ancestral condition to determine which taxa exhibit derived character states. Characters have not been accorded equal weight in this analysis as the polarity of certain characters can be more firmly inferred than can others, and these have been emphasized. This is not an entirely satisfactory procedure, but until developmental and functional aspects of more characters are adequately known it is the pragmatic one. The shared possession of derived characters is explained by recency of common ancestry although the possible confusing effects of parallelism and convergence must always be borne in mind. Alternative ways of grouping the great apes and man on the basis of the shared possession of derived characters have been examined and the most likely set of relationships arrived at. This analysis also provided a definition of the derived features of the ancestral morphotype for each of the clades defined. Fossil hominoids which share derived characters with the great ape and human clade (i.e. derived with respect to the hypothetical common ancestor of the Hominoidea) are assessed on the basis of these characters for evidence of their relationships to extant forms.

A number of the characters which I have employed have become key characters in interpreting fossil hominoids and three of these are discussed below.

Enamel thickness. A major problem in the study of great ape and human phylogeny has been the preoccupation with human characters, which have almost always been assumed to be derived. This has been a particular problem with regard to fossil taxa and the interpretation of the evolu-

Figure 2. This shows the starting position adopted for the present work, that the great apes and man are a clade, but that relationships within that clade are unknown. Morphological characters which vary between great apes/humans and either gibbons and/or Old World monkeys were analysed by outgroup comparisons, and in the light of developmental and functional evidence to assess their probable conditions in the common ancestor of the great ape and human clade (node 1). The characters which were found to be derived with respect to the common ancestor of the Hominoidea define the great ape and human clade (see Table 1). Other characters which vary among great apes and humans, and which therefore provide evidence about relationships within their clade, were also included and their conditions with respect to the common ancestor of the Hominoidea and of the Catarrhini assessed (see Table 1).

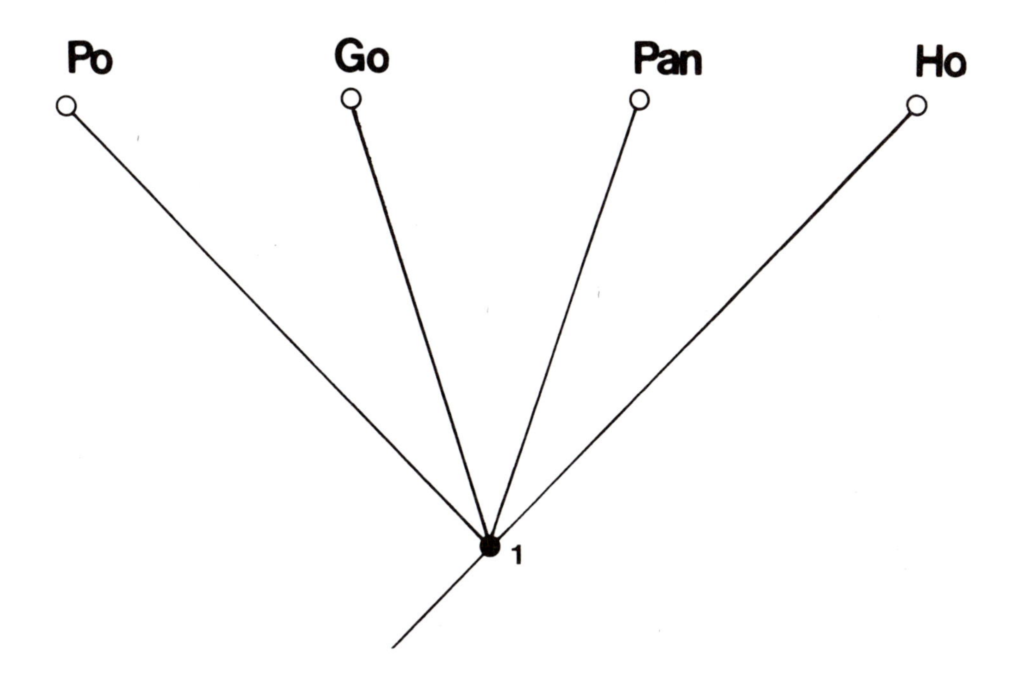

Table 1: Morphological characters employed in this analysis

Character states in the common ancestor of the great ape and human clade:
d = derived with respect to the common ancestor of the Hominoidea;
ph = character retained from the common ancestor of the Hominoidea;
pc = character retained from the common ancestor of the Catarrhini (also retained by the common ancestor of the Hominoidea).

Cranial features:

1) Orbits higher than broad (d).
2) Inferior margin of orbits does not overlap superior portion of the nasal aperture (d).
3) Nasal bones relatively long (d).
4) Nasal bones taper towards frontal (pc), flat (pc), and narrow (d).
5) Broad interorbital pillar (pc).
6) Prominent glabella (pc).
7) Small to medium supraorbital tori, divided at glabella (pc).
8) No post supraorbital sulcus (pc).
9) No fronto-ethmoidal sinus (pc).
10) Deep zygomatic (d).
11) Flattened zygomatic, with inferior inclination, and not flaring (pc).
12) No malar notch in inferolateral surface of zygomatic (pc).
13) Zygomatic foramina small and few in number, below level of the inferior margin of the orbit (pc).
14) Increased alveolar prognathism (lengthened face)(d).
15) Maxillary sinus enlarged (d).
16) Maxillary body lightly built (pc).
17) Palatine makes relatively small contribution to hard palate (d).
18) Palatine foramina large and oval (ph).
19) Abrupt transition from premaxilla to maxilla resulting in a stepped floor of nasal chamber (d).
20) True incisive canal (d).
21) Incisive foramina small (d).
22) Incisive foramina double (pc).
23) Thin cranial vault bones (pc).
24) Skull breadth greatest at supramastoid crests (pc).
25) Moderate postorbital constriction (pc).
26) Low rounded frontal bone (pc).
27) Parietal bone longer than frontal bone, bregma forward placed (d).
28) Sagittal crest may extend to frontal bone (pc).
29) Low temporal bone, inferior edge horizontal (ph).
30) Occipital crest often present (pc).
31) Mastoid process distinct (d).
32) Foramen magnum on base of skull, but posterior (pc).
33) No digastric groove (pc).
34) Large, open spheno-palatine foramen (pc).

Dental and gnathic:

35) $I\underline{1}$ larger cf. molars (d).
36) $I\underline{2}$ rounded, unicuspid (pc).

Table 1: contd.

37) $I\underline{1}$ > $I\underline{2}$ mesio-distal (pc).
38) $I\underline{1}$ broad, as broad as high (ph).
39) $I\underline{1}$ without lingual pillar (ph).
40) Canines robust (d).
41) Canines long relative to height (low cf length)(d).
42) Reduced canine honing (d).
43) Canines sexually dimorphic (male canines large, female canines medium)(pc).
44) Lower P3 not bilaterally compressed, more robust and broadened (d).
45) Lower P4 lengthened (d).
46) Lower P3 low crowned relative to length (ph).
47) Lower P4 talonid almost equal height to cusps (d).
48) Upper premolars elongated mesio-distally (d).
49) Upper premolar heteromorphy very much reduced (d).
50) P4 smaller than P3 (pc).
51) Developmental period of molar crown lengthened cf gibbons providing the potential to form thick enamel (d). Each of the M1-M3 crowns form during about the same length of time in all great apes and humans.
52) Molar enamel all forms quickly (pc).
53) Molar enamel thick (d).
54) Fast formation of tooth roots results in early eruption of teeth (ph, ?pc).
55) Cheek teeth with relatively smooth enamel (pc).
56) Cusps relatively high (pc).
57) High dentine horns (pc).
58) Upper molars with enlarged hypocone which is not linked to lingual cingulum (d).
59) Upper molars with no cingulum or poorly developed (d).
60) $M\underline{1}$ < $M\underline{2}$ > $M\underline{3}$ (pc), $M\underline{2}$ only slightly longer than $M\underline{1}$ (d).
61) Lower molars with five cusps, small hypoconulid (pc).
62) Lower molars with weakly developed cingula (ph).
63) Lower M2 largest tooth (d). Lower molar sizes: M2 > M3 > M1.
64) Tooth rows wider apart anteriorly and more parallel (d).
65) Palate deepened posteriorly (ph), remains shallow anteriorly (pc).
66) Palate still longer than broad but broadened anteriorly cf catarrhine ancestral condition (ph).
67) Palate U-shaped (ph).
68) Deep mandibular symphysis (d).
69) Inferior torus larger than superior (ph).
70) Inferior torus placed very low down symphysis (d).
71) Mandibular corpus robust (if associated with increased enamel thickness, see text and Martin & Andrews, 1984)(d).

Postcrania:

72) Os centrale separate from scaphoid (pc).
73) Metacarpals and phalanges short and curved (pc).
74) Metacarpals with broad distal ends (ph).

Table 1: contd.

75) Development of an intro-articular cartilaginous meniscus fully interposed between the greatly reduced ulnar styloid process and pisiform, resulting in the total exclusion of ulnar carpal articulations (d).
76) Extensive distal radio-ulnar articulation (ph).
77) Proximal ulnar with segmented trochlear notch and U-shaped deep radial notch (ph).
78) Ulnar coronoid process much higher than (reduced) olecranon process (d).
79) Ulna shaft bowed (ph).
80) Radial head rounded and not tilted (ph).
81) Lateral edge of the trochlea that separates trochlea and capitulum
82) anteriorly, wraps around distally to meet with olecranon fossa (ph).
82) Distal humerus with prominent trochlear keels, with deep sulci on either side of the lateral trochlear keel (d).
83) Trochlear width similar to capitulum width, i.e. broadened, with mid portion constricted (d).
84) Capitulum rounded, but not on distal surface (pc).
85) Olecranon fossa deep and well defined with a well developed lateral ridge (pc).
86) Brachialis flange moderately developed (pc).
87) Humeral head rounded, larger than femoral head (ph).
88) Humeral head more medially oriented (ph).
89) Scapula with dorsally positioned glenoid fossa directed more cranio-laterally (ph).
90) Scapula with elongated vertebral border, large robust acrominon, large coracoid (ph).
91) Clavicle elongated (ph).
92) Sternum broad and short, reduced number of sternal elements. (ph).
93) Increased diameter and length of vertebral centra (ph).
94) Thoracic vertebrae still 13 in number, but derived in increased length and in protruding into the thoracic cavity (ph).
95) Short dorsal spines of cervical vertebrae (?)
96) Lumbar vertebrae reduced to 5 (ph).
97) Loss of accessory articular processes of lumbar vertebrae. (ph).
98) Iliac blade broadened (ph).
99) Ischial tuberosities absent (d).
100) Hindlimbs reduced in length so high intermembral index (d).
101) Femoral head small, rounded (pc).
102) Femoral neck long and at a high angle (pc).
103) Femur neck lacks tubercle (ph).
104) Femur with deep and restricted trochanteric fossa (ph).
105) Broad patella groove (pc).
106) Femur with assymetrical condyles (ph).
107) Hallux long and abductable (pc).
108) Not brachiator or knuckle walker (pc).
109) Metatarsals and phalanges fairly straight (ph).

Soft tissue:

110) Origin of pectoralis minor short (d).

Table 1: contd.

111) Development of vermiform appendix (ph)
112) Development of pelvic diaphragm (ph).
113) Low deltoid insertion on humerus (ph), deltoid enlarged (d).
114) Trapezius inserts on clavicle (ph), thickest fibres cranially (d).
115) Dorso epitrochlearis insertion on medial epicondyle (ph).
116) Tendinous origin of teres major (not fleshy)(ph).
117) No knuckle pads over dorsal aspects of middle phalanges (pc).
118) Moderately developed flexor digitorum superficialis (pc).
119) Presence of anterior belly of digastric (pc).
120) Pectoralis abdominis not developed (d).
121) Medial pterygoid as large as lateral pterygoids (d).
122) Tensor palati muscles originate from temporal bone (pc).

References used in compiling table of characters and for my determinations of character polarity where based on literature descriptions rather than direct observations: Aiello, 1981; Andrews, 1978, 1985; Andrews & Cronin, 1982; Andrews & Groves, 1976; Andrews & Tekkaya, 1980; Ankel, 1972; Boyde & Martin, 1982, 1983, 1984a, 1984b; Cave, 1967, 1973; Cave & Haines, 1940; Ciochon, 1983; Clark, 1959; Corruccini, 1975; Dean, 1982; Dean & Wood, 1981; Delson, 1977; Delson & Andrews, 1975; Delson et al., 1977; Feldesman, 1982; Fleagle, 1983; Gleiser & Hunt, 1955; Groves, 1972; Harrison, 1982; Kay, 1982; Kay & Simons, 1983; Kluge, 1983; Lewis, 1969, 1972; Lipson & Pilbeam, 1982; McHenry et al., 1980; Martin, 1983, 1985; Martin & Andrews, 1982, 1984; Martin & Boyde, 1984; Moorees et al., 1963; Morbeck, 1983; Napier & Napier, 1967; Oxnard, 1963; Pilbeam, 1969; Preuss, 1982; Rak, 1983; Rose, 1975, 1983; Schultz, 1930, 1936, 1938, 1944, 1961; Schwartz, 1984; Schwartz et al., 1978; Szalay & Delson, 1979; Tuttle, 1967, 1969, 1970, 1974, 1975; Tuttle & Rogers, 1966; Walker & Pickford, 1983; Ward et al., 1983; Ward & Pilbeam, 1983; Washburn, 1963; Zapfe, 1960.

tion of hominoid enamel and has prevented objective assessment of the relationships of fossil taxa (e.g. '*Ramapithecus*' and *Kenyapithecus africanus*). A major component in the arguments in favour of the hominid status of '*Ramapithecus*' (Simons, 1961, 1976; Simons & Pilbeam, 1972) was the inferred shared possession of a derived character, 'thick enamel' on the molar teeth. Similarly, Schwartz's (1984) case for the orang utan being more closely related to man than are the African apes is also dependent on the interpretation of the thicker enamel seen in these forms as being a shared derived feature. Equally, support for the continued hominid status of '*Ramapithecus*' (now included in *Sivapithecus*) rests largely on the shared possession of thick enamel (Kay, 1981, 1982; Kay & Simons, 1983). It has recently been shown that the common ancestor of the great ape and human clade had thick, fast formed enamel and that this condition has been primitively retained in the hominine line (Martin, 1983, 1985). Both *Pan* and *Gorilla* have secondarily reduced enamel, from thick to thin, and the orang utan also has reduced enamel thickness, from thick to intermediate/thick. Thick enamel in a fossil catarrhine indicates affinities with the great ape and human clade, but does not indicate relationships within that clade. Rather, it is the reduction of enamel thickness in the African apes and in the orang utan which has taxonomic valence within the great ape and human clade. The secondary reduction of enamel thickness in the two African apes appears to be homologous as it occurs via the same developmental route of slowing of average daily ameloblastic secretory activity. This is one line of evidence that Pan and Gorilla have shared a unique common ancestor not shared with Pongo or Homo. The secondary reduction of enamel thickness in the orang utan is not homologous with that in the African apes as the slowing down of ameloblastic secretion is of a different pattern, as well as extent, in the orang utan (Martin, 1983, 1985; Martin & Boyde, 1984). The ability of enamel thickness and formation rate data to resolve relationships within the great ape and human clade is shown in Figure 3.

Mandibular robusticity. Many of the fossil hominoids with thick enamel (e.g. Sivapithecus, Kenyapithecus, Gigantopithecus, Paranthropus and Australopithecus) have robust mandibles, and the possibility that these two characters are interdependent has been discussed by Martin & Andrews (1984). Others, such as Kay & Simons (1983) have argued that robust mandibles are a uniquely hominine attribute. Given the distribution of this character among fossil hominoids, this latter interpretation appears unlikely. Whether the common ancestors of the great ape and human clade had robust or gracile mandibles, a robust mandible is a character of limited value for taxonomy within that clade as it is likely either to be primitive or to have evolved in parallel (Martin & Andrews, 1984). I favour the interpretation that a robust mandible was present in the common ancestor of the great ape and human clade.

Subnasal morphology. This region has been comprehensively reviewed by Ward & Pilbeam (1983) and Ward et al. (1983), but my interpretation of the data differs slightly from theirs. They have argued that three distinct patterns of maxilla/premaxilla relationships exist among hominoids. Firstly, a pattern seen in Proconsul and in gibbons in which no true incisive canal is formed (which is probably the ancestral condition for the Hominoidea). Secondly, an Asian pattern unique to the orang utan among living hominoids, and thirdly, a pattern seen in the African apes and in man in which a true incisive canal is formed and the floor of the nasal cavity is stepped. Ward & Pilbeam (1983) suggest that the African ape/human pattern is a shared derived feature linking these, while I feel that it is more likely that this pattern would also have been present in the common ancestor of the great ape and human clade, with the orang utan being further derived from that condition.

DISCUSSION

The great ape and human clade can be defined by 43 derived morph logical characters with respect to the common ancestor of the Hominoidea (Table 1). Some of these characters were found to be shared with two Middle Miocene hominoids, Dryopithecus from France, Spain, Austria and Hungary, and Kenyapithecus from Kenya. None of the early Miocene taxa from East Africa belonging to the Proconsul group could be shown to share any derived characters with the great ape and human clade (a view also adopted by Fleagle, this volume). Neither Dryopithecus or Kenyapithecus share all derived characters for the great ape and human clade, in known parts, and do not therefore belong within that clade. Although less well known than *Dryopithecus*, *Kenyapithecus* shares two characters with the great ape and human clade which are not present in *Dryopithecus* (see Table 2) and may therefore be more closely related to it than is *Dryopithecus* (the view shown in Figure 5). In addition to craniodental morphology, there is strong evidence from the postcranial skeleton, particularly the distal humerus, that *Dryopithecus*, as well as *Sivapithecus*, has achieved the great ape grade of organisation (Morbeck, 1983; Rose, 1983). *Kenyapithecus* is also derived postcranially but is autapomorphic in many features, although primarily great-ape-like (Morbeck, 1983; Rose, 1983).

Relationships within the great ape and human clade produced the greatest problems. The orang utan is highly derived with respect to the common ancestor of the clade but in almost all respects these specializations are unique to it. Without the support of increased enamel thickness as a derived character there was little evidence from this analysis to support a human/orang clade as had been argued by Schwartz (1984). However, the evidence for an African ape/human clade cannot be considered to be at all robust either (see Figure 5 caption), but with the addition of the molecular data (see Andrews, this volume) this clade does appear to be the most likely. There are some derived characters which are shared among all three living great ape genera as Kluge (1983) has pointed out. Many of these appear to be functionally correlated with large body size and closed hand walking and do not, in my view, outweigh the combined morphological and molecular evidence for an African ape/human clade (see Table 3, and Andrews, this volume). Within the African ape/human clade there is only one character (a spatulate upper lateral incisor) supporting a chimp/human clade, and no evidence for a gorilla/human clade. The shared derived possession of secondarily reduced enamel thickness, in addition to shared specializations relating to knuckle-walking, is convincing evidence for an African ape clade, and this is the position adopted here (see Table 3 and Figure 4). The hominine clade is well defined by derived characters, many of which are shared by australopithecines, and these are considered by Wood & Chamberlain, Grine, and Dean (all in this volume).

In all known parts, *Sivapithecus* (especially as exemplified by the best known species *S. sivalensis*, and to a lesser extent *S.meteai*, *S.punjabicus*, *S.alpani*, and *S.darwini*) displays the derived conditions for great ape and human characters. By contrast, a recently described species '*Sivapithecus*' *simonsi* (Kay, 1982) appears to share no derived features with the great ape and human clade, or even with the Hominoidea (Martin, 1983). In a considerable number of characters, *Sivapithecus* is derived with respect to the common ancestor of the great ape and human clade (Table 3) and most of these are shared only with *Pongo*. This evidence therefore supports the interpretation of *Sivapithecus* (certainly the species *sivalensis* (formerly '*indicus*', see Martin & Andrews, 1984; Martin, 1983) and *meteai*, and less clearly the species *punjabicus* (formerly '*Ramapithecus*'), *alpani*, and *darwini*) as an early member of the orang utan clade (see Andrews & Tekkaya, 1980; Andrews & Cronin, 1982; and Ward & Pilbeam, 1983). The evidence for a *Sivapithecus*/human clade (Kay, 1982; Kay & Simons, 1983) was not found to be convincing in the light of current interpretations of the evolution of hominoid enamel (Martin, 1985).

It is interesting, and probably significant, that the two clades within the great ape and human clade which are best defined by derived characters, i.e. the *Homo* clade and the *Pongo* clade, are also the two to which we are currently able to attach fossil relatives, *Australopithecus* and *Paranthropus*; and *Sivapithecus* respectively. The identification of fossil members of these clades provides evidence about the environments with which character changes are associated and also about the times at which the changes took place and the sequence of events in mosaic evolution. The

Figure 3: The evidence from enamel thickness and formation rates. Node 1 is derived with respect to the common ancestor of the Hominoidea in having thick enamel, all formed quickly. Node 2 represents a shared derived condition between *Pan* and *Gorilla* of having enamel that is secondarily reduced to thin enamel as a result of a single step slowing down of average daily ameloblastic secretory rate. This means that the thin enamel seen in African apes is not homologous with that seen in gibbons or Old World monkeys. Node 3 shows an autapomorphic (uniquely derived condition) condition in *Pongo* in which enamel is secondarily reduced to intermediate/thick, but by a different slowing down process to that seen in the African apes. Enamel thickness/formation data are unable to resolve the trichotomy among the *Homo* clade, the *Pongo* clade, and the African ape clade (Martin, 1985). The shared presence of thick or of thickened enamel can no longer be used to argue relationships within the great ape and human clade.

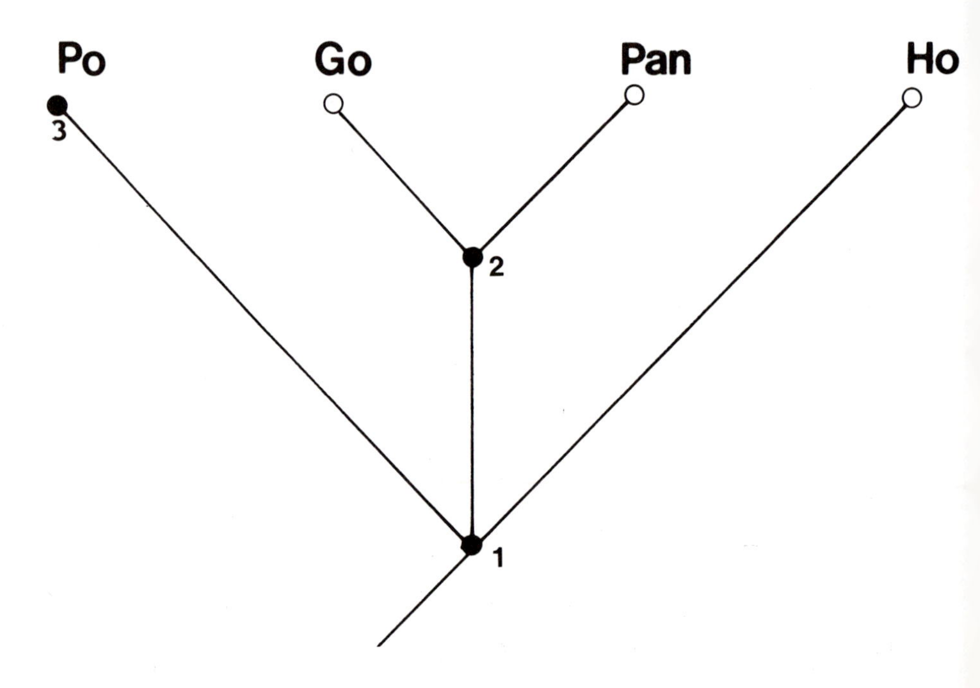

great ape and human clade was distinct from the gibbon clade by the early middle Miocene (as evidenced by early *Kenyapithecus* and *Dryopithecus*). This clade had separated into at least two branches, a pongine clade and an African ape/human clade, by about 12 million years ago (as evidenced by the presence of *Sivapithecus* at that time). There is no fossil evidence currently known relating to the African apes, but the African ape/human clade had already split into an African ape branch and a hominine branch by the middle Pliocene (as evidenced by the existence of hominines at about 4 million BP).

SUMMARY AND CONCLUSIONS

For a considerable portion of this century, the great apes have been considered to be more similar to one another than any was to *Homo*. This view was challenged during the 1960s on the basis of molecular studies which indicated that the African apes were more similar to humans than was the orang utan. Both of these views were based on 'similarity', which was considered, in some unspecified way, to be equivalent to phylogenetic affinity. In general, scientists working from a molecular data base have continued to rely on similarity to reconstruct phylogeny and some have argued for a chimp/human clade within the African ape/human clade on that basis. The molecular data do not lend themselves to phylogeny reconstruction which is best achieved on the basis of the shared possession of derived characters. The molecular data have been analysed in a cladistic framework by Andrews (this volume), who found that they support an African ape/human clade, but were not definitive concerning relationships within that clade. A further problem with the molecular data is that they show rather low levels of variability among members of the great ape and human clade, and especially among the African ape/human clade. In view of the greater known morphological variability within these clades it seemed likely that morphology would provide a better basis for resolving great ape and human relationships at the present time. Morphological data also have the major advantage that they can often be applied to the fossil record.

During the last 20 years a number of cladistic analyses of morphological evidence for great ape and human relationships have been produced. These have had three major conclusions. Firstly, the commonest view has been that the African apes and man form a clade, with *Pan* and *Gorilla* as sister genera within that clade. Secondly, it has been proposed that the great apes do form a clade, to the exclusion of man (Kluge, 1983). Finally, Schwartz (1984) has proposed that the orang utan is more closely related to man than either is to the African apes.

In this chapter, 123 morphological characters have been analysed and their polarities assessed. A few characters that have been of critical importance in interpreting Miocene hominoids have been discussed in more detail in the text. The great apes and man form a clade which became distinct from the gibbon clade before 16 my. *Dryopithecus* and *Kenyapithecus* are early relatives of the great ape and human clade, although these Miocene genera are less closely related to extant great apes and humans than these are to one another (Figure 5). It is possible that *Kenyapithecus* is more closely related to extant great apes and humans than is *Dryopithecus*. The great

Table 2: Derived characters shared among living and fossil great apes and humans. Characters shown as derived for the great ape and human clade (Table 1) are listed here where the condition is known for either *Dryopithecus* or for *Kenyapithecus*. D = derived character present in the genus; P = primitive hominoid or catarrhine condition retained; - = character not known for the genus; V = character variable within genus

Derived character (see Table 1).	*Dryopithecus*	*Kenyapithecus*
15	D	D
16	D	-
19	D	-
20	D	-
21	D	-
40	D	D
41	D	D
42	D	D
44	D	D
45	D	D
47	P	D
48	D	D
49	D	D
53	P	D
58	D	D
59	D	V
60	P	P
63	D	-
68	D	D
70	D	D
71	P	D
82	D	D
83	D	D
113	D	-

ape clade initially diverged into a pongine branch and an African ape/ human branch prior to 12 my BP, and *Sivapithecus* belongs to the pongine clade. The two African ape genera are sister taxa within the African ape/ human clade, but the only fossils which bear on divergences within the African ape/human clade comes from the fossil record for hominines at about 4 my BP.

As the great apes do not represent a monophyletic group, there is no phyletic reason to maintain the family Pongidae for them. As the relationships are of greater importance than the naming of groups, I have minimised nomenclatorial discussion. One possible scheme, favoured by many who work on human evolution, is to retain Hominidae for the uniquely human line, with Gorillidae for the African apes and Pongidae for the orang utan and

Table 3: In this table are listed characters which are derived with respect to the characters listed in Table 1, and on the right hand side are shown the presence/absence of these characters in six groups of hominoids.

Character numbers refer to ancestral conditions for characters listed in Table 1. D = derived character state present in that genus; A = ancestral great ape and human condition retained in that genus (refer to Table 1); V = character variable within the genus (especially relevant to fossil and modern Homo; - = character status unknown or uncertain. Where more than one derived condition exists with respect to the ancestral condition for the great ape and human clade (see Table 1) all have been listed sequentially. Au = australopithecines; Go = Gorilla; Ho = Homo; Pa = Pan; Po = Pongo; Si = Sivapithecus. This analysis has emphasised cranio-dental characters, and also those characters important for determining relationships among extant great apes and humans. Many additional derived characters exist for the hominines, but as these are considered in detail by e.g. Wood & Chamberlain, Dean, and Grine (this volume) they have been omitted here for the sake of brevity.

Characters that have changed from the ancestral great ape and human condition	Au	Go	Ho	Pa	Po	Si
1. Orbits much higher than broad.		P		P	D	D
Orbits secondarily as broad or broader than high	D		D			
2. Orbits overlap upper nasal margin	P	P	D	P	P	P
4. Nasal bones project	P	P	D	P	P	P
5. Interorbital pillar narrow	P	P	P	P	D	D
6. Glabella reduced	P	P	V	P	D	-
7. Supraorbital torus enlarged	D	D		D	P	P
Supraorbital torus continuous	D	D		D	P	P
Supraorbital torus secondarily reduced			D			
8. Postorbital sulcus developed	D	D	V	D	P	-
9. Fronto-ethmoidal sinus present	D	D	D	D	P	-
11. Lateral flare to zygomatic	D	P	D	P	D	D
12. Malar notch in zygomatic	P	P	P	P	D	D
13. Zygomatic foramina above level of inferior margin of orbit, numberous	P	P	P	P	D	D
16. Maxillary body robust	D	P	D	P	P	D
18. Palatine foramina small and slit-like	P	P	P	P	D	D
19. Nasal floor unstepped	P	P	P	P	D	D
21. Incisive foramen large	P	P	D	P	P	P
22. Incisive foramen single	-	P	D	P	D	D
23. Thick cranial bones	D	P	D	P	P	P
24. Maximum skull breadth high up cranium	P	P	D	P	P	P
25. Reduced postorbital constriction	P	P	D	P	P	P
26. Frontal bone flat	P	P	D	P	P	-
Frontal bone more vertical	P	P	D	P	D	-

Table 3 contd.	Au	Go	Ho	Pa	Po	Si
28. Sagittal crest very reduced	P	P	D	P	P	-
29. Inferior edge temporal slopes superiorly towards posterior		P		P	D	-
Inferior edge of temporal slopes inferiorly towards posterior	D		D			-
30. No occipital crest	P	P	D	P	P	-
31. Mastoid process very prominent laterally	P	P	P	P	D	-
Mastoid process enlarged	D	P	D	P	P	-
32. Foramen magnum placed anteriorly	D	P	D	P	P	-
33. Digastric groove present	V	P	D	P	P	-
34. Small, slit-like spheno-palatine fossa	P	P	P	P	D	-
36. I$\underline{2}$ spatulate	D	V	D	D	P	P
37. I$\underline{1}$ much larger than I$\underline{2}$	P	V	P	P	D	D
38. I$\underline{1}$ higher than broad	D	P	D	P	P	P
40. Reduced canine robusticity	V	P	D	P	P	P
41. Very reduced canine crown height	V	P	D	P	P	P
42. No canine honing	D	P	D	P	P	P
43. Reduced canine sexual dimorphism, male and female both small	D	P	D	P	P	P
44. Lower P3 with very broad talonid	P	P		P	D	V
Lower P3 bicuspid	D	P	D	P	P	P
45. Lower P4 shortened	V	P	D	P	P	P
46. Lower P3 crown height level with cheek tooth row	D	P	D	P	P	P
47. Lower P3 talonid height equal to cusp height	D	P	D	P	P	P
48. Upper premolars reduced mesiodistally	V	P	D	P	P	P
49. Upper premolar heteromorphy almost absent	D	P	D	P	P	P
50. P$\underline{4}$ < or = P$\underline{3}$	V	P	D	P	P	P
52. Molar enamel slow formed (1 step)	P	D	P	D		P
Molar enamel slow formed (2 steps)	P		P		D	P
53. Molar enamel thin	P	D	P	D		P
Molar enamel intermediate/thick	P		P		D	P
54. Slow formation of tooth roots	-	P	D	P	P	-
55. Wrinkled occlusal enamel	P	P	P	D		P
Deeply crenulated occlusal enamel	P	P	P		D	P
56. Cusps relatively low	D	P	D	P	D	D
57. Low dentine horns	P	P	P	P	D	D
59. Upper molar cingula absent	D	P	D	P	D	P
60. M$\underline{2}$ < M$\underline{1}$	P	P	D	P	P	P
M$\underline{3}$ reduced in size and morphology	P	P	D	V	V	P
61. Lower M2 with only four cusps	P	P	D	P	P	P
62. Lower molars without cingula	D	P	D	P	D	P
63. Lower M2 < M1	P	P	D	P	P	P
Lower M3 reduced in size	P	P	D	V	V	P
64. Tooth rows parabolic	D	P	D	P	P	P
65. Palate deep anteriorly	P	P	D	P	P	P

Table 3 contd.	Au	Go	Ho	Pa	Po	Si
66. Palate as broad as long	D	P	D	P	P	P
67. Palate parabolic	D	P	D	P	P	P
68. Shallow mandibular symphysis	P	P	D	P	P	P
69. No inferior transverse torus	P	P	D	P	P	P
70. Presence of mental eminence (chin)	P	P	D	P	P	P
71. Mandibular corpus moderate/robust	P		D		D	P
Mandibular corpus gracile	P	D		D		P
72. Os centrale fused to scaphoid	-	D	D	D	V	-
74. Prominent transverse ridge at base of dorsal articular surface of metacarpal heads	P	D	P	D	P	-
Pronounced extension of the articular surface onto the dorsal aspect of the metacarpal heads	P	D	P	D	P	-
76. Prominent bony ridge on dorsodistal aspect of radial articular surface and on distal surface of scaphoid	P	D	P	D	P	-
Volar and ulnar inclination of concave articular surface of the distal radius	P	D	P	D	P	-
85. Extremely deep and well defined olecranon fossa	P	D	P	D	P	-
87. Early fusion of proximal epiphysis of humerus	-	D	P	D	D	-
95. Long dorsal spines of cervical vertebrae	-	D	P	D	D	-
96. Lumbar vertebrae reduced to 4	P	D	P	D	D	-
107. Hallux dramatically reduced, often resulting in the absence of a distal phalanx	P	D	P	D	P	-
109. Metatarsals and proximal and middle phalangeal bones of digits II-V possess marked degree of curvature (important in powerful grasping)	P	P	P	P	D	-
117. Presence of knuckle pads over dorsal aspects of the middle phalanges	-	D	P	D	P	-
118. Extremely strong development of the flexor digitorum superficialis	-	D	P	D	P	-
119. No anterior belly of digastric, posterior belly inserts into mandibular angle below the insertion of the medial pterygoid resulting in a different anatomy of the floor of the mouth	-	P	P	P	D	-
122. Tensor palati muscle originates from the spine of the sphenoid	-	P	D	P	P	-

Sivapithecus. A second possibility is to use Hominidae for the African ape/ human clade with Pongidae for the orang utans. A third possibility, which I currently favour is to use Hominidae for all living and fossil great apes and humans, with subfamilial and tribal distinctions for the various branches within that. However, it is important that phylogenetic conclusions are not obscured by nomenclatorial debate. A more important result of the present analysis is that, when we seek the last common ancestor of man with another primate, we are looking for the common ancestor with both chimpanzees and gorillas. In terms of the commonest fossil recoveries (i.e. teeth and jaws), this common ancestor would probably have teeth with thick enamel and robust jaws and would resemble more closely hominids than modern apes. In the later Miocene, and the Pliocene, it is not so much 'dental apes' which present a problem as 'dental hominids'!

In terms of future aims, we already know a considerable amount about the fossil record of hominine evolution and we know something about orang utan evolution also. The major gaps in our knowledge of hominoid evolution result from the lack of fossils relating to either African ape or to gibbon evolution. This fact should be increasingly emphasised in field

Figure 4. The most parsimonious interpretations of relationships within the great ape and human clade based on molecular as well as morphological evidence. Nodes 1, 2 & 3 are all defined by derived characters that are listed in Tables 1, 2 & 3 and that are discussed in the text.

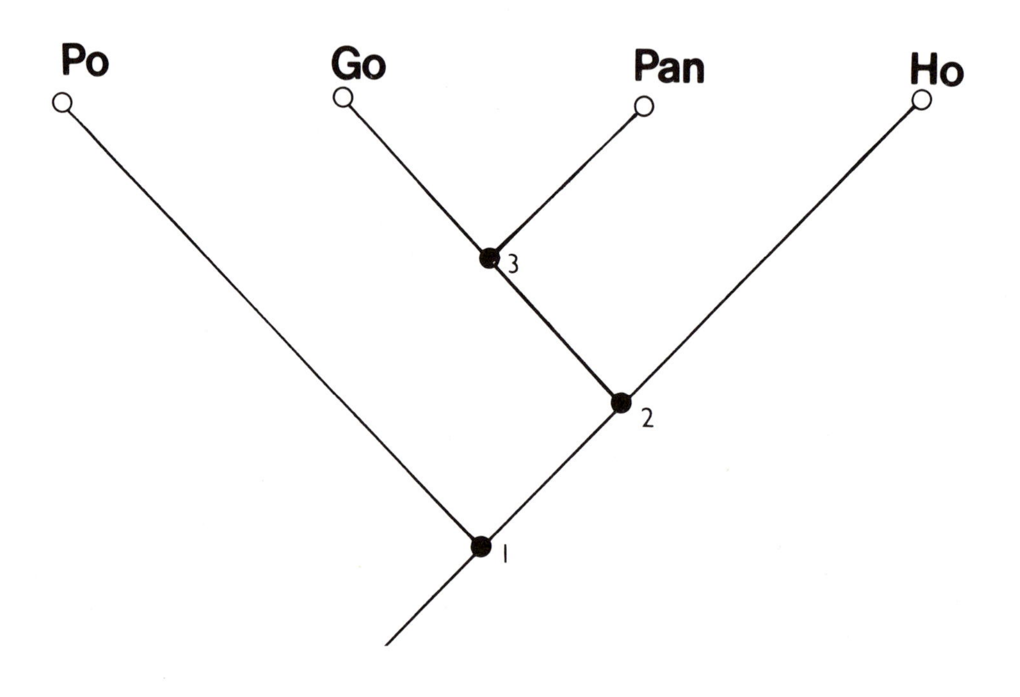

research while continuing to build on the existing record for hominines and pongines. The later Miocene offers great potential for documenting the divergence, and radiation, of the great ape and human clade from the gibbon clade. An increased knowledge and understanding of this period should also be a major goal. It is undeniable that we can learn a great deal about hominoid phylogeny from both morphological and molecular studies of extant

Figure 5: Relationships among extant and extinct great apes and humans. Dry = *Dryopithecus* (*fontani* and *laietanus*); Ken = *Kenyapithecus* (from Fort Ternan and Maboko); and Si = *Sivapithecus* (*sivalensis*, *meteal*, *punjabicus*, *alpani*, and *darwini*). Node 1 is defined by 20 derived characters (see Tables 1 & 2); node 2 is defined by only two further characters (Tables 1 and 2) and may not therefore be considered as definitive. Node 3 is defined by a further 21 derived characters with respect to node 2 (see Tables 1 & 2), at least some of which are known not to be present in either *Dryopithecus* or *Kenyapithecus*. Node 4 is defined by only 6 morphological characters but is strongly supported by the molecular evidence. Node 5 is defined by at least 11 morphological characters and may therefore be considered one of the best known relationships between a fossil and a living hominoid Node 6 is defined by 14 morphological characters mainly associated with knuckle-walking (see Table 3). The possibility that these features evolved in parallel rather than by common descent is made less likely by the shared possession of secondarily thin enamel which provides evidence, from a completely different functional complex, supporting an African ape clade.

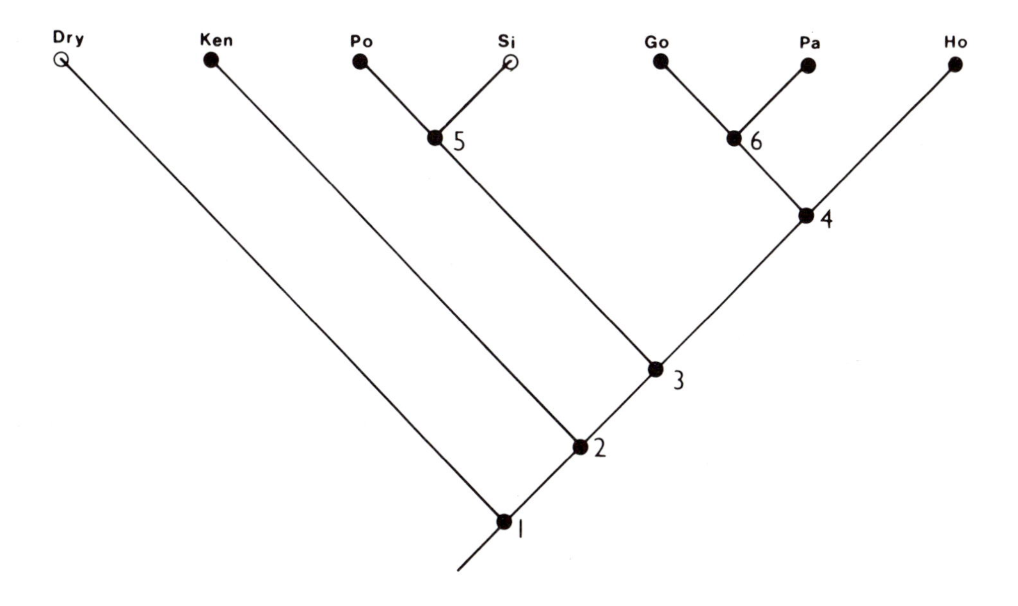

species, and the morphological analyses improve our ability to interpret fossils as and when we recover them. In addition to continuing studies of great apes and humans, and the expansion of the molecular data base, a profitable endeavour would therefore be a concentrated effort to reconstruct the common ancestral morphotype for the gibbons, which would, in turn, facilitate the accurate reconstruction of the common ancestor of the Hominoidea. Without this accomplishment our understanding of hominoid evolution will be limited.

Acknowledgements

I have greatly benefitted from discussing the ideas in this paper with Peter Andrews, Chris Dean, Eric Delson, Fred Grine, Colin Groves, Terry Harrison, David Pilbeam, and Bernard Wood although they are not to be held responsible for any failings in this work. Wendy Martin and Roy Radcliffe produced the figures and the work was supported by the Medical Research Council; the Boise Fund, Oxford; and the Central Research Fund, London. To all of the above, I am grateful.

References

Key References

Andrews, P.J. (1985). Family group systematics and evolution among catarrhine primates. *In* Ancestors: the hard evidence, ed. E. Delson, pp. 14 - 22. New York: Alan Liss.

Andrews, P.J. & Cronin, J.E. (1982). The relationship of Sivapithecus and Ramapithecus and the evolution of the orang-utan. Nature, *297*, 541-46.

Ciochon, R.L. (1983). Hominoid cladistics and the Ancestry of Modern Apes and Humans: A Summary Statement. *In* New Interpretations of Ape and Human Ancestry, eds. R.L. Ciochon and R.S. Corruccini, pp. 783-843. New York: Plenum.

Delson, E., Eldredge, N. & Tattersall, I. (1977). Reconstruction of hominid phylogeny: A testable framework based on cladistic analysis. J. Human Evol., *6*, 263-78.

Martin, L. (1985). Significance of enamel thickness in hominoid evolution. Nature, *314*, 260-63.

Schwartz, J.H. (1984). The evolutionary relationships of man and orang-utans. Nature, *308*, 501-05.

Ward, S.C. & Pilbeam, D.R. (1983). Maxillofacial morphology of Miocene hominoids from Africa and Indo-Pakistan. *In* New Interpretations of Ape and Human Ancestry, eds. R.L. Ciochon & R.S. Corruccini, pp. 211-38. New York: Plenum.

Main References

Aiello, L.C. (1981). Locomotion in the Miocene Hominoidea. *In* Aspects of Human Evolution, ed. C.B. Stringer, pp. 63-97. London: Taylor & Francis.

Andrews, P.J. (1978). A revision of the Miocene Hominoidea of East Africa. Bull. Br. Mus. Nat. Hist. (Geol.), *30(2)*, 85-224.

Andrews, P.J. (1985). Family group systematics and evolution among catarrhine primates. *In* Ancestors: the hard evidence, ed. E. Delson, pp. 14 - 22. New York: Alan Liss.

Andrews, P.J. & Cronin, J.E. (1982). The relationship of Sivapithecus and Ramapithecus and the evolution of the orang-utan. Nature, *297*, 541-46.

Andrews, P.J. & Groves, C.P. (1976). Gibbons and brachiation. Gibbon and Siamang, *4*, 167-218.

Andrews, P. & Tekkaya, I. (1980). A revision of the Turkish Miocene hominoid Sivapithecus meteai. Palaeontology, *23*, 85-95.

Ankel, F. (1972). Vertebral morphology of fossil and extant primates. *In* The functional and evolutionary biology of primates, ed. R.H. Tuttle, pp. 223-40. Chicago: Aldine.

Beneviste, R.E. & Todaro, G.J. (1976). Evolution of Type C viral genes: Evidence for an Asian origin of man. Nature, *261*, 101-08.

Boyde, A. & Martin, L. (1982). Enamel microstructure determination in hominoid and cercopithecoid primates. Anatomy and Embryology, *165*, 193-212.

Boyde, A. & Martin, L. (1983). Enamel microstructure and development in the Hominoidea. J. Anatomy, *136*, 638-40.

Boyde, A. & Martin, L. (1984a). The microstructure of primate dental enamel. *In* Food Acquisition and Processing in Primates, eds D.J. Chivers, B.A. Wood & A.Bilsborough, pp. 341-67. New York: Plenum.

Boyde, A. & Martin, L. (1984b). A non-destructive survey of prism packing patterns in primate enamels. *In* Tooth Enamel IV, eds R.W. Fearnhead & S. Suga, pp. 417-21. Amsterdam: Elsevier.

Brown, W.M., Prager, E.M., Wang, A. & Wilson, A.C. (1982). Mitochondrial DNA sequences of primates: tempo and mode of evolution. J. Molec. Evol., *18*, 225-39.

Cave, A.J. (1967). Observations on the platyrrhine nasal fossa. Amer. J. Phys. Anthropol., *26*, 277-88.

Cave, A.J. (1973). The primate nasal fossa. Biol. J. Linn. Soc., *5*, 377-87.

Cave, A.J. & Haines, R.W. (1940). The paranasal sinuses of the anthropoid apes. J. Anat., *74*, 493-523.

Ciochon, R.L. (1983). Hominoid cladistics and the Ancestry of Modern Apes and Humans: A Summary Statement. *In* New Interpretations of Ape and Human Ancestry, eds. R.L. Ciochon and R.S. Corruccini, pp. 783-843. New York: Plenum.

Clark, W.E. Le Gros (1959). The Antecedents of Man. Edinburgh: Edinburgh University Press.

Corruccini, R.S. (1975). Morphometric affinities in the forelimb of anthropoid primates. Z. Morph. Anthrop., *67*, 19-31.

Darwin, C. (1871). The Ascent of Man. London: John Murray.

Dean, M.C. (1982). The comparative anatomy of the hominoid cranial base. Univ. London PhD thesis.

Dean, M.C. & Wood, B.A. (1981). Developing pongid dentition and its use for ageing individual crania in comparative cross-sectional growth studies. Folia primatol., *36*, 111-27.

Delson, E. (1977). Catarrhine Phylogeny and Classification: Principles, Methods and Comments. J. Human Evol., *6*, 433-59.

Delson, E. & Andrews, P. (1975). Evolution and interrelationships of the catarrhine primates. *In* Phylogeny of the Primates, eds W.P. Luckett & F.S. Szalay, pp. 405-446. New York: Plenum.

Delson, E., Eldredge, N. & Tattersall, I. (1977). Reconstruction of hominid phylogeny: A testable framework based on cladistic analysis. J. Human Evol., *6*, 263-78.

Feldesman, M.R. (1982). Morphometric analysis of the distal humerus of some Cenozoic catarrhines: The late divergence hypothesis revisited. Am. J. Phys. Anthropol., *59*, 73-95.

Fleagle, J.G. (1983). Locomotor adaptations of Oligocene and Miocene hominoids and their phyletic implications. *In* New Interpretations of Ape and Human Ancestry, eds R.L. Ciochon & R.S. Corruccini, pp. 301-24. New York: Plenum.

Geoffroy-Saint-Hillaire, I. (1851). Note sur le Gorille. 3rd series. Annales des Sciences naturelles, *16*, 154-58.

Gleiser, I. & Hunt, E.E. (1955). The permanent mandibular first molar; its calcification, eruption and decay. Am. J. Phys. Anthropol., *13*, 253-84.

Goodman, M. (1963). Man's place in the phylogeny of the primates as reflected in serum proteins. *In* Classification and Human Evolution, ed. S.L. Washburn, pp. 204-34. Chicago: Aldine.

Goodman, M., Braunitzer, G., Stangl, A. & Schank, B. (1983). Evidence on human origins from haemoglobins of African apes. Nature, *303*, 546-48.

Gregory, W.K. (1927). The origin of man from the anthropoid stem - when and where? Proc. Am. Phil. Soc., *66*, 439-63.

Groves, C.P. (1972). Systematics and phylogeny of gibbons. Gibbons and Siamang, *1*, 1-89.

Haekel, E. (1866). Generelle morphologie der oranismen: Allgemeine Grundzuge der organischen Formen-Wissenschaft, mechanisch begrundet durch die von Charles Darwin feformiste Descendenz-Theorie, 2 vols. Berlin: Georg Reisner.

Harrison, T. (1982). Small-bodied apes from the Miocene of East Africa. Univ. London PhD thesis.

Hasegawa, M., Yano, T. & Kishino, H. (1984). A new molecular clock of mtDNA and the evolution of hominoids. Proc. Japan Acad. Sci., *60*, 95-8.

Hennig, W. (1966). Phylogenetic systematics. Urbana: Univ. Illinois Press.

Huxley, T.H. (1863). Evidence as to man's place in nature. London: Williams & Norgate.

Kay, R.F. (1981). The Nut-Crackers - A new theory of the Adaptations of the Ramapithecinae. Am. J. Phys. Anthrop., *55*, 141-52.

Kay, R.F. (1982). Sivapithecus simonsi, a new species of Miocene hominoid, with comments on the phylogenetic status of the Ramapithecinae. Int. J. Primatol., *3*, 113-73.

Kay, R.F. & Simons, E.L. (1983). A reassessment of the relationship between later Miocene and subsequent Hominoidea. *In* New Interpretations of Ape and Human Ancestry, eds R.L. Ciochon & R.S. Corruccini, pp. 577-624. New York: Plenum.

Keith, A. (1910). Description of a new craniometer and of certain age changes in the anthropoid skull. J. Anat. Physiol., Lond., *44*, 251-70.

Kluge, A.G. (1983). Cladistics and the Classification of the Great Apes. *In* New Interpretations of Ape and Human Ancestry, eds R.L. Ciochon and R.S. Corruccini, pp. 151-77. New York: Plenum.

Lewis, O.J. (1969). The hominoid wrist joint. Am. J. Phys. Anthropol., *30*, 251-68.

Lewis, O.J. (1972). Osteological features characterizing the wrists of monkeys and apes, with a reconsideration of this region in Dryopithecus (Proconsul) africanus. Am. J. Phys. Anthropol., *36*, 45-58.

Lipson, S. & Pilbeam, D. (1982). Ramapithecus and Hominoid Evolution. J. Human Evol., *11*, 545-48.

McHenry, H.M., Andrews, P. & Corruccini, R.S. (1980). Miocene Hominoid Palatofacial Morphology. Folia primatol, *33*, 241-52.

Martin, L.B. (1983). The relationships of the later Miocene Hominoidea. Univ. London PhD thesis.

Martin, L. (1985). Significance of enamel thickness in hominoid evolution. Nature, *314*, 260-63.

Martin, L. & Andrews, P. (1982). New ideas on the relationships of the Miocene hominoids. Primate Eye, *18*, 4-7.

Martin, L. & Andrews, P. (1984). The phyletic position of Graecopithecus freybergi KOENIGSWALD. Courier Forschungsinstitut Senckenberg, *69*, 25-40.

Martin, L. & Boyde, A. (1984). Rates of enamel formation in relation to enamel thickness in hominoid primates. *In* Tooth Enamel IV, eds R.W. Fearnhead & S. Suga, pp. 447-51. Amsterdam: Elsevier.

Moorees, C.F.A., Fanning, E.A. & Hunt, E.E. (1963). Age variation of formation stages for ten permanent teeth. J. Dent. Res., *42*, 1490-1502.

Morbeck, M.E. (1983). Miocene Hominoid Discoveries from Rudabanya: Implications from the Postcranial Skeleton. *In* New Interpretations of Ape and Human Ancestry, eds R.L. Ciochon & R.S. Corruccini, pp. 369-404. New York: Plenum.

Napier, J.R. & Napier, P.H. (1967). A handbook of living Primates. London: Academic Press.

Nuttall, G.H.F. (1904). Blood Immunity and Blood Relationships. Cambridge: Cambridge University Press.

Owen, R. (1848). On a new species of chimpanzee (Troglodytes savagei). Proc. Zool. Soc., Lond., *16*, 27-35.

Owen, R. (1851). Osteological contributions to the natural history of the chimpanzees (Troglodytes) and orangs (Pithecus). No.IV Trans. Zool. Soc., Lond., *4*, 75-88.

Oxnard, C.E. (1963). Locomotor adaptations in the primate forelimb. Symp. Zool. Soc. London, *10*, 165-82.

Pilbeam, D.R. (1969). Tertiary Pongidae of East Africa: Evolutionary relationships and taxonomy. Bull. Peabody Mus. Nat. Hist., New Haven, *31*, 1-185.

Pilgrim, G.E. (1927). A Sivapithecus palate and other primate fossils from India. Palaeont. Ind. N.S., *14*, 1-24.

Preuss, T.M. (1982). The face of Sivapithecus indicus: Description of New, Relatively Complete Specimen from the Siwaliks of Pakistan. Folia primatol., *38*, 141-57.

Rak, Y. (1983). The Australopithecine Face. New York: Academic Press.

Rose, M.D. (1975). Functional proportions of primate lumbar vertebral bodies. J. Human Evol., *4*, 21-38.

Rose, M.D. (1983). Miocene hominoid postcranial morphology: Monkey-like, ape-like, neither, or both? *In* New Interpretations of Ape and Human Ancestry, eds R.L. Ciochon & R.S. Corruccini, pp. 405-17. New York: Plenum.

Sarich, V.M. & Cronin, J.E. (1976). Molecular systematics of the primates. *In* Molecular Anthropology, ed. M. Goodman & R.E. Tashian, pp. 141-70. New York: Plenum.

Sarich, V.M. & Wilson, A.C. (1967). Immunological time scale for hominoid evolution. Science, *158*, 1200-03.

Savage, T.S. & Wyman, J. (1847a). Notice of the external characters and habits of Troglodytes gorilla, a new species of orang from the Gaboon river. Proc. Bost. Soc. Nat. Hist., *2*, 245-47.

Savage, T.S. & Wyman, J. (1847b). The first scientific identification of the gorilla as a new and hitherto unclassified species of man-like ape. Proc. Bost. Soc. Nat. Hist., *5*, 417-43.

Schultz, A.H. (1930). The skeleton of the trunk and limbs of the higher Primates. Human Biology, *2*, 303-48.

Schultz, A.H. (1936). Characters common to higher primates and characters specific to man. Q. Rev. Biol., *11*, 259-83, 425-55.

Schultz, A.H. (1938). The relative length of the regions of the spinal column in Old World primates. Am. J. Phys. Anthrop., *24*, 1-22.

Schultz, A.H. (1944). Age changes and variability in gibbons; A morphological study of a population sample of a man-like ape. Am. J. Phys. Anthrop., *2*, 1-129.

Schultz, A.H. (1961). Vertebral column and thorax. Primatologia, *4*, 1-66.

Schwartz, J.H. (1984). On the evolutionary relationships of humans and orang-utans. Nature *308*, 501-05.

Schwartz, J.H., Tattersall, I. & Eldredge, N. (1978). Phylogeny and Classification of the Primates Revisited. Yrbk. Phys. Anthrop., *21*, 95-133.

Sibley, C.G. & Ahlquist, J.E. (1984). The phylogeny of the hominoid primates, as indicated by DNA-DNA hybridization. J. Molec. Evol., *20*, 2-15.

Simons, E.L. (1961). The phyletic position of Ramapithecus. Postilla, *57*, 1-9.

Simons, E.L. (1976). The nature of the Transition in the Dental Mechanism from Pongids to Hominids. J. Human Evol., *5*, 511-28.

Simons, E.L. & Pilbeam, D.R. (1972). Hominoid Paleoprimatology. *In* The Functional and Evolutionary Biology of the Primates, ed. R.H. Tuttle, pp. 36-62. Chicago: Aldine.

Szalay, F.S. & Delson, E. (1979). Evolutionary History of the Primates. New York: Academic Press.

Templeton, A.R. (1984). Phylogenetic inference from the restriction endonuclease cleavage maps with particular reference to the evolution of humans and the apes. Evolution, *37*, 221-44.

Tuttle, R.H. (1967). Knuckle-walking and the evolution of hominoid hands. Am. J. Phys. Anthrop., *26*, 171-206.

Tuttle, R.H. (1969). Knuckle-walking and the problem of human origins. Science, *166*, 953-61.

Tuttle, R.H. (1970). Postural, propulsive and prehensile capabilities in the cheirida of chimpanzees and other Great Apes. *In* The Chimpanzee, volume 2, ed. G.H. Bourne, pp. 167-253. Basel: Karger.

Tuttle, R.H. (1974). Darwin's apes, dental apes, and the descent of man: Normal science in evolutionary anthropology. Curr. Anthropol., *15*, 389-98.

Tuttle, R.H. (1975). Parallelism, brachiation and hominoid phylogeny. *In* Phylogeny of the Primates, eds W.P. Luckett & F.S. Szalay, pp. 447-80. New York: Plenum.

Tuttle, R.H. & Rogers, C.M. (1966). Genetic and selective factors in reduction of the hallux in Pongo pygmaeus. Am. J. Phys. Anthrop. *24*, 191-98.

Walker, A.C. & Pickford, M. (1983). New postcranial fossils of Proconsul africanus and Proconsul nyanzae. *In* New Interpretations of Ape and Human Ancestry, eds R.L. Ciochon & R.S. Corruccini, pp. 325-51. New York: Plenum.

Ward, S.C., Kimbel, W.H. & Pilbeam, D. (1983). Subnasal alveolar morphology and the systematic position of Sivapithecus. Am. J. Phys. Anthrop., *61*, 157-71.

Ward, S.C. & Pilbeam, D.R. (1983). Maxillofacial morphology of Miocene hominoids from Africa and Indo-Pakistan. *In* New Interpretations of Ape and Human Ancestry, eds R.L. Ciochon & R.S. Corruccini, pp. 211-38. New York: Plenum.

Washburn, S.L. (1963). Behavior and human evolution. *In* Classification and Human Evolution, ed S.L. Washburn, pp. 190-203. Chicago: Aldine.

Zapfe, H. (1960). Die Primatenfunde aus der miozanen Spaltenfullung van Neudorf an der March (Devinska Nova Ves), Tschechoslowakei. Schweiz. Palaeont. Abh., *78*, 1-293.

11 BIPEDALISM: PRESSURES, ORIGINS AND MODES

M.H. Day,
Division of Anatomy,
United Medical and Dental Schools
of Guy's and St Thomas's Hospitals,
London SE1 7EH.

INTRODUCTION

Upright posture and bipedal gait in man and his ancestors have been a matter of concern to anatomists and anthropologists for over a century. The recognition that this unique form of posture and gait had an evolutionary history within the primate Order has led to speculation upon its causes, its mechanisms and its earliest appearance in the fossil record of the Hominidae. It is the history of these aspects of bipedalism that are reviewed in this contribution.

The word 'pressures' in the title is intended to mean evolutionary pressures, or selective pressures. These can perhaps be best expressed as the question "What survival advantages could have been conferred on a primate by its assumption of upright posture and a bipedal gait?"

The word 'origins' can be expressed as another question "How did this extraordinary posture and means of progression arise in the primate Order?" The word 'modes' can be interpreted as an appraisal of the skeletal modes, or patterns of morphology that can be recognised from the fossil record of hominid evolution. This review will be limited to the femoro-pelvic complexes of the australopithecines and of Homo erectus.

PRESSURES

If it is accepted as a premise that pressure for the evolution of bipedalism will most likely lie in survival advantage over a former locomotor method, then this pressure could act in one, two, or all three of the following ways:-

1) Improved food acquisition
2) Improved predator avoidance
3) Improved reproductive success.

The possibility that the evolution of upright stance and bipedalism might enhance food acquisition and thus confer survival advantage was suggested over twenty years ago. Hewes (1961) argued strongly that freeing of the hands for food carriage in bipeds would have had advantages in an open savanna environment. Napier (1964) suggested that this may have been more important in woodland savanna since this was a transitional environment which would eventually lead to open country which would, in turn, favour the full development of what he termed 'striding bipedalism'.

The work of Ripley (1979) has shown the evolutionary links between Prost's concept of 'positional behaviour' (Prost, 1965) and food acquisition. Recently this has been explored further by Rose (1984) in terms of bipedalism. He cites three factors in relation to bipedalism and food acquisition, 1) the importance of the arboreal/terrestrial habitat shift, 2) the 'endurance-hunting' hypothesis, and 3) vertical climbing abilities, as some of the many influences that would be operating in a slowly changing pattern of food acquisition behaviour. This is in contradiction to Lovejoy's (1981) 'single event' ancestral biped hypothesis. Recently Shipman (1983, 1984) has argued, from the evidence of cut marks on bones from Olduvai Gorge, that scavenging was the most likely food acquisition strategy of early man and that bipedalism is "consistent with the hypothesis that two million years ago hominids were scavengers rather than accomplished hunters".

The second pressure, predator avoidance by the evolution of upright stance and bipedal gait, could well have been enhanced, in my view, by the combined ability to run and climb trees. Most large predatory carnivores are very fast over short distances while man is not. Upright stance and tree climbing ability both combine to increase the visual horizon for early warning of danger so that an early retreat to a place of safety becomes a viable strategy. An alternative course of action for the biped on the ground has been advocated by Kortlandt (1980). This is the use of the freed hands to grasp and brandish thorn bush branches in order to keep large carnivores at bay - an imaginative but seemingly hazardous procedure that might have been a last resort.

The third pressure for survival advantage that could stem from upright posture and bipedalism, is improved reproductive success. The need to nurse, protect and travel in search of food or safety with a helpless infant that has a prolonged infant dependency may well require freed hands for food and infant carriage. It may also require some male parental investment in a monogamous nuclear family structure - this according to Lovejoy (1981) is a powerful evolutionary pressure towards bipedalism, and thus enhanced reproductive fitness and survival advantage.

There is also the question of energetics which affects at least two of these aspects of survival advantage. In a food acquisition strategy the energy expended in obtaining the food clearly should not exceed that which can be obtained from the food. Standing upright and picking fruit is economical since experiments have shown that human standing is little more expensive in energy costs than lying down (Joseph, 1960). Human bipedal walking is no more expensive in energy terms than mammalian quadrupedal walking. This may be because the gain of potential energy by the rise in the centre of gravity at each stride roughly equals the loss of kinetic energy in the descent (Alexander, 1976). A running man expends twice as much energy as a mammalian quadruped of the same size, but he can endure for much longer. 'Endurance hunting' in man is known from many parts of the world. In a dramatic demonstration, Louis Leakey himself ran down an antelope as a young man, killed it and skinned it with a stone tool.

Recently Carrier (1984) has suggested that hypothetical early hominids were "diurnal endurance hunters" capable of losing metabolic heat through cholinergic sweating and lack of body hair rather than panting, the dissipation of heat being a limiting factor in endurance running. Hunters can afford to run only if the energy balance of such an activity shows a high energy reward for their efforts. The risks of failure are also great since energy invested in a failed pursuit is a complete loss. Dogged pursuit would seem to have a higher success rate than the stalk and rush strategy. The combination of bipedal running ability, cholinergic sweating and lack of hair on the body may have given the early hominids an opportunity to exploit a vacant ecological niche - daylight hunting.

In summary, I believe that pressures for the evolution of upright stance and bipedal gait can be explained in terms of improved survival advantage through enhanced food acquisition (new options in positional behaviour, food carriage and endurance hunting), predator avoidance (running and climbing, early warning of danger and freed hands for threat displays and defence) and improved reproductive success (two freed hands for safer infant carriage and for nursing at pectoral mammae in the light of a prolonged infant dependency).

ORIGINS

The origin of bipedalism has long been believed to have been one of the keys to our understanding of human evolution, and competing hypotheses have been debated since the turn of the century. Keith (1912-34) supported by Gregory (1916-49) argued on comparative grounds that 'the apes' habit of brachiation played an important role in the development of upright posture in man. Wood Jones (1940) disagreed and favoured the evolutionary much earlier postural erectness seen in arborealism. Morton (1922-35) favoured "arboreal bipedalism" in a gibbon-like creature as a hominid precursor. The 1940s saw the virtual collapse of the brachiationist position when Keith retracted his view and Wood Jones (1940) and Straus (1940, 1949) pressed home the vision of a monkey-like hominid quadrupedal precursor. Washburn (1950) favoured a chimpanzee-like model and Tuttle (1967) later put forward the knuckle walking hypothesis. Straus (1962) drew attention to the observation that truncal erectness occurs in all primates at some point or other in their locomotor repertoire, and Napier & Walker (1967) extended this view and proposed that the basic primate locomotor pattern was vertical clinging and leaping and that truncal erectness was thus a basic primate feature that Napier later added to Le Gros Clark's 'trends' as a primate characteristic (Napier & Napier, 1967). Their study of the *Proconsul africanus* forelimb led Napier & Davis (1959) to the conclusion that it belonged to a semi-brachiating, primitive, pongid with a quadrupedal arboreal heritage. More recently Lewis (1971) has suggested, on the basis of wrist anatomy, that man, chimpanzee and gorilla share a close common ancestry by virtue of their brachiating adaptations. This contention has been criticised by Morbeck (1972, 1975), Schon & Ziemer (1973), and the position weakened by the experimental work of Jenkins (1981) and the statistical analyses of McHenry & Corruccini (1983). I have little doubt that Lewis will respond to this debate in due time.

Quadrumanal climbing has been cited as a link between brachiation and bipedalism (Fleagle *et al.* 1981) on the basis of electromyographic and kinetic studies of primates. They suggested a possible human ancestor with forelimb morphology associated with brachiation, and a hind limb morphology adapted for bipedalism. The climbing hypothesis also gained some support from a recent review (Aiello & Day, 1982).

Even this brief survey of the literature shows that we really have no clear idea of what form of locomotion, in what creature, preceded, or was immediately pre-adaptive for, upright posture and bipedal gait. The situation has been examined by Prost (1980) who believed that it is doubtful whether human bipedalism was ever derived from non-human bipedalism. The whole position has been reviewed by Preuschoft (1978) who concluded that "there was no full agreement about the motor habits from which bipedality evolved". Perhaps with Tuttle (1974) we should not be looking to models based on living forms and, with Fleagle et al (1981), hope that if and "when we find such an ancestor we will be able to interpret its fossils".

MODES

It is a matter of some irony that the earliest clear evidence of hominid locomotion are not fossil bones at all but trails of bipedal footprints. The now famous Laetoli Site G footprint trails from Tanzania were found in 1978 (Leakey & Hay, 1979). Two sequential trails of three individuals of differing foot sizes were uncovered in solidified volcanic ash securely dated at 3.6 million years BP. The tracks run for over 70 ft and there are some dozens of footprints available for study. The trails disclose one single track on the left and two overprinted tracks on the right. The stride length, pace length and the dynamic angles of gait are known and some estimates have been given of standing height (White, 1980) and walking speed (Charteris et al., 1981).

Photogrammetric analysis of experimental human soft substrate footprints were made and compared with the Laetoli fossil footprints (Day & Wickens, 1980). The comparison revealed seven points of similarity between the experimental human and fossil footprints including the heel impression deeper on the medial side, oblique ball impressions, lateral weight transference, a medial arch and deep hallux impressions. Further observations taken recently on other prints from the series have disclosed a consistently varus hallux position, hallucial drag at 'toe-off', a 'barefoot gap' between the hallux and the remaining toes, features commonly found in the feet and footprints of the modern human habitually unshod (Day, 1985). The evidence of the Laetoli footprints is clear and unequivocal and establishes upright bipedalism of an apparently human kind, at least in terms of weight and force transference from the foot to the ground, the position of the hallux and the spread of the unconfined forefoot as far back as 3.6 million years B.P.

If we now turn to the patterns of skeletal morphology known from the early hominid fossil record, I have long held the view (Day, 1973), that there are two skeletal morphologies (other than that of Homo sapiens) within the Family Hominidae. These can be characterised by groups, or constellations, of features which when fully developed may be diagnostic of the genus Australopithecus, or of the species Homo erectus. On this occasion

I will restrict myself to two particular areas to illustrate this contention - the pelvis and the femur, or the femoropelvic complex. There are other areas with other features, and groups of features that may prove to be equally revealing. The two femoropelvic morphological patterns that I recognise are the australopithecine and Homo erectus types.

Australopithecine pattern

Broom's and Robinson's recovery of the Sts 14 pelvis and partial skeleton of a 'gracile' australopithecine from Sterkfontein was the first evidence of an australopithecine femoropelvic complex and showed a series of features that betrayed its bipedal functional adaptation; such as a broad, backwardly-directed, S-shaped iliac blade, a small acetabulum and a short ischium. Those who recovered and described the fossils as well as Le Gros Clark (1955) and Napier (1964), concluded that the australopithecines were bipedal, a conclusion regarded as premature by Zuckerman (1966, 1967). Later, Zuckerman and his colleagues (1973) accepted that their results on the pelvis were conflicting in terms of muscle pull and weight transmission and in the summary to their paper they conceded that "it is impossible to advance an unequivocal interpretation of these findings" and again "The possibility is not excluded that the overall locomotor pattern of Australopithecus included in addition to bipedalism (italics mine) components yet undefined." On reflection I doubt whether Le Gros Clark, had he lived, would have dissented from the latter statement, neither for that matter would Napier now. Lovejoy et al. (1973) did disagree, however, and contended then that no locomotor differences exist between the gait of man and the gait patterns of either the large or the small form of australopithecine.

Examination of the Al 288 innominate bone discloses numerous similarities with that from Sterkfontein including the forward rake and shape of the ilium, the small acetabulum, the short ischium and the position of the auricular surface. It seems clear that the innominate morphological pattern characterised by the distinctive group of features shown by Sts 14, as well as in other specimens such as Sts 65 and SK 3155(b), has been repeated in Al 288. Yet the latter specimen was recovered from a site 2,000 miles away and from deposits perhaps as much as a million years older (Fig. 1). Two reconstructions of the Sts 13 and Al 288 pelves, one by Robinson and the other by Walker (in Shipman, 1981) are remarkably similar.

If we turn now to the upper end of the femur, it was Napier (1964) who recognized SK 82 and SK 97 from Swartkrans in South Africa as hominid and it was he who pointed out their combination of features. They possess small heads, relatively long, flat, necks, they lack greater trochanter flare, they have a posterior position of the lesser trochanter and no intertrochanteric line. In 1969 some five years later, Dr Mary Leakey recovered from the museum in Nairobi a proximal femoral fragment from Olduvai. Despite the lack of articular surfaces, her unerring eye had spotted the similarities between the Swartkrans femora and that from Olduvai. In describing this specimen (Olduvai Hominid 20) it was possible to add to the complex the presence of an obturator externus groove on the back of the neck and subsequently the presence of a psoas major groove on the front was noted (Day, 1969).

These features confirmed that the australopithecine hip was habitually held in an extended, or even hyperextended, position during upright stance and bipedal gait. Since 1969 there are now approximately 14 similar australopithecine femoral necks known - 5 from Koobi Fora (East Rudolf) 6 from Hadar (Ethiopia) and one from the Middle Awash, Ethiopia in addition to those already mentioned. The latest from Middle Awash (White, 1984) is dated at 4 million years BP which makes it the oldest fossil evidence of the australopithecine lower limb skeleton from anywhere in the world.

The internal anatomy of the australopithecine femoral neck is of interest and can be examined by means of X-rays. The modern human femoral neck shows a 'calcar femorale' and a pillar of trabeculae that transfer weight from the head to the neck and shaft. These features are absent in the gorilla and chimpanzee. The australopithecine femoral neck shows a well marked buttress on the inner aspect of the neck that compares with the human calcar femorale and it also shows some trabecular thickening towards the

Fig. 1. Three hip bones showing a group of similar anatomical features including a large acetabulum, a stout acetabulo-cristal buttress and a rotated ischium. Left: Arago XLIV; Centre: Olduvai Hominid 28; Right: KNM-ER 3228.

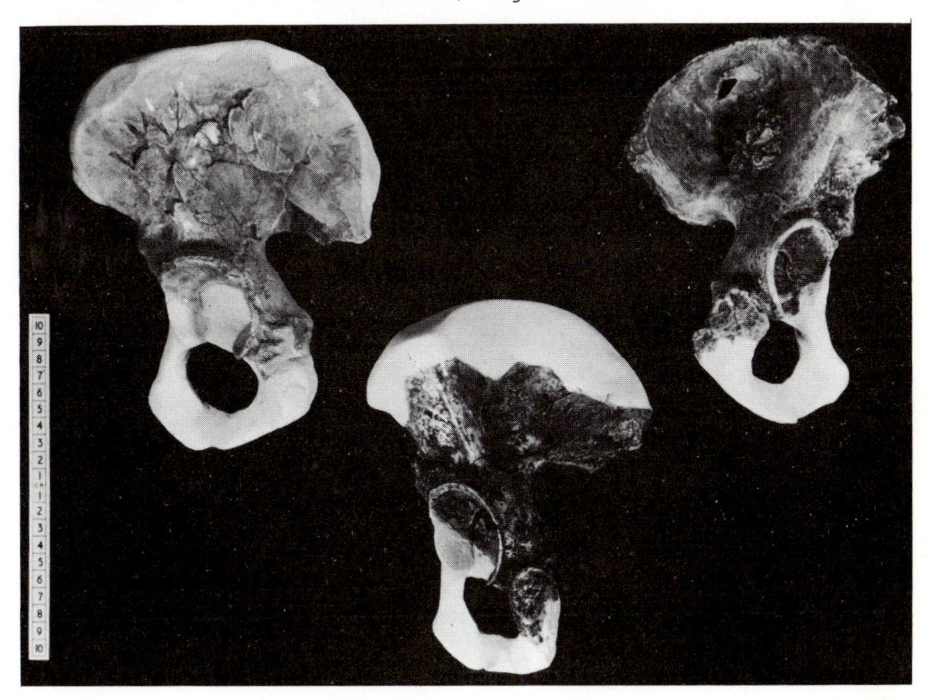

head, in effect an incipient pillar (Figs. 2-4).

These finds, and the investigations performed on them, have shown the essential similarity of the morphology of the known australopithecine pelvic remains, and also the similarities between the known femoral remains. What of the morphological differences between the so-called 'gracile' and 'robust' australopithecines, that is, between Australopithecus africanus and Australopithecus robustus (Paranthropus robustus to some taxonomists)?

Figs. 2, 3 & 4. X-ray photographs and drawings of three femoral necks (Fig. 2 Olduvai Hominid 20, Fig. 3 Swartkrans 97 and Fig. 4 Swartkrans 82) that show medial buttressing and radiating trabeculae that intersect and support the neck in upright stance and bipedal gait.

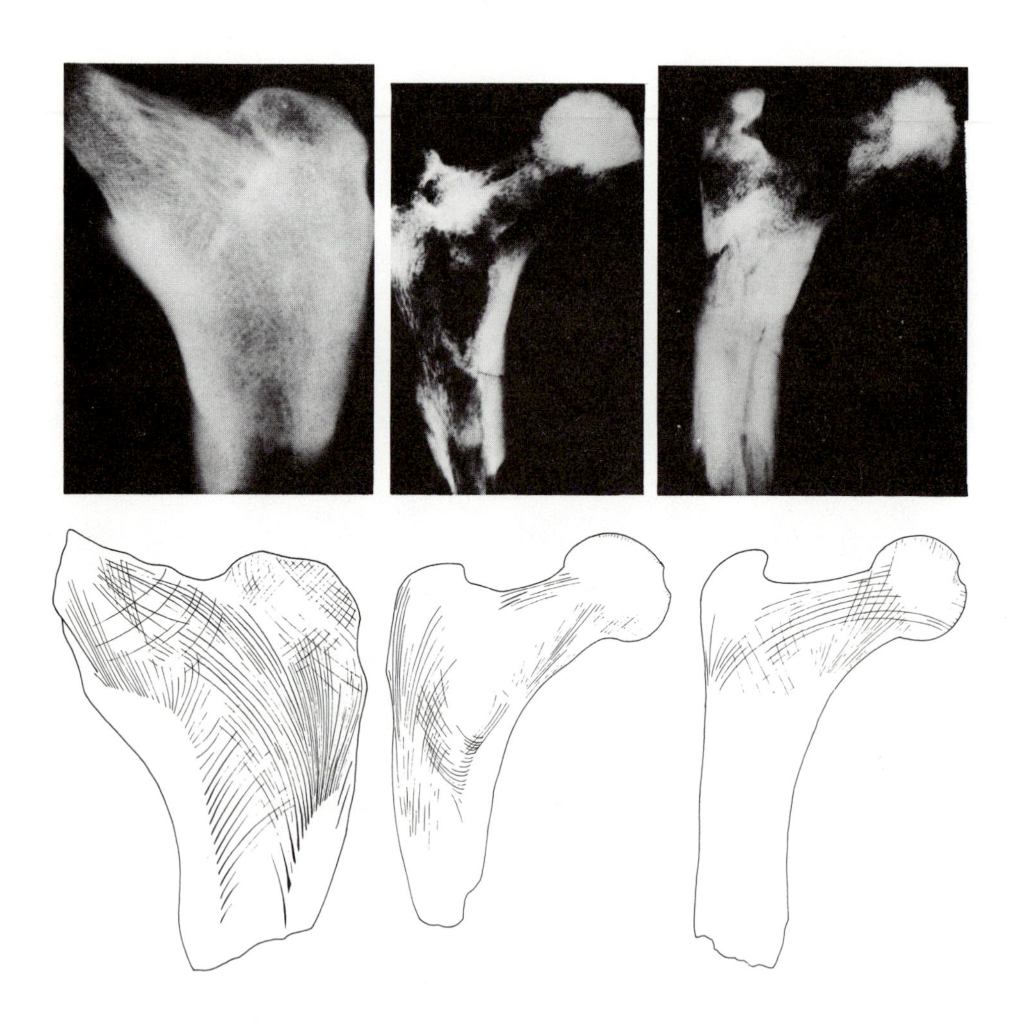

The femora that we have range in size from small to large and there is no clear indication of two morphological types, or of two size groups. The only so-called 'robust' or Paranthropus pelvis known is the SK 50 specimen from Swartkrans and that is so distorted as to be unreliable as evidence. Such features that it does display seem to recall the 'gracile' pelves from Sterkfontein and Hadar. It seems, therefore, that the femoropelvic complex has little to contribute to the discussion of the relationship between the two species (or genera) of australopithecines and from this evidence alone there seems little reason to believe that the larger form was more arboreally adapted than the smaller, or vice versa.

Homo erectus pattern

The earliest postcranial bone that has been attributed to Homo erectus is the famous femur recovered by Dubois from Java (Dubois, 1894). From the beginning, the Trinil femur has always been controversial in terms of its provenance, its dating and its pathology. Only one feature has been the subject of universal agreement - that is its functional anatomy. Although Dubois drew attention to some details that he considered distinctive he was in no doubt that its owner was capable of bipedal gait. He emphasised this by naming the calotte and femur from Trinil Pithecanthropus erectus - later Homo erectus. Successive anatomists agreed with him. Manouvrier (1895), Hepburn (1897), Weinert (1928), Weidenreich (1941), Le Gros Clark (1964) all agreed that there was no feature or combination of features that marks this femur as distinguishable from that of modern man. In 1973, with Theya Molleson, I came to the same conclusion (Day & Molleson, 1973) yet on analytical grounds we could not disprove its antiquity or contemporaneity with the Java calotte.

While Java produced the first finds of Homo erectus, China in the 1920s produced the best material including femora nearly certainly associated with the Peking Homo erectus skulls. Careful description of the Peking femora (Weidenreich, 1941) revealed a complex of features unique and distinct from that of Homo sapiens. Any one of these features can probably be matched from a large collection of modern human femora both anatomically and metrically. But the combination of features has not been matched yet in a modern human series. This complex included platymeria, a medial shaft ridge and convexity, a low narrow point, thick cortical walls and lack of A/P bowing.

The recovery in 1970 of the Olduvai Hominid 28 femur and pelvic fragment confirmed Weidenreich's description of this complex of femoral features. The detailed resemblance of the OH 28 femur to the Peking femora was so striking that there was no hesitation in attributing it, and its closely associated pelvis, to Homo erectus (Day, 1971). Indeed the pelvic fragment was also shown to possess a distinctive combination of anatomical features, unmatched as a group with modern human pelves. The most striking feature of this complex is a prominent, or even massive, acetabulocristal buttress, as well as a large acetabulum, a medially-rotated ischium and a small, obliquely placed, auricular surface. The case for a femoropelvic complex associated with Homo erectus was beginning to build and the conventional view, supported by Le Gros Clark (1955) that the postcranial skeleton of Homo erectus did not differ significantly from that of modern man, was beginning to look less tenable.

In the early 1970s the Koobi Fora site in North Kenya produced numerous australopithecine finds as well as specimens such as the ER 730 mandible and the ER 737 femur which were cautiously attributed to the genus Homo (Day & Leakey, 1973). In 1975 the recovery of the ER 3733 skull from deposits dated at 1.5 my BP led to its attribution to Homo erectus by Leakey & Walker (1976) and the realisation that Homo erectus was earlier on the scene than had been believed. At about the same time from even older deposits, dated at not less than 1.9 my BP, a pelvic fragment was recovered. This find was astonishing in that its features paralleled the OH 28 specimen both anatomically and metrically and yet it was over one million years older. The same suite of features is present in ER 3228 as in the Olduvai Hominid 28 specimen (Rose, 1984)(Fig. 1).

The Koobi Fora site has also produced two almost complete femora ER 1472 and 1481 from deposits not less than 1.9 my old that show some features in common with the Peking and Olduvai Hominid 28 femora. Indeed Kennedy (1983) has gone so far as to attribute these femora to Homo erectus. If this should prove to be a correct attribution then they are the earliest evidence of Homo erectus from Africa or from anywhere in the world. This attribution has not, however, gone uncontested (Trinkaus, 1984).

The presence or absence of Homo erectus in the European fossil record is a matter of current active debate (Howell, 1982, pers. comm.). Among the candidates for this status are the hominid remains from Arago found by Henri and Marie-Antoinette de Lumley that include 3 femoral fragments (Arago 48, 51 and 53) and a pelvic fragment (Arago 44). The site is dated at about 450,000 my BP. Comparison of the Arago pelvic fragment with the fossil hip bones known from Africa (OH 28 and ER 3228) shows remarkable similarities. The complex of features described as peculiar to OH 28 and ER 3228 can be clearly identified in Arago 44; that is a large acetabulum, stout acetabulocristal buttress, a small oblique auricular surface, a forwardly raked anterior superior iliac spine, a stout horizontal buttress and lack of demarcation between the iliac fossa and the true pelvis. Linear discriminant function analysis, using nine variables and 5 comparative samples, showed that the fossil bones, when entered as a group, had close metrical similarities to each other and to the group drawn from Homo sapiens. It seems that the mode of locomotion employed by the fossil group was more closely allied to that of upright bipedal Homo sapiens than the locomotor mode used by the other groups in the comparison (Day, 1982, 1984). Comparison of the Arago femora by Sigmon (1982) has shown the pattern of traits recognised for Peking and OH 28 and ER 737 is present in large part in the Arago femora. It also appears that the femoropelvic complex of Peking, OH 28 and ER 3228 can be clearly identified in Arago 44, 48, 51 and 53. If this femoropelvic complex, in fully developed form, is shown to be specific for Homo erectus then this species can be described now in both cranial and postcranial terms over as long a period as 1.5 million years.

What then, of the Trinil femur which is modern in form, yet alleged to be of Middle Pleistocene antiquity? Last year, ten years after Theya Molleson and I were unable to prove or disprove the antiquity of this fossil, Bartstra (1983) has claimed, from new investigations into the sedimentary geology, stratigraphy and prehistory of the region of Trinil, that Dubois' excavations at Trinil overlooked a Solo River High Terrace sediment and that a younger layer overlaid the Kabuh deposits. Bartstra states "...at the spot where the skull cap and thigh bones of Pithecanthropus erectus were found, both High Terrace and Kabuh sediments are present. Dubois excavated through both units!!" Serious doubts arise therefore as to the provenance of the skull cap and the femur - to add to the anatomical doubts I have already mentioned. A possible means of distinguishing between fossils of the older Kabuh Beds and the younger High Terrace sediments has presented itself. Energy dispersive microanalysis (Goodhew & Chescoe, 1980) can detect the heavy elemental content of particular fossil material both qualitatively and quantitatively. Comparison of the figures obtained for the relative amounts of elements present in the calotte, the femora, and several other specimens from the deposits, has shown, by discriminant function analysis, that Trinil Femur I is highly significantly different from the other material in its composition (Fig. 5). There are several possible reasons for this, one of which is that it is much younger than believed and not contemporary with the calotte - an answer that would explain its modern morphology and remove this anomaly from the morphological record of Homo erectus postcrania. This work is at an early stage but I hope that with new material, collected under proper stratigraphic control, to be able to contribute further to this problem in the near future.

DISCUSSION

Two skeletal modes, or morphologies, other than that of modern man have been characterised from femoropelvic remains from Africa, Asia and Europe in terms of Homo erectus and Africa alone with regard to Australopithecus. What is the meaning of these differences in functional and taxonomic terms? Nobody now seriously disputes the upright stance and bipedalism of the australopithecines, or of Homo erectus.

The morphological opinions of Le Gros Clark (1955) and Napier (1964) with regard to the australopithecines and bipedalism have been amply vindicated by the Laetoli footprint discoveries and by subsequent fossil finds and analyses. The only debate that is left centres around the so-called 'arboreal' features of early hominid remains and their meaning in terms of behaviour. The Homo erectus femoropelvic complex is also clearly bipedal in its functional anatomy and may well shade downwards into Homo habilis and upwards into the erectus/sapiens transitionals of a later period. Much work remains to be done, not least by those who are skilled in the methods of allometry. The advent of two complete Homo erectus skeletons, one reported from West Turkana (WT 15000) and one from Yinkou County, China, may give new opportunities for research in this area and provide new insights into the significance of this particular skeletal mode.

Man's upright posture and bipedal gait has been considered an outstanding feature of his humanity by a long list of authors including Haeckel, Darwin,

Engels, Hooton, Freud, Weidenreich, Keith, Schultz, Washburn, Mayr, Le Gros Clark, Piveteau, Delmas, Vallois, Howells, Dart, Hewes and Napier. Its advantages have been speculated upon by almost as many other writers; with freeing the hand for many purposes leading the field, followed by a widened range of vision as a close second. Scavenging was cited in 1953 by Eiseley and also Bartholomew & Birdsell (1953) as of importance since "carrion robbery" requires freed hands (30 years before its revival on other evidence by Shipman (1983, 1984)). In this field of speculation very

Fig. 5. Canonical variate plot of the relative quantities of heavy elements detected by X-ray microanalysis in fossil material from Java including the Pithecanthropus I calotte, the Trinil femora (I-VI), the Solo remains and a sample of Elephas from Kedung Brubus (KD). Trinil femur I is highly significantly different from all of the other material sampled.

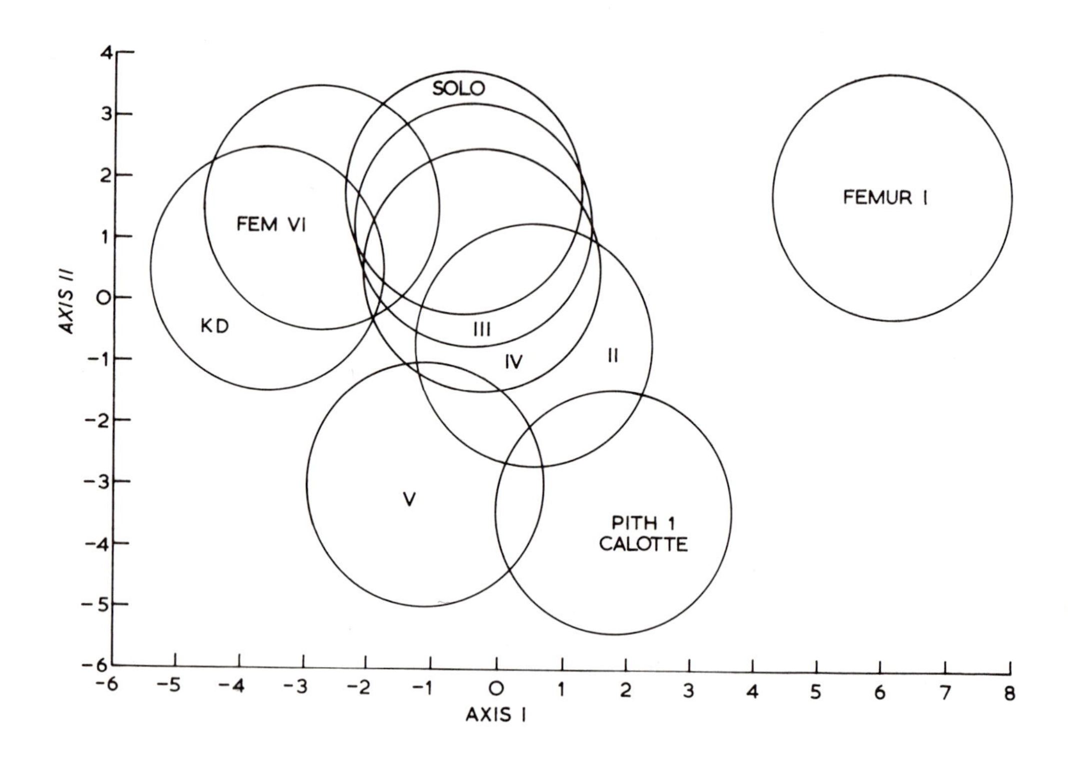

little is new. The search for hominid ancestors with a clear preadaptation to bipedalism must still go on with known extant locomotor models looking less and less likely to fit the bill. Only in the field of fossil discovery does the "hard evidence" continue to accumulate through the tireless efforts of field researchers. Fossil discoveries demand explanations that fit the facts from those of us who try to describe and to analyse.

ACKNOWLEDGEMENTS

I thank the Anatomical Society of Great Britain and Ireland for their invitation to deliver the 5th Anatomical Society Review Lecture, which was coupled with the Louis Leakey Memorial Lecture for 1984.

This paper is dedicated to the memory of Louis Seymour Bazett Leakey.

REFERENCES

Key references

Day, M.H. (1985). Hominid locomotion - from Taung to the Laetoli footprints. *In*: Past, Present and Future of Hominid Evolution. Alan Liss Inc., New York.

Napier, J.R. (1964). The evolution of bipedal walking in the hominids. Arch. Biol. Liège, LXXV: 673-708.

Preuschoft, H. (1978). Recent results concerning the biomechanics of man's acquisition of bipedality. *In*: Recent Advances in Primatology, *3*: Evolution. (Eds) D.J. Chivers & K.A. Joysey, Acad. Press, New York, 435-58.

Sigmon, B.A. (1982). Comparative morphology of the locomotor skeleton of Homo erectus and the other fossil hominids, with special reference to the Tautavel innominate and femora. Prem. Cong. Internat. Paleo. Hum. UISPP Nice: Pretirage.

Tuttle, R. (1974). Darwin's Apes, Dental Apes, and the Descent of Man: Normal Science in Evolutionary Anthropology. Curr. Anthrop. *15*: 389-426.

Main references

Aiello, L.C. & Day, M.H. (1982). The evolution of locomotion in the early Hominidae. *In*: Progress in Anatomy 2 (Eds) R.J. Harrison & V. Navaratnam, p. 81-97, Cambridge University Press.

Alexander, R. McN. (1976). Mechanics of bipedal locomotion. *In*: Perspectives in experimental biology: Zoology 1 (Ed.) S.P. Davies, Oxford, Pergamon.

Bartholomew, G.A. & Birdsell, J.B. (1953). Ecology and the Protohominids. Amer. Anthrop. *55(4)*: 481-98.

Bartstra, G.-J. (1983). The fauna from Trinil, type locality of Homo erectus: A Reinterpretation. Geologie en Mijnbouw, 62:329-36.

Carrier, D.R. (1984). The energetic paradox of human running and hominid evolution. Curr. Anthrop. 25(4): 483-95.

Charteris, J., Wall, J.C. & Nottrodt, J.W. (1981). Functional reconstruction of gait from the Pliocene hominid footprints at Laetoli, Northern Tanzania. Nature, 290: 496-98.

Clark, W.E., Le Gros (1955). The Fossil Evidence for Human Evolution. Chicago, Univ. of Chicago Press.

Clark, W.E., Le Gros (1964). The Fossil evidence for Human Evolution. Chicago, Univ. of Chicago Press.

Day, M.H. (1969). Femoral fragment of a robust australopithecine from Olduvai Gorge, Tanzania. Nature, *221*:230-33.

Day, M.H. (1971). Postcranial remains of Homo erectus from Bed IV, Olduvai Gorge, Tanzania. Nature, *232*:383-87.

Day, M.H. (1973). Locomotor features of the lower limb in hominids. Symp. Zool. Soc. Lond. *33*:29-51.

Day, M.H. (1982). The Homo erectus pelvis: Punctuation or gradualism? Prem. Cong. Internat. Paleo. Hum. UISPP Nice: Pretirage: 411-21.

Day, M.H. (1984). The postcranial remains of Homo erectus from Africa, Asia and possibly Europe. Cour. Forsch. Inst. Senckenberg *69*: 113-21.

Day, M.H. (1985). Hominid locomotion - from Taung to the Laetoli footprints. *In*: Past, Present, and Future of Hominid Evolution. Alan Liss Inc. New York.

Day, M.H. & Leakey, R.E.F. (1973). New evidence of the genus Homo from East Rudolf, Kenya. I. Am. J. Phys. Anthrop. *39*:341-54.

Day, M.H. & Molleson, T.I. (1973). The Trinil Femora. *In*: Human Evolution (Ed.) M.H. Day, Symp. SSHB 11 London: Taylor & Francis Ltd: 127-54.

Day, M.H. & Wickens, E.H. (1980). Laetoli Pliocene hominid footprints and bipedalism. Nature, *286*:385-87.

Dubois, E. (1894). Pithecanthropus erectus, eine menschenaehnliche Ubergangsform aus Java. Batavia: Landesdruckerei.

Eiseley, L.C. (1953). Fossil man. Scient. Am. *189*(*6*):65-72.

Fleagle, J.G., Stern, J.T. Jr., Jungers, W.L., Susman, R.L., Vangor, A.K., Wells, J.P. (1981). Climbing: A biochemical link with brachiation and with bipedalism. *In*: Vertebrate Locomotion (Ed.) M.H. Day, Symp. Zool. Soc. Lond. *48*:359-75.

Gregory, W.K. (1916-1949). *In*: Tuttle, R. Darwin's Apes, Dental Apes, and the Descent of Man: Normal Science in Evolutionary Anthropology. Curr. Anthrop., *15*:389-426.

Goodhew, P.J. & Chescoe, D. (1980). Microanalysis in the transmission electron microscope. Micron, *11*:153-181.

Hepburn, D. (1897). The Trinil Femur (Pithecanthropus erectus) contrasted with the femora of various savage and civilised races. J. Anat. Physiol. *31*:1-17.

Hewes, G.W. (1961). Food transport and the origin of hominid bipedalism. Amer. Anthrop.: 687-710.

Howell, F.C. (1982). Pers. Comm. to delegates UISPP, Nice.

Jenkins, F.A. Jr. (1981). Wrist rotation in primates: A critical adaptation for brachiators. Symp. Zool. Soc. Lond. :429-51.

Jones, F.W. (1940). Attainment of the upright posture of man. Nature *146*:26-7.

Joseph, J. (1960). Man's Posture. Springfield, III, C.C. Thomas.

Keith, A. (1912-1934). *In:* Tuttle, R. Darwin's Apes, Dental Apes, and the Descent of Man: Normal Science in Evolutionary Anthropology. Curr. Anthrop., *15*:389-426.

Kennedy, G.E. (1983). A morphometric and taxonomic assessment of a hominine femur from the Lower Member, Koobi Fora, Lake Turkana. Am. J. Phys. Anthrop. *61*: 429-36.

Kortlandt, A. (1980). How might hominids have defended themselves against large predators and food competitors? J. Hum. Evol. *9*: 79-112.

Leakey, M.D. & Hay, R.L. (1979). Pliocene footprints in the Laetolil Beds at Laetoli, Northern Tanzania. Nature, *278*:317-23.

Leakey, R.E.F. & Walker, A. (1976). Australopithecus, Homo erectus and the single species hypothesis. Nature *261*:572-74.

Lewis, O.J. (1971). Brachiation and the early evolution of the Hominoidea. Nature, *230*:577-78.

Lovejoy, C.O. (1981). The Origin of Man. Science, *211*:341-50.

Lovejoy, C.O., Heiple, K.G. & Burstein, A.H. (1973). The gait of Australopithecus. Am. J. Phys. Anthrop. *38*:757-79.

Manouvrier, L. (1895). Discussion dur "Pithecanthropus erectus" comme precurseur presume de l'homme. Bull. Soc. Anthrop. Paris 6: 12-47.

McHenry, H.M. & Corruccini, R.S. (1983). The wrist of Proconsul africanus and the origin of hominoid postcranial adaptations. *In*: New Interpretations of Ape and Human Ancestry (Eds) R.L. Ciochon & R.S. Corruccini, Plenum Press, New York and London.

Morbeck, M.E. (1972). A re-examination of the forelimb of the Miocene hominoidea. Unpub. PhD thesis, Univ. California, Berkeley.

Morbeck, M.E. (1975). Dryopithecus africanus forelimb. J. Hum. Evol. *4*: 39-46.

Morton, D.J. (1922-1935). *In*: Tuttle, R. Darwin's Apes, Dental Apes, and the Descent of Man: Normal Science in Evolutionary Anthropology. Curr. Anthrop. *15*:389-426.

Napier, J.R. (1964). The evolution of bipedal walking in the hominids. Arch. Biol. Liege, LXXV:673-708.

Napier, J.R. & Davis, P.R. (1959). The fore-limb skeleton and associated remains of Proconsul africanus. Fossil Mammals of Africa 16, Brit. Mus. (Nat. Hist.).

Napier, J.R. & Napier, P. (1967). A Handbook of LIving Primates. Academic Press, London.

Napier, J.R. & Walker, A.C. (1967). Vertical clinging and leaping - a newly recognized category of locomotor behaviour of primates. Folia primat. *6*:204-19.

Preuschoft, H. (1978). Recent results concerning the biomechanics of man's acquisition of bipedality. *In*: Recent Advances in Primatology, 3: Evolution. (Eds) D.J. Chivers & K.A. Joysey, Acad. Press, New York, 435-58.

Prost, J. (1965). A definitional system for the classification of primate locomotion. Amer. Anthrop. *67*:1198-214.

Prost, J.H. (1980). Origin of Bipedalism. Am. J. Phys. Anthrop. *52*:175-89.

Ripley, S. (1979). Environmental grain, niche diversification, and positional behaviour in Neogene primates: An evolutionary hypothesis. *In*: Environment, Behaviour and Morphology (Eds) M.E. Morbeck, H. Preuschoft, N. Gomberg, Gustav Fischer, New York and Stuggart.

Rose, M.D. (1984). A hominine hip bone, KNM-ER 3228, from East Turkana, Kenya. Am. J. Phys. Anthrop. *63*:371-78.

Schon, M.A. & Ziemer, L.K. (1973). Wrist mechanism and locomotor behaviour of Dryopithecus (Proconsul) africanus. Folia Primat. *20*:1-11.

Shipman, P. (1981). Life History of a Fossil: An Introduction to Vertebrate Taphonomy and Paleoecology. Cambridge, Mass: Harvard Univ. Press.

Shipman, P. (1983). Early hominid lifestyles: Hunting and gathering or foraging and scavenging? *In*: Animals and Archaeology (Eds) J. Clutton-Brock & C. Grigson, BAR International Series 163.

Shipman, P. (1984). Scavenger hunt. Nat. Hist. April 1984.

Sigmon, B.A. (1982). Comparative morphology of the locomotor skeleton of Homo erectus and the other fossil hominids, with special reference to the Tautavel innominate and femora. Prem. Cong. Internat. Paleo. Hum. UISPP Nice: Prétirage.

Straus, W.L. Jr. (1940). The posture of the great ape hand in locomotion, and its phylogenetic implications. Am. J. Phys. Anthrop. 27:199-207.

Straus, W.L. Jr. (1949). The riddle of man's ancestry. Quart. Rev. Biol. 24:200-23.

Straus, W.L. Jr. (1962). Fossil evidence of the evolution of the erect, bipedal posture. Clin. Orthopaed. 25:9-19.

Trinkaus, E. (1984). Does KNM-ER 1481A establish Homo erectus at 2.0 myr BP? Am. J. Phys. Anthrop. 64:137-39.

Tuttle, R.H. (1967). Knuckle walking and the evolution of hominid hands. Am. J. Phys. Anthrop. 26:171-206.

Tuttle, R. (1974). Darwin's Apes, Dental Apes, and the Descent of Man: Normal Science in Evolutionary Anthropology. Curr. Anthrop. 15:389-426.

Washburn, S.L. (1950). The analysis of primate evolution with particular reference to the origin of man. Cold Spring Harbor Symp. Qunat. Biol. 15:67-78.

Weidenriech, F. (1941). The extremity bones of Sinanthropus pekinensis. Palaeont. sin. New Ser. D. 5:1-150.

Weinert, H. (1928). Pithecanthrupus erectus. Z. Ges. Anat. 87:429-547.

White, T.D. (1980). Evolutionary implications of Pliocene hominid footprints. Science 208:175-76.

White, T.D. (1984). Pliocene hominids from the Middle Awash, Ethiopia. Cour. Forsch. Inst. Senckenberg 69:57-68.

Zuckerman, S. (1966). Myths and methods in anatomy. J. Roy. Coll. Surg. Ed. 11:87-114.

Zuckerman, S. (1967). The functional significance of certain features of the innominate bone in living and fossil primates. J. Anat. Lond. 101:608.

Zuckerman, S. (1973). Some locomotor features of the pelvic girdle in primates. Symp. Zool. Soc. Lond. 33:71-165.

12 DENTAL TRENDS IN THE AUSTRALOPITHECINES: THE ALLOMETRY OF MANDIBULAR MOLAR DIMENSIONS

W.L. Jungers,
Department of Anatomical Science,
State University of New York, Stony Brook,
N.Y. 11794, U.S.A.

F.E. Grine,
Departments of Anthropology and Anatomical Sciences,
State University of New York, Stony Brook,
N.Y. 11794, U.S.A.

INTRODUCTION

To many palaeoanthropologists the australopithecines represent an evolutionary grade, made up of the comparatively small-brained, albeit rather large-toothed, hominids from the Pliocene and early Pleistocene of Africa. The first of these creatures to be described consisted of an incomplete juvenile skull from the Buxton Limeworks Quarry at Taung, a specimen designated sixty years ago by Dart (1925) as the holotype of a novel genus and species, *Australopithecus africanus*. The hominid fossils that have been subsequently extracted from the Sterkfontein Member 4 and Makapansgat Member 3 and 4 breccias have been accepted by most authorities as being attributable to the hypodigm of *A. africanus* (Robinson 1954a, 1968; Tobias 1967; Clarke 1977; White *et al*. 1981; Rak, 1983). The recently discovered Pliocene hominid remains from the Laetolil Beds and the Hadar Formation sediments have been assigned to a separate species of *Australopithecus*, *A. afarensis*, by Johanson *et al*. (1978). While Tobias (1980) and Boaz (1983) have contended that the Laetoli and Hadar fossils are referrable to *A. africanus*, White *et al*. (1981), Rak (1983), White (1985) and Kimbel *et al*. (1985) have argued that a suite of features serves to distinguish *A. afarensis* from *A. africanus* and other early hominid taxa.

Apart from *A. afarensis* and *A. africanus*, which are commonly referred to by the informal, and quite misleading epithet 'gracile', most students of hominid evolution recognize a second group, the so-called 'robust' australopithecines. The first of these 'robust' forms, discovered at the site of Kromdraai, was attributed by Broom (1938) to a distinct genus and species, *Paranthropus robustus*, while the subsequently discovered australopithecine remains from Swartkrans were considered by him (Broom, 1949) to represent a second species of *Paranthropus*, *P. crassidens*. Robinson (1954a), however, maintained that the Kromdraai and Swartkrans australopithecines could be accommodated in a single species (*P. robustus*), and while the majority of contemporary opinion seems to hold with this position, Howell (1978) has recently proposed that a species-specific distinction between these samples is, after all, warranted. Elsewhere, one of us (F.E.G.) has cited both deciduous and permanent odontological differences between the Kromdraai and Swartkrans australopithecines which are consistent with a taxonomic distinction (Grine, 1982, 1984a, 1985).

The presence of 'robust' australopithecines in eastern Africa was established with the discovery in 1959 of a massively constructed cranium in Bed I of Olduvai Gorge. This specimen was attributed by L.S.B. Leakey (1959) to a novel genus and species, Zinjanthropus boisei, whereas Tobias (1967) concluded that it should be recognized as representing a distinct species of Australopithecus, viz. A. boisei. Numerous additional specimens of 'hyper-robust' australopithecines have been recovered more recently from Beds I and II of Olduvai Gorge, the Humbu Formation at Peninj, Chemoigut Formation at Chesowanja, Members D through G and Member L of the Shungura Formation in the Omo, and from the Lower and especially the Upper and Ileret Members of the Koobi Fora Formation (Howell, 1978; Grine, 1981).

Tobias has suggested that these eastern African australopithecines, together with those from the southern African sites of Kromdraai and Swartkrans, might be regarded as representing 'a superspecies, A. robustus comprising two semi-species, A. robustus and A. boisei' (1975:293). Some workers (e.g. Clarke, 1977; Bilsborough, 1978) postulate an even closer relationship amongst these forms, by viewing them as representing two subspecies, whilst others (e.g. Howell, 1978; Rak, 1973) consider the eastern African 'robust' specimens to be representative of a distinct species. This latter view has been supported by analyses of the deciduous dentitions attributable to the southern and eastern 'robust' australopithecines (Grine, 1984a, 1985).

Not only has the level of taxonomic distinction amongst the various 'robust' australopithecine samples been the subject of different opinion, there also has been (and continues to be) a notable lack of consensus over the distinctiveness of the 'robust' australopithecines as a group. Thus, although the case for a 'robust' australopithecine lineage is widely accepted by students of hominid palaeontology, the taxonomic and, indeed, the functional implications of the characters utilized in the recognition of this lineage have been the subject of vigorous debate since Robinson's (1954b) proposal of the 'dietary hypothesis'. The observable morphological differences between the 'gracile' and 'robust' australopithecines were perceived by Robinson (1954b) as having been related to profoundly different trophic adaptations, and they were interpreted by him as validating a generic distinction between Australopithecus and Paranthropus. While the generic distinctiveness of Paranthropus has been questioned by numerous workers, only Tobias (1967) and Wolpoff (1974) have actually analysed in any detail Broom's and Robinson's morphological criteria for this genus. Wolpoff (1974) argued that the differences between the 'gracile' and 'robust' australopithecines are minimal in number and importance, and that the differences which occur are, in large part, related to concomitant differences in body size.

Allometric scaling

The problem of sorting those features that are requisite consequences of allometric scaling from those that are evidence of ecologically significant functional differences is of fundamental importance in any attempt to interpret the course of australopithecine evolution. Unfortunately, we have few a priori expectations about the relation between morphology and size change. The observation has also been made that highly correlated, size-related changes may themselves reflect major adaptive differences along a size gradient of individuals or taxa (Kay, 1975; Smith, 1980; Jungers, 1984). For these reasons, the question of size and allometric scaling has been the subject of considerable discussion for the last decade. Because trophic factors are central to the construction of models of early hominid ecology and evolution, much of the debate has focused upon dental variables, with considerable attention having been paid to relative canine and cheek-tooth sizes (Pilbeam & Gould, 1974; Kay, 1975; Wolpoff, 1978, 1982; Wood, 1978, 1984; Wood & Stack, 1980). For example, relative canine reduction in the 'robust' australopithecines may be related to a shift in canine function from shearing to apical crushing (Wallace, 1978; Grine, 1981, 1984b). Moreover, it would appear that the comparatively larger cheek teeth of the 'robust' australopithecines were not related solely to the maintenance of "functional equivalence" in larger bodies (Grine, 1981). In comparison to their 'gracile' relatives, the 'robust' australopithecines did not require relatively larger post-canine teeth merely in order to process greater amounts of qualitatively similar foods per chewing cycle in order to maintain metabolic equivalence.

Recent work on the allometric scaling of australopithecine teeth has focused on the relationship between overall crown size and relative cuspal size in the mandibular molars. Such considerations would seem to be potentially relevant to questions concerning early hominid tooth function and diet.

The observation that the lower molars of the 'robust' australopithecines are larger, on average, than those of the 'gracile' Plio-Pleistocene hominids has been documented extensively (e.g. White et al. 1981; Wood, 1981; Wood & Abbott, 1983). Moreover, several studies have indicated that the comparatively large molars of the 'robust' australopithecines appear to be distinctive in that they possess relatively large talonids together with relatively reduced trigonids (McHenry & Corruccini, 1980; Wood et al. 1983; Hills et al. 1983; Wood, 1984).

On the basis of a canonical variates analysis of crown measurements, including the 'basal diameters' of the five cusps, occlusal length, diagonal crown length, cervical length and breadth and the breadths of the trigonid and talonid, McHenry & Corruccini (1980) reported that the 'robust' australopithecines appear to be characterized by the combination of a large hypoconulid, small metaconid, protoconid and entoconid, and a broad talonid. Apart from the fact that the separation of the 'robust' australopithecines reported by McHenry & Corruccini (1980) involved a number of other variables (e.g. the breadth and depth of the mandibular corpus and crown heights), some of the measurements of the tooth crown (e.g. the 'basal' diameters of the cusps and diagonal measurements) employed in their canonical variates analysis are of questionable utility and are subject to modification by attrition of the occlusal surface. The studies by Wood and his colleagues, on the other hand, utilized comparisons of cuspal areas with total occlusal areas, as determined by planimetry of enlarged photographs of first and second permanent mandibular molars (Wood et al. 1983, Hills et al. 1983; Wood, 1984). The study by

Wood *et al.* (1983), which entailed univariate and principal components analyses of cuspal areas of early hominid molars, recorded that 'robust' australopithecine crowns displayed relative reduction of trigonid cusps (protoconid and metaconid) and concomitant enlargement of the hypoconid and entoconid. Hills *et al.* (1983) expanded this early hominid data base to include the analysis of comparative cuspal areas in extant humans, gorillas and orangutans. Intraspecific and interspecific allometric relationships between cuspal and crown base areas were assessed by examining the correlations between these areas and by calculating the slope of the regression line of the log of cusp area on the log of basal crown area.

On the basis of these comparisons it was reported that, on balance, there are no grounds for regarding the reduction of the trigonid and the enlargement of the talonid in the 'robust' australopithecines as a simple (i.e. one that is pervasive among hominoids) allometric effect (Hills *et al.* 1983; Wood, 1984). In these studies, however, molar crown base area, which was employed as the measure of overall size, was determined as the sum of the individual cuspal areas comprising each specimen. Thus, since the summation of variables selected for analysis (i.e. individual cuspal areas) constitute the X axis parameter (i.e. overall crown base area), the question of dependent versus independent variables is rendered moot, and the X variable necessarily contains the potential accrual of mensurational errors for each of the five cusps used in its calculation. Least squares regression seems inadvisable under such circumstances (Kuhry & Marcus 1977; Harvey & Mace, 1982).

Before any metric trend can be identified as an 'allometric phenomenon', the expected size-related changes in hominoid molar dimensions must be phrased precisely. One appropriate way to phrase this question has been offered by Hills *et al.* (1983): is the nature of size-related dental change similar within various hominoid taxa (a criterion of commonality), and does inter-specific scaling amongst extant hominoids predict the direction of change in a fossil taxon? In complementary fashion, it might also be asked whether a given intraspecific trend can be extrapolated to other taxa within an evolutionary lineage. For example, does an extension of the allometric trends evinced by *A. africanus* successfully predict the molar shapes of the southern African 'robust' australopithecines?

In order to pursue these issues further, we have analyzed the relationships between trigonid, talonid and crown length dimensions in a series of australopithecine and extant anthropoid specimens. While it is recognized that discrete cuspal sizes probably approximate more closely to functionally relevant parameters than do the gross buccolingual dimensions of the trigonid and talonid, the latter have been chosen for examination for two reasons. Firstly, the buccolingual diameters of the trigonid and talonid are not necessarily dependent upon, and nor do they contribute to, the determination of the mesiodistal length of the crown. Secondly, the sample of Plio-Pleistocene hominid specimens for which these gross dimensions can be recorded is notably larger than the sample presently available for individual cuspal sizes because of the rather large number of crowns on which occlusal wear has obliterated definitive cuspal boundaries. In the present instance we have restricted the allometric analysis of crown dimensions to the first permanent mandibular molars, not only because this tooth represents the most common element available for the various australopithecine samples, but also because it is generally the most stable of the molars, in respect to metrical as well as non-metrical features, within extant anthropoid species (Gingerich & Schoeninger, 1979). Amongst the extant non-human anthropoids, Pan troglodytes and Gorilla gorilla were selected for comparative analysis since the size ranges of their molars slightly overlap those between the smallest and largest early hominid teeth, because these closely related taxa can be analogized to an 'evolutionary lineage' (Shea, 1985; Jungers & Susman, 1984), and because size-related dental trends shared by these taxa may have been present in the last common ancestor of the African ape and hominid (sensu stricto) clades.

MATERIALS AND METHODS

The buccolingual (BL) diameters of the trigonid and talonid and the mesiodistal (MD) diameter of the first permanent mandibular molar were recorded for combined sex samples of Pan troglodytes (N = 36) and Gorilla gorilla (N = 24). The samples for both taxa comprised approximately equivalent proportions of male and female individuals, and only a single tooth, usually from the left side, was measured for each specimen. Only those crowns that displayed no appreciable interproximal attrition were chosen for measurement.

A total of 40 specimens from southern and eastern Africa constitute the australopithecine sample under consideration. For purposes of analysis, some five sub-samples, each of which corresponds to a purported taxon, were recognized from within the total sample. Australopithecus afarensis is represented by four specimens from the Laetolil Beds (LH 2, LH 3, LH 4, LH 16) and by seven specimens from the Hadar Formation (AL 128-23, AL 266-1, AL 288-1, AL 333 w-1, AL 333 w-32, AL 333-74, AL 400-1a). Eight specimens, comprising the single individual from Taung, five from Sterkfontein (Sts 9, Sts 24, Sts 52, Stw 107, Stw 127), and two from Makapansgat (MLD 2, MLD 24) are attributable to A. africanus. Amongst the 'robust' australopithecines, three specimens from Kromdraai (TM 1517, TM 1536, KB 5223) comprise the Paranthropus robustus sample, while twelve individuals from Swartkrans (SK 6, SK 23, SK 25, SK 55, SK 61, SK 63, SK 828, SK 838b, SK 843, SK 1587, SK 1588, SK 3974) constitute the sample attributed to P. crassidens. Finally, some six specimens, including four from the Koobi Fora Formation (KNM-ER 1509, KNM-ER 1820, KNM-ER 3230, KNM-ER 3890), one from Olduvai Gorge (OH 30) and the single mandible from Peninj, compose the sample referred to P. boisei.

Measurements of the A. africanus, P. robustus and P. crassidens specimens were recorded by one of us (F.E.G.), and in each instance only a single molar of any one individual was utilized. The measurements were taken according to the definitions by Grine (1984a), and in those instances where the crown had suffered from slight interproximal reduction, the MD diameter was corrected so that the value recorded approximated that for the pristine condition of the molar. Metrical data pertaining to the P. bosei teeth were kindly supplied by Wood, while those for the Laetoli and Hadar specimens of A. afarensis were kindly supplied by White and

Johanson. The dimensional values recorded by Wood and White for the eastern African fossils were corroborated by us on casts, and the fact that these values are comparable to those recorded by Grine for the southern African australopithecines is substantiated by the similarity of values obtained by Wood, White and Grine for the fossils from the Transvaal caves (Grine, 1984a).

The allometric relationships between the BL diameter of the trigonid (Y_1) and the MD crown diameter (X), and those between the BL diameter of the talonid (Y_2) and the MD crown diameter were assessed through the major axis line-fitting procedure, where natural logs of the original data were utilized. The major axis version of Model II regression analysis was employed because neither the X nor the Y term is error-free, and because the assumption of comparable error variance in these parameters seems reasonable in this instance (Kuhry & Marcus, 1977a; Sokal & Rohlf, 1981). The slope of this line represents the estimate of the exponent in Huxley's well-known power formula, $Y = bX^k$. For two linear variables, the null hypothesis of geometric similarity (isometry) is represented by a slope of 1.0; values of $k < 1.0$ and values of $k > 1.0$ represent negative and positive allometry, respectively.

ALLOMETRY OF MOLAR DIMENSIONS

The means of the three dimensions obtained for the M_1s of the various australopithecine samples are recorded in Table 1. It is readily apparent from this table that the first molar tends to increase in size, not only in MD length, but also in BL breadth across both the trigonid and talonid from *Australopithecus afarensis* to *Paranthropus boisei*. Moreover, whilst the mean values for the *P. crassidens* (i.e. Swartkrans) sample are significantly larger than those of *A. africanus* in all three dimensions, the averages for the small Kromdraai (i.e. *P. robustus*) sample are consistently a little smaller than the corresponding *A. africanus* sample means.

If one compares the BL diameters of the trigonid and talonid portions of the crown by a simple index (Table 1), it is evident that whereas the trigonid tends to be relatively broader in the 'gracile' australopithecines, the talonid tends to be buccolingually dominant in the 'robust' forms. Thus, there is a clear trend for the talonid to display an increase in relative breadth from *A. afarensis* to *P. boisei*. The allometric basis of this trend is considered below. The Kromdraai fossils fall between *A. africanus* and the Swartkrans 'robust' australopithecines along this particular metrical 'cline'. While the talonid tends to increase in relative breadth over the trigonid in the 'robust' australopithecines, what must be evaluated is whether this tendency is pervasive enough to be expected with any overall increase in the size, or length of the molar.

Comparisons of the BL diameters of the trigonid and talonid with the MD diameter of the M_1 for pooled samples of comparatively small-toothed common chimpanzees (*Pan troglodytes*) and large-toothed gorillas (*Gorilla gorilla*) reveal that, within these African apes, the major axis slopes for both BL dimensions do not differ significantly from isometry (Fig. 1).

Indeed, if anything, the interspecific slope for trigonid diameters (1.07) is slightly higher than that for the talonid diameters (0.98). Within the smaller crowned Pan sample, however, the talonid increases in size relatively faster than the trigonid (slopes of 1.25 and 0.90 respectively), while in the Gorilla sample the BL diameters of both crown moieties are positively allometric with respect to MD length.

With regard to the total australopithecine sample (Fig. 2), talonid

Table 1: Mean metrical values of Australopithecine mandibular first molars

MESIODISTAL DIAMETER	N	$\overline{X}$	SD	SE
A. afarensis	11	13.04	0.88	0.27
A. africanus	8	13.96	0.97	0.34
P. robustus	3	13.87	1.02	0.59
P. crassidens	12	14.83	0.55	0.16
P. boisei	6	16.13	0.82	0.34
BUCCOLINGUAL TRIGONID				
A. afarensis	11	12.49	0.88	0.27
A. africanus	8	12.60	1.12	0.40
P. robustus	3	12.30	0.72	0.42
P. crassidens	12	13.37	0.71	0.21
P. boisei	6	13.98	0.67	0.27
BUCCOLINGUAL TALONID				
A. afarensis	11	12.24	0.95	0.29
A. africanus	8	12.46	1.06	0.37
P. robustus	3	12.43	0.65	0.38
P. crassidens	12	13.78	0.79	0.23
P. boisei	6	14.65	0.60	0.24
TALONID/TRIGONID INDEX				
A. afarensis	11	97.96	3.27	0.99
A. africanus	8	98.96	1.96	0.69
P. robustus	3	101.12	1.75	1.01
P. crassidens	12	103.11	1.70	0.49
P. boisei	6	104.81	1.97	0.81

Figure 1. Major axis plots of the buccolingual diameters of the trigonid and talonid compared to the mesiodistal diameters of mandibular first permanent molars of combined sex samples of <u>Pan troglodytes</u> (C) and <u>Gorilla gorilla</u> (G).

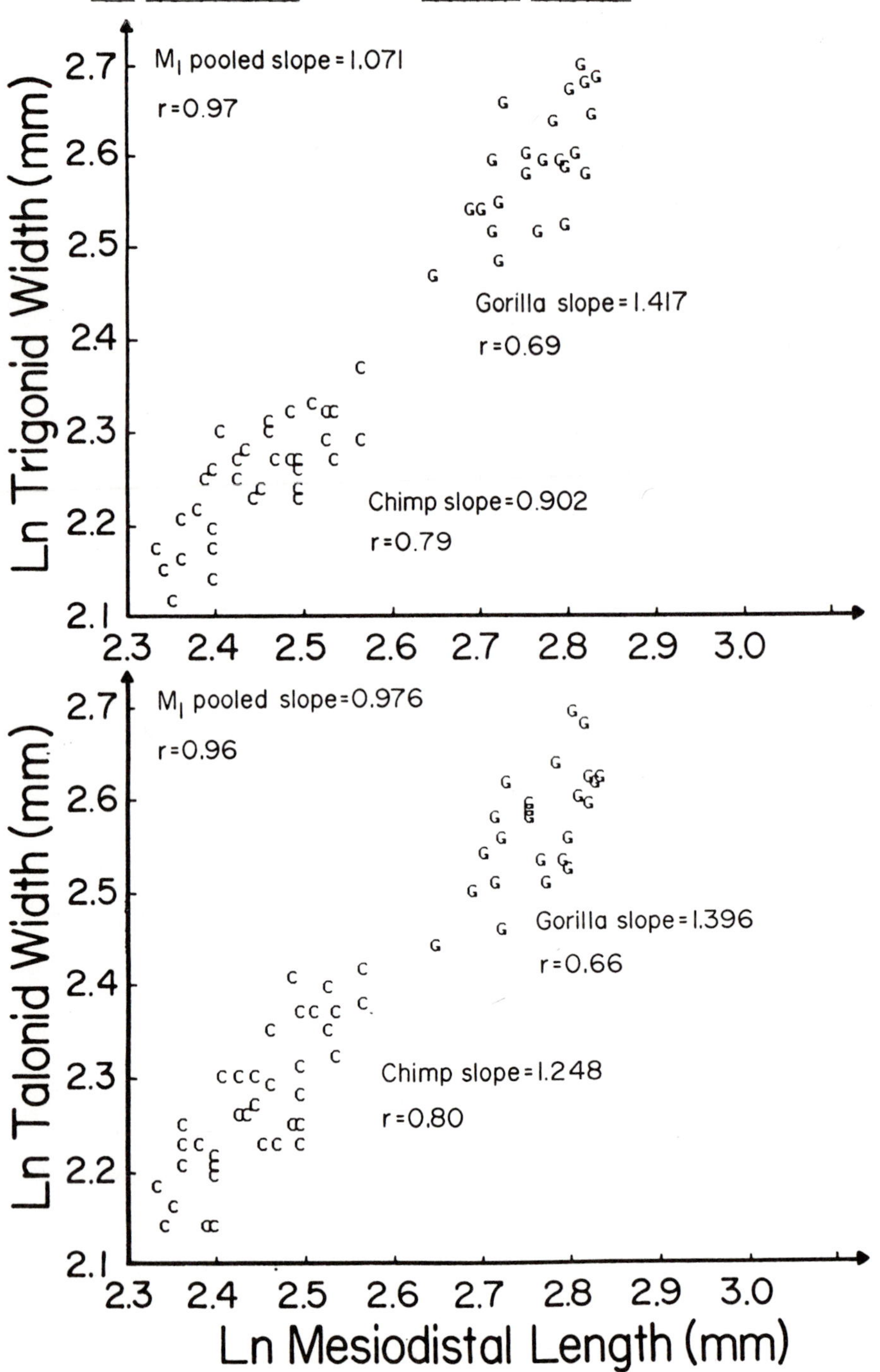

breadth tends to scale positively (k = 1.10) with molar length, whereas the BL diameter of the trigonid tends to exhibit negative allometric scaling (k = 0.87). As was noted for the interspecific African ape samples, the confidence limits to both slopes for the pooled australopithecine samples include isometry. Nevertheless, in comparison to MD length, the trigonid tends to enlarge somewhat faster in the pooled ape sample, whereas amongst the australopithecines the talonid exhibits a slightly greater relative increase in breadth.

When the total australopithecine sample is divided into its constituent 'gracile' and 'robust' components, however, the allometric basis of the tendency for the talonid to increase at the expense of the trigonid appears to be more sample-specific (Table 2). Thus, within both the 'gracile' and 'robust' samples the talonid slopes are slightly steeper than those of the trigonid, but the allometric coefficient for the talonid is higher for the 'gracile' australopithecines (slope = 1.19) than for the 'robust' sample (slope = 1.09). Comparison of the intraspecific trigonid and talonid slopes recorded for the *Pan* and *Gorilla* samples (Fig. 1) with those obtained for the separate 'gracile' and 'robust' australopithecine samples does not appear to support the notion that the 'robust' Plio-Pleistocene hominids, in general, evince a uniquely distinctive reduction of the trigonid and concomitant enlargement of the talonid of the M_1. Indeed, this tendency is more pronounced in chimpanzees than in 'robust' australopithecines. Moreover, the talonid slopes of both australopithecine samples are lower than those for either African ape sample.

If, however, the particular samples comprising the 'gracile' and 'robust' samples are considered individually, some rather interesting features emerge. In both *Australopithecus afarensis* and *A. africanus* the trigonid as well as the talonid display positive allometric increase (Table 2). The slope of the trigonid relationship is but only a little higher in *A. africanus* (value of 1.31) than in the *A. afarensis* (value of 1.22) sample, and while the *A. africanus* talonid slope value (1.24) is slightly lower than the corresponding trigonid value, the slope for the talonid of the *A. afarensis* sample (1.42) is higher than both the corresponding trigonid slope and the talonid slope for the *A. africanus* sample. In addition, the Laetoli and Hadar specimens attributed to *A. afarensis* tend to fall above *A. africanus* specimens in both the trigonid and talonid plots (Fig. 2); *A. afarensis* M_1s therefore tend to be relatively broader than *A. africanus* homologues for their length.

As noted above, the trigonid and talonid slopes determined for the 'robust' australopithecine sample are both very close to isometry. Plotted by themselves, however, the 'robust' specimens describe a rather sigmoidal relationship for the BL diameters of both the trigonid and talonid (Fig. 3). While the allometric values pertaining to the BL dimensions of the Swartkrans (*Paranthropus crassidens*) molars are strongly positive (1.76 for trigonid and 1.87 for talonid), the corresponding slopes for the comparatively small *P. boisei* sample of six individuals (0.88 and 0.71 respectively) are suggestive of negative scaling (Table 2). Although we are reluctant to attach biological meaning to the confidence limits of the

Figure 2. Major axis plots of the buccolingual diameters of the trigonid and talonid compared to the mesiodistal diameters of mandibular first permanent molars of australopithecine specimens. Pooled slopes are those obtained for the total australopithecine sample. Laetoli and Hadar specimens of A. afarensis (L), Sterkfontein and Makapansgat specimens of A. africanus (A), Taung A. africanus specimen (T), Kromdraai and Swartkrans 'robust' australopithecine specimens (P), eastern African specimens attributed to P. boisei (B).

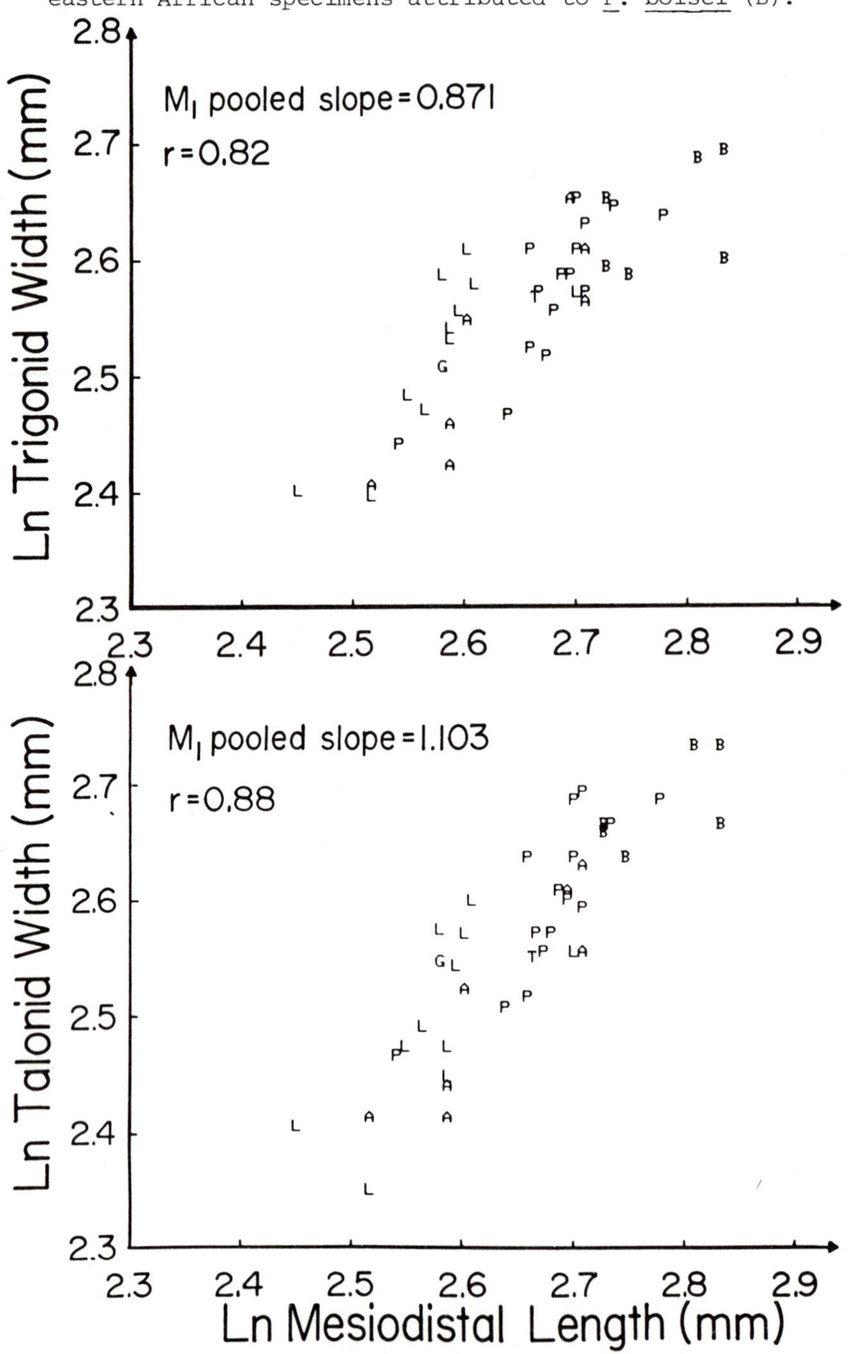

slopes of the australopithecine samples, given their small sizes and the sensitivity of confidence limits to sample size, it is perhaps noteworthy that the only instance in which these limits do not include isometry pertain to the talonid diameters of the Swartkrans molars. At the same time the three smallest 'robust' australopithecine molars from Kromdraai may obtain comparably flat slopes with regard to the BL diameters of both the trigonid and talonid (Figure 3). It is noteworthy that the 'robust' australopithecine specimens from Kromdraai fall at the inflection between the slopes defined by the _A. africanus_ and Swartkrans (_P. crassidens_) samples. That is, utilizing the major axis regressions for the _A. africanus_ trigonid and talonid dimensions, the Kromdraai molar diameters are within about 3 percent of the values predicted for their MD lengths. Similarly, according to the corresponding Swartkrans sample slopes of 1.76 and 1.87,

Table 2: Allometric relationships of first molar dimensions in Australopithecine samples

BL TRIGONID ON MD DIAMETER

	N	Slope	r	$\log_e$ Intercept
Total australopithecine sample	40	0.87	0.82	0.24411
Australopithecus afarensis	11	1.22	0.79	-0.60954
Australopithecus africanus	8	1.31	0.89	-0.91091
Combined 'gracile' sample	19	1.14	0.78	-0.44963
Paranthropus crassidens	12	1.76	0.70	-2.15615
Paranthropus boisei	6	0.88	0.46	0.20514
Combined 'robust' sample	21	0.96	0.79	0.00381

BL TALONID ON MD DIAMETER

	N	Slope	r	$\log_e$ Intercept
Total australopithecine sample	40	1.10	0.88	-0.36348
Australopithecus afarensis	11	1.42	0.70	-1.14754
Australopithecus africanus	8	1.24	0.91	-0.73425
Combined 'gracile' sample	19	1.19	0.77	-0.59140
Paranthropus crassidens	12	1.87*	0.73	-2.40600
Paranthropus boisei	6	0.71	0.66	0.71448
Combined 'robust' sample	21	1.09	0.84	-0.32994

*The confidence limits of this slope value do not include isometry.

Figure 3. Major axis plots of the buccolingual diameters of the trigonid and talonid compared to the mesiodistal diameters of mandibular first permanent molars of 'robust' australopithecine specimens. Slopes are those obtained for the pooled 'robust' australopithecine sample. Kromdraai and Swartkrans specimens (P), P. boisei specimens (B).

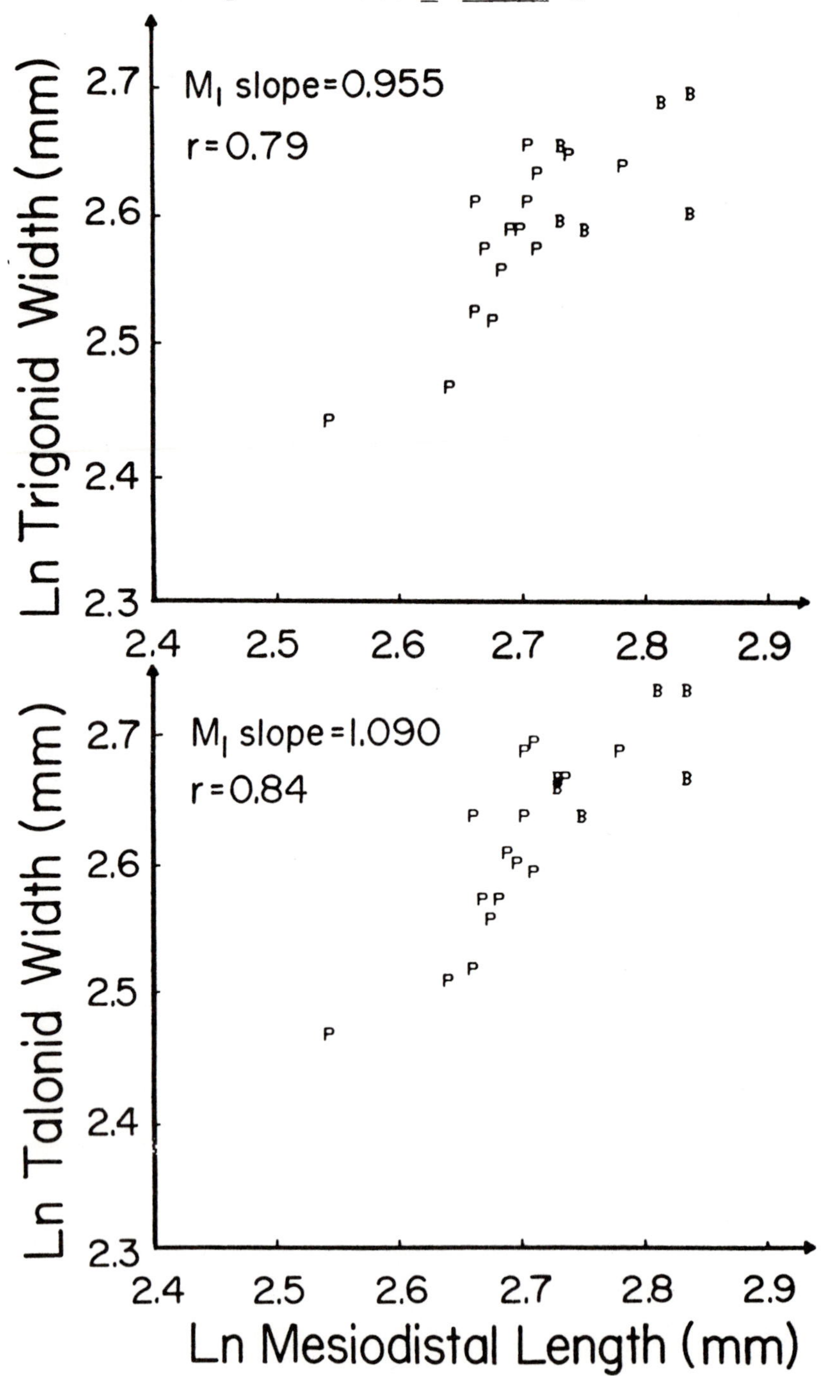

the observed dimensions of the Kromdraai crowns are within approximately 3 percent of the values interpolated for their MD diameters.

Thus, within the P. crassidens (Swartkrans) sample, the first mandibular molars display pronounced buccolingual broadening for length, but the relative increase in the talonid diameter occurs only slightly faster than the relative expansion of the trigonid. Within the P. boisei (and possibly the P. robustus) samples, on the other hand, the homologous crowns do not show an allometrically positive increase in either of the two BL diameters relative to their MD elongation. Thus, statements to the effect that 'robust' australopithecine mandibular molars tend to display relative BL broadening of the talonid and relative BL reduction of the trigonid appear to be correct in reference to the Swartkrans (P. crassidens) sample, as well as from a pooled, interspecific perspective, although the differences in the trends between the mesial and distal moieties are quite small in either case. Within-group analyses of these crown dimensions, however, appear to distinguish between P. crassidens and P. boisei trends. The M_1s of P. crassidens show positive allometric scaling of the BL diameters of both the trigonid and talonid; that is, the Swartkrans 'robust' australopithecine crowns tend to become broader as length increases. On the other hand, the larger molars of P. boisei, and probably also the smaller Kromdraai homologues, display negative allometric increase in these BL diameters. The M_1s of P. boisei are notable (and distinctive in their own right) for their tendency toward relative MD elongation, a finding that would be obscured by a consideration of only the mean talonid/trigonid index values.

SUMMARY AND CONCLUSION

Since Robinson's (1954b) proposal that Australopithecus and Paranthropus occupied dissimilar 'adaptive zones', the functional and taxonomic implications of the morphological differences between the 'gracile' and 'robust' australopithecines have been the subject of vigorous debate. Because trophic factors are central to the understanding of australopithecine ecology and evolution, considerable attention has been paid to dental parameters, in an attempt to sort those which are evidence of significant functional differences from those which may be related to the effects of allometric scaling.

Recently, several studies have focused upon the relationship between individual cuspal size and overall crown size in the mandibular molars of the Plio-Pleistocene hominids (McHenry & Corruccini, 1980; Wood et al., 1983; Hills et al., 1983; Wood, 1984). On the basis of these analyses it has been posited that the 'robust' australopithecines differ from their 'gracile' relatives in the unique possession of molars with a relatively small trigonid and large talonid.

We have investigated the allometric relationships amongst buccolingual and mesiodistal crown dimensions for the first permanent mandibular molars of australopithecine samples utilizing gross BL determinations of trigonid and talonid breadth and of MD length. The total, pooled, australopithecine sample considered here tends to scale molar breadth with molar length such that the BL diameter of the trigonid exhibits negative allometry, while the

BL diameter of the talonid increases in a positive allometric fashion. The higher mean talonid/trigonid index values obtained for the 'gracile' australopithecine phena, in comparison with those recorded for 'robust' australopithecine samples, reflect this trend. However, the allometric trends recorded for the separate 'gracile' and 'robust' australopithecine samples are not the same. Within the 'gracile' sample the BL diameters of both the trigonid and talonid increase faster than MD length, whereas these breadths are near isometry in the 'robust' australopithecines. Differences among the phena that constitute the 'robust' australopithecine sample are especially noteworthy. Thus, whilst the P. crassidens M_1s from Swartkrans display pronounced trigonid and talonid broadening as MD length increases, these BL diameters are negatively allometric for the larger-crowned P. boisei specimens, and probably for the smaller-toothed P. robustus individuals from Kromdraai. Whereas the molars of P. crassidens tend to broaden as length increases, P. boisei homologues are notable for their comparative MD elongation.

ACKNOWLEDGEMENTS

We are sincerely grateful to Professor P.V. Tobias and Mr. A.R.Hughes of the University of the Witwatersrand, and to Drs. E.S. Vrba and C.K. Brain of the Transvaal Museum, for permission to study the Plio-Pleistocene fossils in their care. Our gratitude is extended also to Professors B.A. Wood and T.D. White for providing us with measurements for a number of African hominid fossils, and to Dr. Ian Tattersall of the American Museum of Natural History, and Dr. W.H. Kimbel of the Cleveland Museum of Natural History for permitting us access to their collections of African apes. We would like to thank Professor B.A. Wood, Dr. Peter Andrews and Dr. L.B. Martin, the co-organizers of the Symposium on 'Major Topics in Primate Evolution' for their invitation to present this paper at that meeting. Special thanks go to Bernard Wood for the generous hospitality shown F.E.G. throughout the duration of the Symposium. The figures were drawn by Ms. L. Jungers, and we thank Mrs. M. Walker for typing the manuscript. Supported in part by NSF grant BNS-8217635 to W.L.J. and by L.S.B. Leakey Foundation and S.U.N.Y. Biomedical grants to F.E.G.

REFERENCES

Key references

Gingerich, P.D. & Schoeninger, M.J. (1979). Patterns of tooth size variability in the dentition of primates. A. J. Phys. Anthrop., *51*, 457-66.

Grine, F.E. (1985). Australopithecine evolution: the deciduous dental evidence. *In* Ancestors: the Hard Evidence, ed. E. Delson. New York: Alan R. Liss. (In Press).

Harvey, P.H. & Mace, G.M. (1982). Comparisons between taxa and adaptive trends: problems of methodology. *In* Current Problems in Sociobiology, eds. King's College Research Group, pp. 343-61. Cambridge: Cambridge University Press.

Pilbeam, D.R. & Gould, S.J. (1974). Size and scaling in human evolution. Science, *186*, 892-901.

Wood, B.A. (1981). Tooth size and shape and their relevance to studies of hominid evolution. Phil. Trans. R. Soc. Lond., B, *292*, 65-76.

Main references

Bilsborough, A. (1978). Some aspects of mosaic evolution in hominids. *In* Recent Advances in Primatology, vol. 3, Evolution, eds D.J. Chivers & K.A. Joysey, pp. 335-50. New York: Academic.

Boaz, N.T. (1983). Morphological trends and phylogenetic relationships from middle Miocene hominoids to late Pliocene hominids. *In* New Interpretations of Ape and Human Ancestry, eds R.L. Ciochon & R.S. Corruccini, pp. 705-20. New York: Plenum.

Broom, R. (1938). The Pleistocene anthropoid apes of South Africa. Nature, *142*, 377-79.

Broom, R. (1949). Another type of fossil ape-man (Paranthropus crassidens). Nature, *163*, 57.

Clarke, R.J. (1977). The cranium of Swartkrans hominid, SK 847, and its relevance to human origins. Unpublished Ph.D. thesis, University of the Witwatersrand, Johannesburg.

Dart, R.A. (1925). Australopithecus africanus: the man-ape of South Africa. Nature, *115*, 195-99.

Gingerich, P.D. & Schoeninger, M.J. (1979). Patterns of tooth size variability in the dentition of primates. Am. J. Phys. Anthrop. *51*, 457-66.

Grine, F.E. (1981). Trophic differences between 'gracile' and 'robust' australopithecines: a scanning electron microscope analysis of occlusal events. S. Afr. J. Sci., *77*, 203-30.

Grine, F.E. (1982). A new juvenile hominid (Mammalia: Primates) from Member 3, Kromdraai Formation, Transvaal, South Africa. Ann. Tvl. Mus., *33*, 165-239.

Grine, F.E. (1984a). The deciduous dentition of the Kalahari San, the South African Negro and the South African Plio-Pleistocene hominids. Unpublished Ph.D. thesis, University of the Witwatersrand, Johannesburg.

Grine, F.E. (1984b). Deciduous molar wear of South African australopithecines. *In* Food Acquisition and Processing in Primates, eds. D.J. Chivers, B.A. Wood and A. Bilsborough, pp. 525-34. New York: Plenum.

Grine, F.E. (1985). Australopithecine evolution: the deciduous dental evidence. *In* Ancestors: the Hard Evidence, ed. E. Delson. New York: Alan R. Liss. (In Press).

Harvey, P.H. & Mace, G.M. (1982). Comparisons between taxa and adaptive trends: problems of methodology. *In* Current Problems in Sociobiology, eds King's College Research Group, pp. 343-61. Cambridge: Cambridge University Press.

Hills, M., Graham, S.H. & Wood, B.A. (1983). The allometry of relative cusp size in hominoid mandibular molars. Am. J. Phys. Anthrop., *62*, 311-16.

Howell, F.C. (1978). Hominidae. *In* Evolution of African Mammals, eds V.J. Maglio & H.B.S. Cooke, pp. 154-248. Cambridge, Mass.: Harvard University Press.

Johanson, D.C., White, T.D. & Coppens, Y. (1978). A new species of the genus Australopithecus (Primates: Hominidae) from Pliocene of eastern Africa. Kirtlandia, *28*, 1-14.

Jungers, W.L. (1984). Aspects of size and scaling in primate biology with special reference to the locomotor skeleton. Yrbk. Phys. Anthrop., *27*, in press.

Jungers, W.L. & Susman, R.L. (1984). Body size and skeletal allometry in African apes. *In* The Pygmy Chimpanzee: Evolutionary Biology and Behavior, eds. R.L. Susman, pp. 131-77. New York: Plenum.

Kay, R.F. (1975). Allometry and early hominids. Science, *189*, 63.

Kimbel, W.H., White, T.D. & Johanson, D.C. (1985). Craniodental morphology of the hominids from Hadar and Laetoli: evidence of Paranthropus and Homo in the mid-Pliocene of eastern Africa? *In* Ancestors: the Hard Evidence, ed. E. Delson. New York: Alan R. Liss. (In Press)

Kuhry, G. & Marcus, L.F. (1977). Bivariate linear models in biometry. Syst. Zool., *26*, 201-9.

Leakey, L.S.B. (1959). A new fossil skull from Olduvai. Nature, *184*, 491-3.

McHenry, H.M. & Corruccini, R.S. (1980). Late Tertiary hominoids and human origins. Nature, *285*, 397-8.

Pilbeam, D.R. & Gould, S.J. (1974). Size and scaling in human evolution. Science, *186*, 892-901.

Rak, Y. (1983). The Australopithecine Face. New York: Academic.

Robinson, J.T. (1954a). The genera and species of the Australopithecinae. Am. J. Phys. Anthrop., *12*, 181-200.

Robinson, J.T. (1954b). Prehominid dentition and hominid evolution. Evolution, *8*, 324-34.

Robinson, J.T. (1968). The origin and adaptive radiation of the australopithecines. *In* Evolution und Hominization, 2nd ed., ed. G. Kurth, pp. 150-75. Stuttgart: Gustav Fischer.

Shea, B.T. (1985). Ontogenetic allometry and scaling: a discussion based on the growth and form of the skull in African apes. *In* Size and Scaling in Primate Biology, eds. W.L. Jungers, pp. 175-205. New York: Plenum..

Smith, R.J. (1980). Rethinking allometry. J. Theor. Biol., *87*, 97-111.

Sokal, R.R. & Rohlf, F.J. (1981). Biometry (2nd ed.). San Francisco: Freeman.

Tobias, P.V. (1967). The Cranium and Maxillary Dentition of Australopithecus (Zinjanthropus) boisei. Olduvai Gorge, vol. 2. C Cambridge: Cambridge University Press.

Tobias, P.V. (1975). New African evidence on the dating and the phylogeny of the Plio-Pleistocene Hominidae. Trans. R. Soc. New Zealand, *13*, 289-96.

Tobias, P.V. (1980). "Australopithecus afarensis" and A. africanus: critique and an alternative hypothesis. Palaeont. afr., *23*, 1-17.

Wallace, J.A. (1978). Evolutionary trends in the early hominid dentition. *In* Early Hominids of Africa, ed. C. Jolly, pp. 285-310. London: Duckworth.

White, T.D. (1985). The hominids of Hadar and Laetoli: an element-by-element comparison of the dental samples. *In* Ancestors: the Hard Evidence, ed. E. Delson. New York: Alan R. Liss. (In Press).

White, T.D., Johanson, D.C. & Kimbel, W.H. (1981). Australopithecus africanus: its phyletic position reconsidered. S. Afr. J. Sci., *77*, 445-70.
Wolpoff, M.H. (1974). The evidence for two australopithecine lineages in South Africa. Yrbk. Phys. Anthrop., *17*, 113-39.
Wolpoff, M.H. (1978). Some aspects of canine size in the australopithecines. J. Hum. Evol., *7*, 115-26.
Wolpoff, M.H. (1982). Relative canine size. J. Hum. Evol., *10*, 151-58.
Wood, B.A. (1978). Models for assessing relative canine size in fossil hominids. J. Hum. Evol., *8*, 493-502.
Wood, B.A. (1981). Tooth size and shape and their relevance to studies of hominid evolution. Phil. Trans. R. Soc. Lond., B, *292*, 65-76.
Wood, B.A. (1984). Interpreting the dental peculiarities of the 'robust' australopithecines. *In* Food Acquisition and Processing in Primates, eds D.J. Chivers, B.A. Wood and A. Bilsborough, pp. 535-44. New York: Plenum.
Wood, B.A. & Abbott, S.A. (1983). Analysis of the dental morphology of Plio-Pleistocene hominids. I. Mandibular molars - crown area measurements and morphological traits. J. Anat. *136*, 197-219.
Wood, B.A. & Stack, C.G. (1980). Does allometry explain the differences between "gracile" and "robust" australopithecines? Am. J. Phys. Anthrop., *52*, 55-62.
Wood, B.A., Abbott, S.A. & Graham, S.H. (1983). Analysis of the dental morphology of Plio-Pleistocene hominids. II. Mandibular molars - study of cusp area, fissure pattern and cross-sectional shape of the crown. J. Anat., *137*, 287-314.

13 AUSTRALOPITHECUS: GRADE OR CLADE?

B.A. Wood and A.T. Chamberlain,
Department of Anatomy & Biology as Applied to Medicine,
The Middlesex Hospital Medical School,
Cleveland Street, London, W1P 6DB, U.K.

Present address:
Department of Anatomy,
The University of Liverpool,
P.O. Box 147, Liverpool L69 BX3, UK.

INTRODUCTION

This contribution uses a particular technique, cladistic analysis, to examine the relationships between early hominid taxa. In particular, it addresses the problem of what is meant by the genus Australopithecus. Does it refer to all the taxa within a lineage, or clade, or is it used to identify taxa at the same level of evolutionary organisation, or grade?

This analysis was designed to be as objective as possible; the characters were chosen for technical reasons and not because they support one phylogenetic scheme or another. No prior assumptions were made about how the taxa were related beforehand, but the results of the analysis do suggest which of the possible arrangements of the taxa best fit the data. Comparable analyses may be found in Andrews & Franzen (1984) and Delson (1985).

Taxonomic review

Most contemporary classifications of fossil hominids recognise only two genera, Australopithecus and Homo. A third genus, Paranthropus, was introduced when the hominid remains from Kromdraai were published (Broom, 1938), and it subsequently came to include both the Kromdraai remains and many of the hominids recovered from Swartkrans (Broom, 1949). Washburn & Patterson (1951) were probably the first authors to suggest that Paranthropus be sunk into Australopithecus. Leakey et al. (1964) and Howell (1965) suggested that it be reduced to subgeneric rank, but Tobias (1967) went further and formally proposed that Australopithecus and Paranthropus be sunk as subgenera, and that the 'old' Paranthropus material (both P. robustus (Broom, 1938) and P. crassidens (Broom, 1949)) should be included in a single species, Australopithecus robustus (Broom). In this paper, Australopithecus will be taken to include Paranthropus, but readers should refer to the evidence presented in other contributions in this book which suggests that the 'robust' australopithecines may be more distinct than the present classification suggests.

Most recent reviews and references to australopithecines assume that the name embraces a group of hominids with relatively small-brains and large cheek-teeth, and four taxa Australopithecus afarensis, Australopithecus africanus, A. robustus and Australopithecus boisei are usually included within that definition. Homo is conventionally assumed to include three species Homo habilis, Homo erectus and Homo sapiens, the latter including remains previously attributed to Homo neanderthalensis.

'Clades' and 'Grades'

In an evolutionary context in which there is evidence of divergence (or cladogenesis), episodes of progressive change (or anagenesis) and periods of apparent persistence of form (stasigenesis), fossil taxa can be visualised as moving through a matrix, the columns and rows of which represent two very different types of classification. These ideas were reviewed and clarified by Huxley (1958) who suggested a simple terminology (Fig. 1). The 'columns' in the matrix diagram represent tracks of diversification, which pass through successive 'rows' corresponding to levels of biological organisation. Huxley considered these classifications complementary, and named the columns 'clades', and the rows 'grades'. Grades are identifiable because of a combination of relative stasis and parallel, or convergent, evolution. Thus, if we accept this simple model and return to the hominid fossil record, does Australopithecus, as presently defined, correspond to a grade or a clade? This is not an altogether sensible question for a clade may also be a grade e.g. the class Aves, but, to put the same question in a different form, do the characters which define an australopithecine have the same origin?

Consequences of a clade classification

There is already evidence that the various australopithecine taxa have sufficient in common to be regarded as a grade. Scientists may disagree about the details of such a grade definition, but common features would probably include a relatively small brain, and relatively large cheek teeth and a distinctive femoro-pelvic complex. However, to determine whether the four taxa also correspond to a clade, we require an operational definition for such a grouping. We shall follow Huxley (1958) and take a clade to be equivalent to a 'monophyletic unit' (Huxley, 1958:27). Such a unit contains all the known descendants, and nothing but the known descendants, of a common ancestor. How, then, is such a unit to be recognised? In short, a monophyletic unit is defined on the basis of

Figure 1. Simplified diagram of patterns of evolutionary change to show the difference between the concepts of 'clades' and 'grades' (adapted from Huxley, 1958).

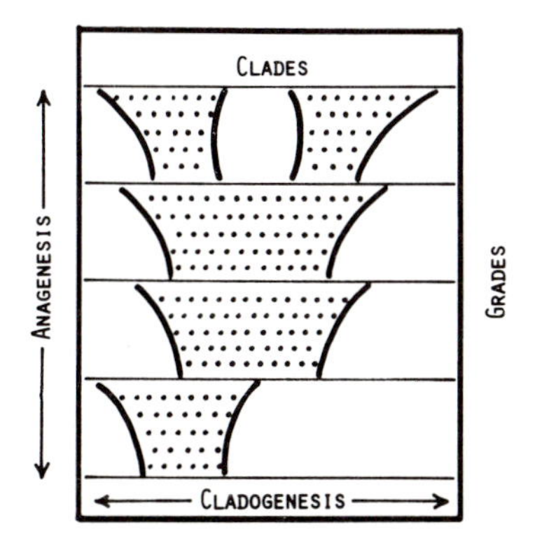

characters which are apparently unique to that unit, and its members are those taxa which possess these characters. We are using 'character' in the sense of a character state; that is a series of different versions of the same homologous structure. Thus, the root number and morphology of the same premolar would form a series of character states, whereas, for example, the presence or absence of a simian shelf as opposed to the possession, or not, of a superior transverse torus, would not constitute alternative states of the same character. The shelf and the torus may have the same effect, that is buttressing the mandibular symphysis, but they are analogous structures, not homologous ones.

A grade may, as we have stated, be coextensive with a monophyletic group, but it may also correspond to other combinations of taxa. In contrast to a monophyletic group, which includes all the descendants of a common ancestor a paraphyletic group contains some, but not all of such descendants. A polyphyletic group is essentially a special case of a paraphyletic group; in effect it is a grouping in which many members of a larger monophyletic unit are missing from the classification. In summary, whereas a monophyletic group can be defined on the basis of inferred shared-derived characters, paraphyletic groups can only be defined on the basis of inferred shared-primitive characters, and polyphyletic groups only on the basis of convergent characters.

What are the implications, if any, of a clade classification for *Australopithecus*? An important consequence is that, no *Australopithecus* species could be ancestral to *Homo*. This must follow a clade classification because any descendants of an australopithecine must either possess the characters that make it an australopithecine in the primitive condition, or in a derived form. In this case *Homo* must have those characters, but having been 'removed' from *Australopithecus*, it must, by definition, leave *Australopithecus* as a paraphyletic group and not a monophyletic one, *Homo* being the missing member. This is not an abstruse point of detail (Tattersall & Eldredge, 1977), for most of the more recent hominid phylogenies which have been suggested in the literature assume that one or other of the known australopithecine taxa is a common ancestor of later hominids, be that taxon *A. africanus* (Tobias, 1973, 1980), or *A. afarensis* (Johanson & White, 1979; White *et al.*, 1981).

Protocol

Intuition, if not actual evidence, suggests that the outcome of a cladistic analysis is likely to be at least partly dependent on the selection of characters on which it is based. Simple logistical considerations have led some authors to use previously published data and observations as the basis for such analyses, but we believe that this practice has limitations which we will outline below. To help counter these and other criticisms which can be levelled at this approach, we have drawn up what amounts to a research protocol, and the five main elements in it (Table 1) will be discussed below.

Taxonomic units. A cladistic analysis should ideally use taxonomic units which are assembled on the basis of phenetic information, with due allowance made for what are believed to be the appropriate comparative analogues for intraspecific variation, including allometrically-based shape differences. While most of the taxa we have used for this study have been determined using methods which approximate to this ideal, we recognise that some have not. Likewise there are genuine disagreements about the limits of taxonomic variation and the relative significance and importance of variation in different parts of the skeleton (e.g. A. afarensis vs A. africanus, and Homo habilis vs A. africanus). However, we have used the taxonomic groups as listed in Table 2 as the units for this analysis, and more details of what each category included are given later in the paper. In general we have adhered to conventional definitions and assumed that all our units are monophyletic ones. It is noteworthy that we have not pooled A. boisei and A. robustus as a single monophyletic unit, nor have we assumed that the taxonomic unit we describe as H. habilis bears any special relationship to H. erectus or H. sapiens.

Outgroups. The identification of monophyletic groups is dependent on the ability to tell which characters are inferred shared-derived, and which are inferred shared-primitive retentions. Patterson (1982) has provided a useful review of tests which seek to discriminate between the two, and these include making reference to the ontogenetic and comparative contexts of the characters. The latter can be explored by reference to 'outgroups' which are usually selected from more, or less, distantly related taxa. The point of making reference to outgroups is that if the character state is widely distributed in such groups, then it is considered more likely that it is a shared-primitive retention. Some cladistic studies have used not only ontogenetic and outgroup data, but also functional interpretations and stratigraphy to support assessments of character polarity. However, we follow Bonde (1977:776-7) in disregarding functional interpretations and stratigraphy when evaluating morphocline polarity. Outgroups used in this study are listed in Table 3.

Table 1: Major elements of the research protocol

1. Taxonomic units
2. Outgroups
3. Sampling
4. Quantification
5. Standardisation for size

Sampling. It is tempting, and indeed pragmatic, to use previously published data as the basis for a cladistic analysis, but such a strategy has problems unless individual data items, and the set as a whole, are assessed against certain criteria. Clearly each data item should be as objective as possible. Subjective assessments of morphology should be avoided, for statements like 'deeper than', or 'more inflated than' are difficult, if not impossible, to judge or to score. Where possible character states should be based on quantitative data. Corruccini & McHenry (1980) were scrupulous in confining their characters to quantitative ones, but their cladistic analysis was based on a relatively narrow set of mostly mandibular dental traits. The data set should also not contain items which are likely to be highly correlated. For example, the length of the canine root and mandibular robusticity appear to be highly correlated, so that mandibles with relatively large canines tend to have lower robusticity indices than those with relatively reduced canines (Wood, 1978; Smith, 1983; Wood & Chamberlain, in press). Clearly, it is preferable that these two characters are not included in the same analysis.

Evidence about character states culled from the literature is inevitably liable to at least two kinds of bias. The first is that their distribution across the regions of the body is likely to be uneven, reflecting as it does the prevailing and past preoccupations of palaeoanthropologists, and not any concerted attempt at a systematic analysis of the skeleton. Secondly, given the evidence for palimpsest, or mosaic, evolution, if traits are selected from a limited region of the skeleton, any resulting

Table 2: Taxonomic units used in the analysis

Australopithecus:	*Homo*:
Australopithecus afarensis	*Homo erectus*
Australopithecus africanus	*Homo habilis*
Australopithecus boisei	*Homo sapiens*
Australopithecus robustus	

Table 3: Outgroups used in the analysis

Proconsul

Sivapithecus

Gorilla

Hylobates

Pan paniscus

Pan troglodytes

Pongo

cladistic hypotheses must be regarded as regional ones, and not necessarily interpreted as reflecting the relationships of the whole animal.

A final consideration is that information about character states should be drawn from consistent fossil hypodigms. Observations in the literature about the condition of a character in H. habilis prior to 1972 are likely to be based on a very different hypodigm than the one that would be used in a comparable study undertaken a decade later. Some of the differences are due to the addition of newly recovered material, but others are due to reinterpretation of existing material; changes of either sort in the reference hypodigm are likely to influence any decision about the distribution of character states. It is not unknown for cladistic studies to perpetuate errors. For example, in several analyses (e.g. Kay (1982)) a derived character state has been inferred for mandibular corpus robusticity in the Ramamorphs, yet these taxa have mandibular robusticity index values which are indistinguishable from those of extant hominoids.

Thus, in response to these perceived difficulties, we devised a sampling strategy, which is summarised below, and a set of criteria for identifying suitable characters, which will be discussed in the following section. The elements of the sampling strategy are as follows:-

1) The analysis is confined to the skull. This was a decision dictated by the nature of the hominid fossil record, in which cranial remains predominate (Tobias, 1972), and by the related fact that most taxonomic definitions are based on cranial remains. The current debate about the likely femoral morphology of Homo habilis (Kennedy, 1983; Trinkaus, 1984) is an example of the present uncertainty which affects the attribution of even virtually complete postcranial remains.

2) The skull was sampled by dividing it into five regions, or functional complexes (Table 4). These were selected to provide a relatively comprehensive coverage of the head. An attempt was made to ensure that each region was represented by an equivalent number of traits, but these attempts were thwarted because many potential traits had to be rejected because they could not be quantified, or because, even if they could be

Table 4: Anatomical and functional regions used in the analysis

1. Vault and endocranium
2. Face
3. Palate, upper jaw and maxillary dentition
4. Cranial base
5. Mandible and mandibular dentition

quantified, the ratio of within to between group variation made them unreliable as characters. Within each region an effort was made to use traits which were drawn from as many morphological and functional units as possible. Particular point was made of avoiding clusters of characters, all of which were likely to be directly related to a single function e.g. mastication.

3) Traits that were evidently likely to be correlated were eliminated where possible. For example, if a trait which related the size of the anterior teeth to the size of the dental row as a whole was included, then one which attempted to do the same for the molar teeth was not. Such a trait would have been strongly negatively correlated with the existing one.

4) The same fossil hypodigms were used throughout the study. This allowed the distribution of the character states in the fossil groups to be directly compared.

Quantification. All but three of the final list of 39 characters were quantified as continuous variables. Two of the three exceptions involve the location of structures relative to the tooth row, these were the locations of the base of the zygomatic process of the maxilla in the upper jaw, and the position of the mental foramen in the lower; both these characters were quantified as discrete variables. The third exception concerns the occipital and marginal venous sinuses which were simply recorded as present (i.e. a visible trace on the occipital) or absent.

Standardisation for size. When one uses metrical data in an analysis of this sort consideration has to be given to the problem of whether the use of 'raw' data will lead to differences in character states which are purely the result of disparities in the overall size of the taxa. This poses particular problems when trying to establish the polarity of a character on the basis of its distribution in outgroups that may be, and which in the case of this study, actually are, very different in overall size. Size affects analyses such as this not only by its direct effect, but also by the influence it has on shape, as described by allometric studies. The latter can be eliminated using empirically-derived coefficients as correction factors. Coefficients can be derived from the comparative samples, but there is, in general, too little fossil evidence to derive coefficients for each taxon. Coefficient values are known to vary from taxon to taxon (Wood, 1975) so, in the absence of allometry coefficients which could be directly derived from the fossil taxa, we decided not to make any allometry corrections. Ratios help to reduce, if not actually eliminate, differences due to absolute size. Despite their acknowledged drawbacks (Atchley et al., 1976; Albrecht, 1978), we have used ratios in this study and nearly all characters have been expressed by relating one linear variable to another.

MATERIALS AND METHODS

Taxonomic units

The taxa used in this study are listed in Table 2 and the details of the sample of each taxon are given below.

A. afarensis

The hypodigm included all the hominid remains from the Laetoli Beds, Tanzania and the Hadar Formation, Ethiopia. Data for the characters were taken from White (1977, 1980), Johanson *et al.* (1982), Kimbel *et al.* (1982), White & Johanson (1982) and Kimbel *et al.* (1984).

A. africanus

The hypodigm consisted of hominid remains recovered from Sterkfontein Mb4 (Partridge, 1978, but see Wilkinson, 1983) and Makapansgat Mb3 and 4 (Partridge, 1979). Data for this taxon came from measurements made by one of the authors (Wood, in preparation) and Dean & Wood (1982).

A. boisei

The hypodigm included all material referred to this taxon from Olduvai, Koobi Fora, Peninj and Chesowanja, with the addition of three mandibles (L7A-125, 18-18, L74A-21) from the Omo. Data based on sources as given for *A. africanus*.

A. robustus

The hypodigm comprises the remains from Swartkrans Mbs 1 and 2 and Kromdraai. The sample excludes those remains which have been attributed to *Homo* cf. (Clarke, 1977). Data based on sources as given for *A. africanus*.

H. erectus

The hypodigm consisted of the site samples from Trinil, Sangiran (Djetis and Kabuh) and Zhoukoudian (Lower Cave). Data are based on measurements taken from Weidenreich (1936, 1937, 1943) and Wood (in preparation).

H. habilis

The hypodigm consists of all material referred to *H. habilis* from (i) Bed I and Lower to Middle Bed II, Olduvai (cf. Howell, 1978; Leakey, 1978), and (ii) Faunal Unit 3 (approximately equivalent to the Lower Member, Koobi Fora Formation) at Koobi Fora. Individual specimens from Koobi Fora included in this taxonomic unit are: KNM-ER 1470, 1483, 1590, 1801, 1802, 1813, 3731, 3732, 3734. Data based on sources as given for *A. africanus*.

H. sapiens

Data were based on (i) measurements taken by the authors on a balanced-sex sample of ten skulls from the Dissecting Room collection of The Middlesex Hospital Medical School, (ii) measurements given in Wood (1975) and (iii) measurements of *H. s. neanderthalensis* taken from Trinkaus (1983).

Outgroups

The outgroups are listed in Table 3, and details of the source of the data are given below.

Fossil outgroups

Dental and mandibular data for Proconsul africanus were based on measurements given by Andrews (1978). Data representing the Ramapithecus/ Sivapithecus group (the Ramamorphs) were based on measurements from Pilbeam et al. (1980). In the event, too few data were available for the fossil outgroups for them to be included in the final cladograms.

Extant outgroups

Much of the dental data for Hylobates, Pongo, Pan and Gorilla were based on measurements taken from Swindler (1976). Crown areas were computed separately for each sex from the mean diameters and then the male and female values were averaged. Cranial data were based on measurements made by the authors on balanced-sex samples of ten crania of each of the species Hylobates agilis, Pongo pygmaeus, Pan troglodytes and Gorilla gorilla from the primate collection of the B.M.(N.H.). Some data for the extant groups were taken from Wood (1975).

All linear cranial and mandibular measurements were taken to the nearest 1 mm and dental measures to the nearest 0.1 mm. Angles were measured to the nearest five degrees.

Characters

The 39 characters from the five regions (Table 4) are listed in Table 5; a detailed definition of each character is given in Appendix A. The metrical data recorded for each character were summarised using the mean value and the range and are provided in Appendix B. The polarity of the character states was assessed using the outgroups, and the inferred shared-primitive condition was taken to be that which appeared with the highest frequency in the outgroups (Table 6). Decisions about the boundary between the primitive and a derived state, or between derived states, were based on consideration of both the mean values and the ranges; the samples for the fossil taxa were judged to be too small to rely solely on the results of techniques such as Students' 't' test. As a rule of thumb, the state was judged to be derived if the range overlap with the primitive state fell to less than 50%. In some examples the boundaries between the states were determined empirically by recognising apparent clustering of the mean values. When either the outgroups or the taxonomic units included values which suggested more than one derived trend, then the two trends were recognised separately and a forward prime designation was given to the derived condition with the wider distribution, and a backward prime superscript marked the trend with the more limited distribution.

Cladogram construction and testing

Cladograms for each region were constructed making no assumption other than the polarity of the characters. The 'best fit' cladogram for each region was the arrangement which resulted in the most intact synapomorphies for that region. We regard two taxa as synapomorphous for

Table 5: List of Characters

Vault	
V1	Postorbital constriction
V2	Parietal shape
V3	Parietal-occipital arc ratio
V4	Occipital sagittal curvature
V5	Relative vault height
V6	Relative vault length
V7	Frontal angle
V8	Encephalisation Quotient
V9	Occipito-marginal sinus
V10	Relative biparietal width
V11	Relative posterior cranial length
Face	
F1	Position of zygomatic
F2	Relative facial width
F3	Facial shape
F4	Alveolar prognathism
F5	Nasal bone shape
F6	Subnasal projection
F7	Malar projection
Palate	
P1*	Incisor size heteromorphy
P2*	Premolar size heteromorphy
P3	Premolar root number
P4	Relative length of molar row
P5	Maxillo-alveolar index
P6	Relative palate depth
P7	Alveolar tooth row divergence
Base	
B1	Petro-tympanic angle
B2	Location of basion
B3	Relative length of tympanic
B4	Relative width of mandibular fossa
B5	Angulation of foramen magnum
Mandible	
M1	Corpus robusticity
M2	Position of mental foramen
M3	Relative height of mental foramen
M4	Alveolar tooth row divergence
M5	Relative height of foramen spinosum
M6	Relative length of canine socket
M7	Premolar root number
M8	P^3 crown shape
M9*	Molar size heteromorphy

*These dental characters, when calculated for the hominids, only incorporated data which were available for the same individual.

a character when they are both derived in the same direction from the primitive state. We interpret a taxon which possesses state n of character A as being synapomorphous with other taxa that either possesses n of A or a more derived state than n of A. All such cladograms resulted in broken synapomorphies, and these, along with the intact synapomorphies, are listed beside each cladogram. No attempt was made to weight characters.

The regional cladograms were compared with each other using the mean Consistency Index (CI), which has been adapted from Kluge & Farris (1969). In this simple technique the number of possible states of each character (i.e. the inferred primitive condition and the number of derived states) is divided by the number of times any of these character states actually appear in the cladogram: the resulting value is the CI for the particular character. A CI score of unity indicates that none of the synapomorphies of that character are broken in constructing the cladogram. As more synapomorphies are broken the index takes a lower value. A mean CI was calculated for each regional cladogram, and a similar calculation was made for the summary cladograms. More sophisticated techniques for constructing and comparing cladograms have been described (Estabrook <u>et al</u>., 1977) but they were not available to the authors.

Figure 2. Regional cladogram for the vault and endocranium. Characters and character states are those given in Tables 5 and 6. Synapomorphies at each node are represented by a solid horizontal line crossing both the lines above the bifurcation. Autapomorphies are represented by a horizontal line across the line leading to that taxon. Dashed lines indicate broken synapomorphies.

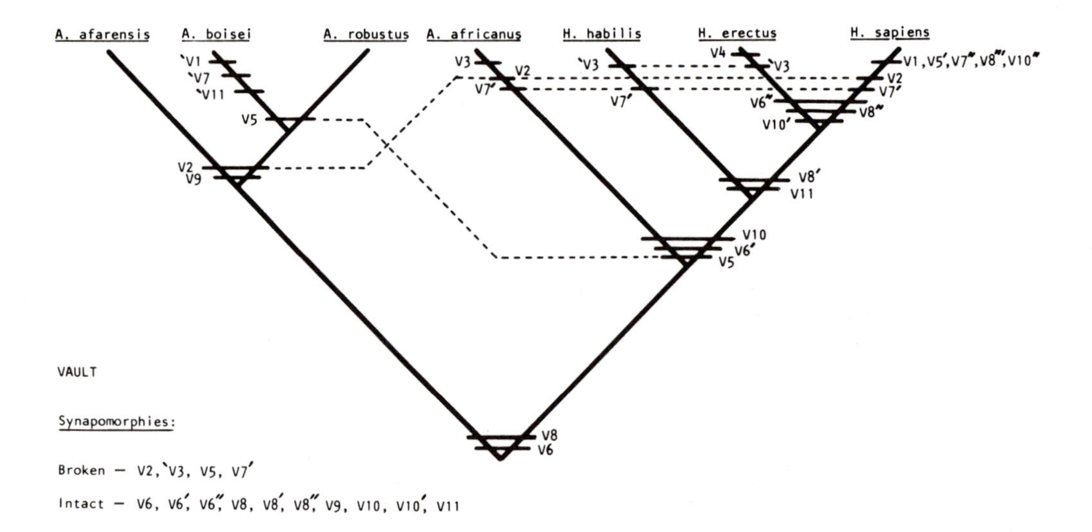

Table 6: Inferred character states for each outgroup and taxon.

	Dryo	Hylo	Pong	Gori	Pan_t	Pan_p	Rama	afar	afri	robu	bois	habi	erec	nean	sapi
M1	0	0	0	0	0	0	0	1	1	1	1	1	0	0	0
M2	-	0	0	0	0	-	0	0	0	0	0	0	0	1	0
M3	-	`1	0	0	0	-	`1	0	0	-	1	0	0	-	0
M4	1	1	0	0	0	-	-	1	1'	1	1'	1	1	1	1'
M5	-	0	0	0	0	-	-	1	1	1	1	-	1	-	1'
M6	-	0	0	0	0	-	-	1	1	1'	1'	1'	1	-	1
M7	0	-	0	0	0	-	-	0	0	0	0	1	1'	-	1'
M8	0	0	0	0	1	1	1'	1'	1''	1''	1''	1'	1'	1''	1'
M9	1''	0	0	0	0	1	1	1'	1	1'	1'	1	`1	1	1
V1	-	0	0	`1	0	-	-	0	0	0	`1	0	0	-	1
V2	-	`1	0	0	0	-	-	1	1	-	1	0	0	-	1
V3	-	`1	0	0	0	-	-	-	1	-	0	`1	`1	0	0
V4	-	0	0	0	0	-	-	0	0	-	0	0	1	0	0
V5	-	0	0	0	0	-	-	0	1	-	1	1	1	-	1'
V6	-	0	`1	0	0	-	-	-	1'	-	1	1'	1''	-	1''
V7	-	0	0	`1	0	-	-	0	1	0	`1	1'	0	-	1''
V8	0	0	0	0	0	0	-	1	1	1	1	1'	1''	-	1'''
V9	0	-	0	0	0	-	-	1	0	1	1	0	0	-	0
V10	-	1	0	0	0	-	-	0	1	-	0	1	1'	-	1''
V11	-	0	0	0	0	-	-	-	0	-	`1	1	1	-	1
P1	0	0	``1	0	0	0	``1	`1	`1	`1	`1	`1	1'	1'	1
P2	0	0	0	0	0	0	0	0	1	1'	1'	0	0	0	0
P3	-	0	0	0	1	-	1	1'	1'	0	0	1'	1''	1'	1''
P4	-	0	0	0	0	-	-	1'	1	1'	1'	1'	-	-	1
P5	-	0	0	`1	0	-	`1	1	1	1'	1'	1''	1''	-	1'''
P6	-	0	0	0	0	-	-	0	1	0	1'	1	1''	-	0
P7	1	0	0	0	0	-	-	1	1	1	1'	1'	-	1'	1''
F1	0	0	0	0	0	-	0	1	1	1'	1'	1	-	-	0
F2	-	1	`1	0	0	-	-	-	0	0	`1	1	0	-	0
F3	-	1''	0	0	0	-	-	-	1	1	0	1'	1'	-	1'
F4	-	1'	`1	0	0	-	-	-	1	1'	1'	1'	1''	1''	1''
F5	-	1'	0	0	0	-	0	-	0	0	0	0	0	1	1
F6	-	0	`1	0	0	-	-	-	0	0	0	1	1'	-	1
F7	-	0	0	0	0	-	-	-	1	1'	1'	0	1	0	1
B1	-	0	0	0	0	-	-	1	0	1''	1'	1	1	-	1'
B2	-	0	0	0	0	-	-	1	1	1	1	1	-	-	0
B3	-	0	0	0	0	-	-	-	1'	1	1	1'	-	-	1'
B4	-	1	0	0	0	-	-	0	0	-	0	0	1	-	1
B5	-	0	-	-	0	0	-	1	0	-	1	1'	1'	-	1'

Figure 3. Regional cladogram for the face. Note that there are too few data for A. afarensis to be included in the cladogram. Legend as for Fig. 2.

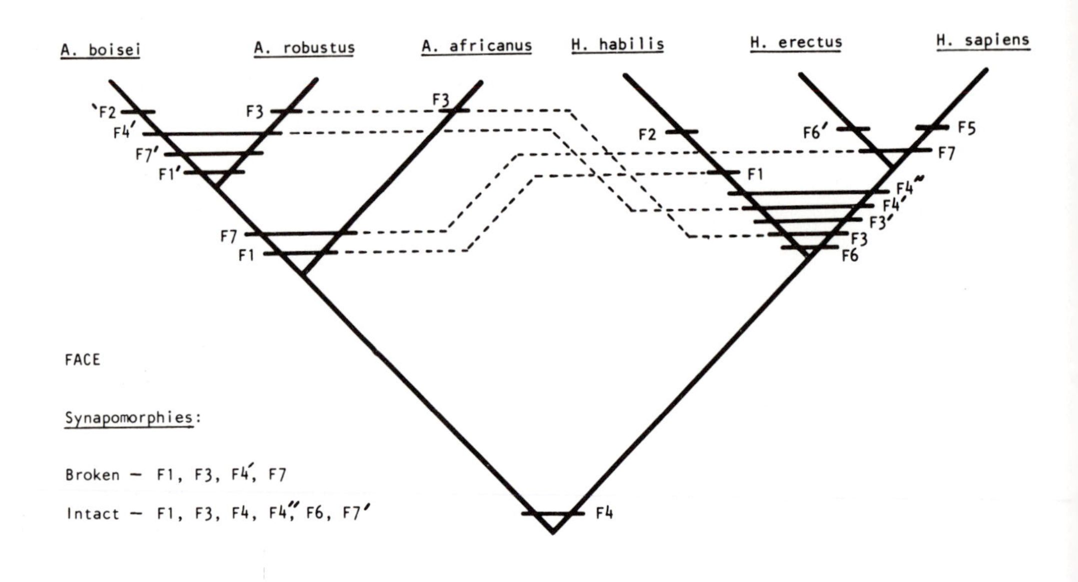

Figure 4. Regional cladogram for the palate, upper jaw and maxillary dentition. Legend as for Fig. 2.

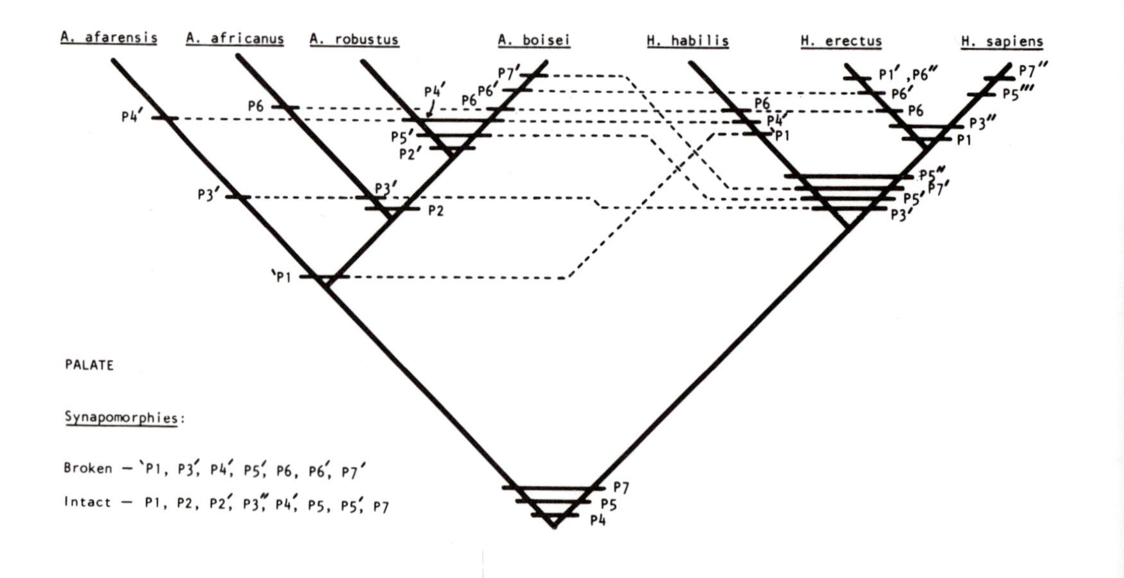

Figure 5. Regional cladogram for the cranial base. Legend as for Fig. 2.

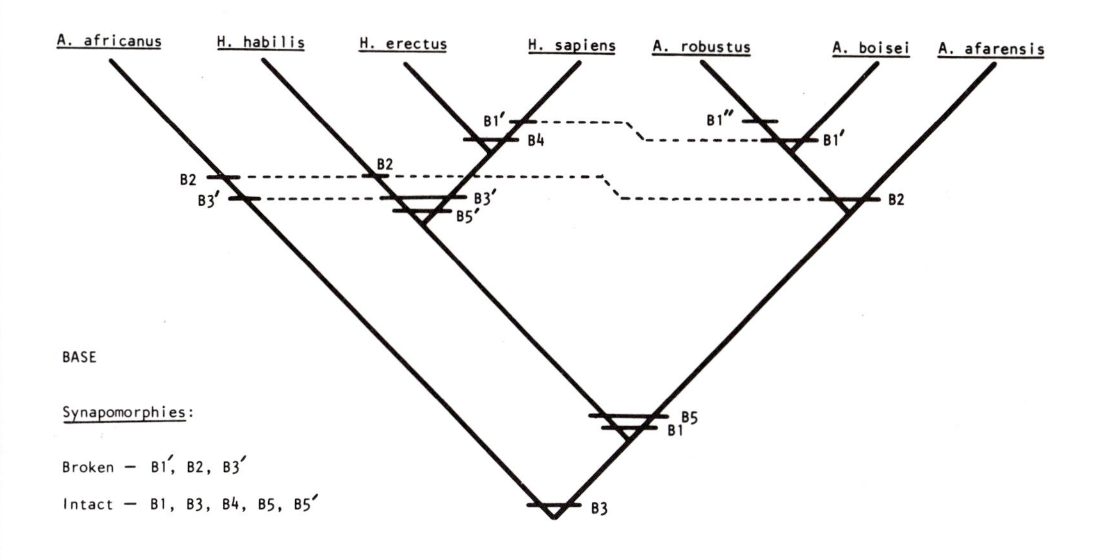

Figure 6. Regional cladogram for the mandible and mandibular dentition. Legend as for Fig. 2.

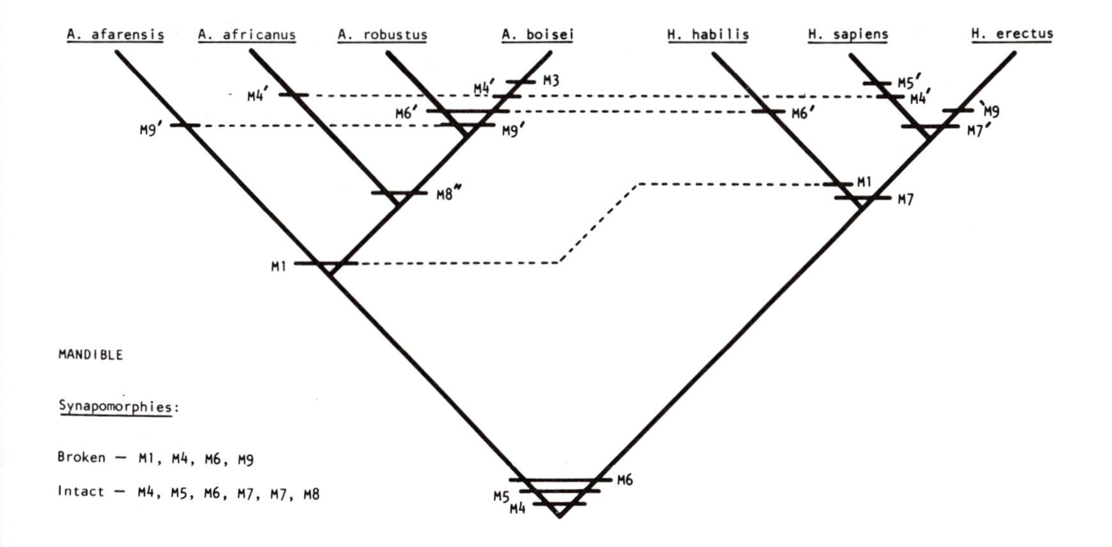

RESULTS

The five 'best-fit' regional cladograms are presented in Figs. 2-6 together with a list of the intact and broken synapomorphies for each cladogram. Each of the regional cladograms have the three *Homo* taxa as a monophyletic group, and this group is referred to as *Homo* in the summary cladograms presented in Fig. 7. Three of the regional cladograms (face, palate and mandible) are consistent with an australopithecine clade (Fig. 7(i)) and the available data on the face are compatible with a second clade scheme (Fig. 7 (ii)). In contrast, the vault and endocranium and the cranial base are consistent with grade classifications (Fig. 7 (iii) and (iv)). Inferred homoplasies, or broken synapomorphies, are common, particularly so in the 'best fit' cladograms for the palate and cranial base. The *Homo* monophyletic group (made up of *H. sapiens*, *H. erectus* and *H. habilis*) is characterised by no fewer than eight synapomorphies (V8, V11, F3', F4", F6', P5", B5' and M7). The two 'robust' australopithecine taxa are also consistently linked together, but only three character states (F1', F7' and P2') are synapomorphic for this clade. Consistency Index scores for all 39 characters (Table 7) suggest that the two clade arrangements and the grade pattern supported by cranial vault and endocranial evidence are the most parsimonious cladistic schemes.

DISCUSSION

This attempt to sample characters widely across the cranium has not resulted in any single, preeminent, cladistic scheme. All the 'best fit' regional cladograms involve broken synapomorphies and thus homoplasy is present in all regions. The CI scores (Table 7) suggest that some regions have 'better fit' cladograms than others, with a range of mean scores from 0.65 for the palate cladogram to 0.84 for the vault cladogram. Inspection of the CI scores for each of the 39 characters (Table 7) shows that just over half (20) of the characters gave similar scores for each of the six summary cladograms that were examined in this way. Thus, the differences in the mean CI scores for the summary cladograms rests on the distribution of the remaining nineteen characters and

Table 7: Consistency Indices for regional and summary cladograms

	CLADE 'A'	CLADE 'B'	GRADE 'A'	GRADE 'B'	GRADE pC'	GRADE 'D'
Vault	0.77	0.79	0.84	0.77	0.77	0.74
Face	0.76	0.76	0.74	0.72	0.72	0.74
Palate	0.65	0.62	0.62	0.61	0.62	0.64
Base	0.60	0.65	0.64	0.73	0.58	0.64
Mandible	0.77	0.74	0.71	0.71	0.74	0.74
Summary	0.72	0.73	0.73	0.71	0.70	0.71

these are not evenly spread across the five regions (i.e. vault = 6; face = 3; palate = 4; base = 3 and mandible = 3).

Of the total of 128 character states for the 39 characters, 22, or 17%, were autapomorphic for one or other of the taxonomic categories of hominid (Table 8) and were thus not relevant to the arrangements of the taxonomic groups. One taxon, A. afarensis, showed no autapomorphies and three taxa, A. africanus, A. robustus and H. habilis were characterised by a single autapomorphic trait. In contrast to this paucity of unique features, A. boisei and H. erectus each show five autapomorphic traits and H. sapiens is distinguished by nine unique character states.

The five regional cladograms are consistent with one or other of the first four summary cladograms which are illustrated in Fig. 7. Three of the 'best fit' regional cladograms are compatible with 'Clade A', and one each with the 'Clade B', 'Grade A' and 'Grade B' schemes. Without further information, it would be tempting to conclude that the first of the two 'clade' arrangements ('Clade A') was the most parsimonious cladogram, but reference to Table 7 shows that the mean CI score for 'Clade A' is actually less (but only marginally so) than the scores for the 'Clade B' and 'Grade A' schemes. This result may reflect on the inadequacy of the CI to assess the cladograms, but we believe that it is as likely to be due to the extent of homoplasies, which may have the effect of making each 'best fit' regional cladogram only marginally more parsimonious than the second, or even the third, ranking schemes for that region.

When the summary cladograms derived from the present study are compared to schemes recently proposed in the literature, there is some correspondence. 'Clade A' is equivalent to the cladogram in Corruccini & McHenry (1980) and White et al. (1981) and the scheme 'Grade A' corresponds to the cladogram put forward by Olson (1981). However, it should be noted that Olson's hypodigms differ from ours (e.g. A. afarensis was restricted to large specimens, with all the 'gracile' Hadar material placed in Homo sp. indet, as the sister taxon of a clade containing A. africanus and Homo). In addition to the cladograms suggested by the present study, two further 'grade' schemes were tested using these data. The first is 'Grade

Table 8: Autapomorphic Character States

TAXA	AUTAPOMORPHIC CHARACTER STATES								
A. afarensis									
A. africanus	V3								
A. robustus	B1								
A. boisei	M3	V1	V7	V11	F2				
H. habilis	F2								
H. erectus	'M9	V4	P1'	P6"	F6'				
H. sapiens	M5'	V1	V5'	V7"	V8"'	V10	P5"'	P7"	F5

Figure 7. Summary of regional cladograms together with two alternative cladograms taken from the literature. In all cases Homo stands for the monophyletic unit consisting of H. sapiens, H. erectus and H. habilis.

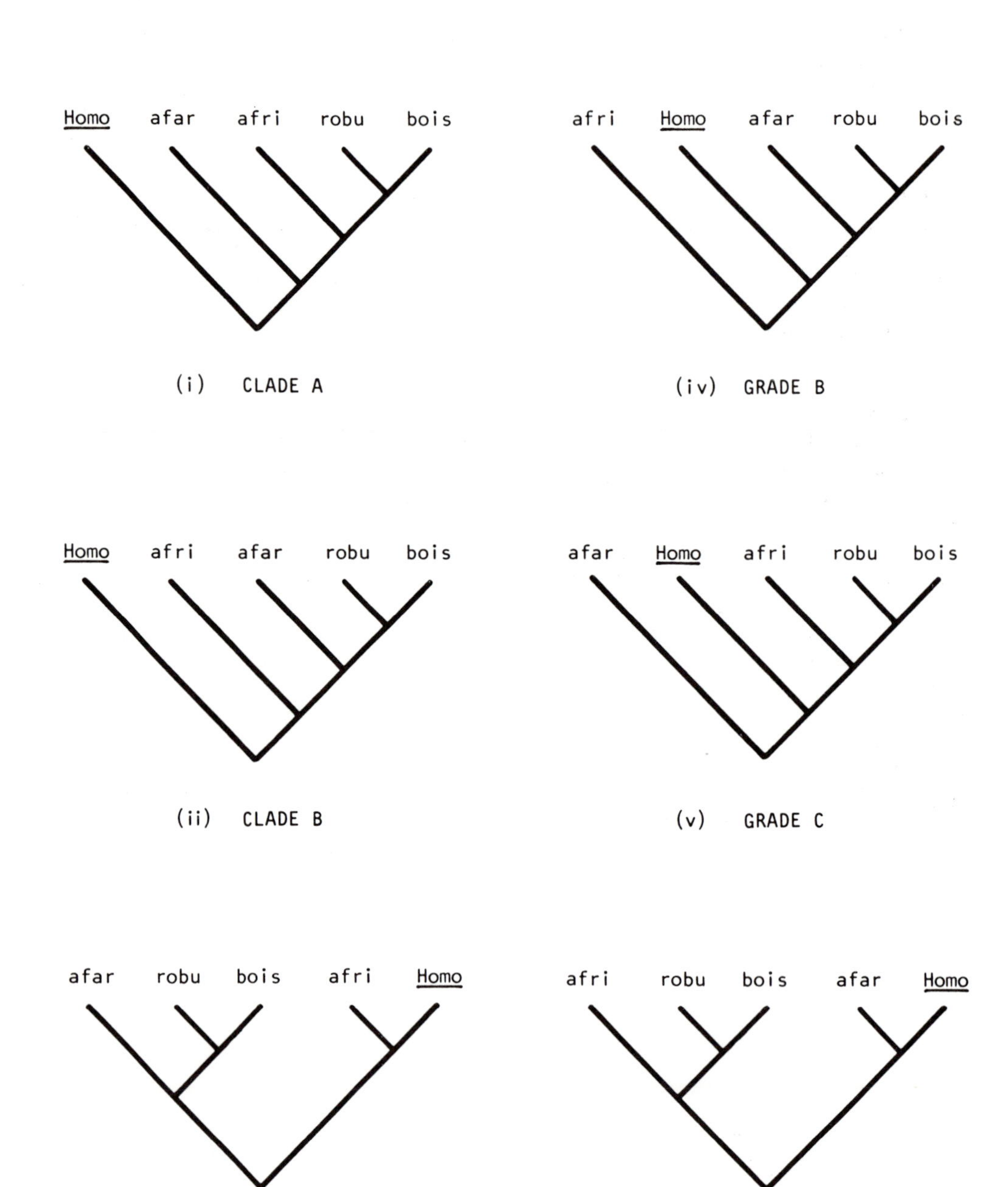

C', which is, we believe, the cladogram consistent with a recent analysis of the sister group relationship of early hominid taxa (Kimbel et al, 1984). The second scheme, 'Grade D', links Homo and A. afarensis in the same clade, as the sister group of the remaining Australopithecus species. However, the CI scores suggest that the last three of the 'grade' cladograms show fewer intact synapomorphies, and more broken synapomorphies, than do the two 'clade' and the first of the 'grade' schemes. Although the present study is based on a different set of data than that used by White et al. (1981), it is interesting that the cladograms for the three regions most involved with mastication - that is those for the face, palate and mandible - are the ones which are congruent with the cladogram of White et al. (1981) which was "rooted in a functional-adaptive analysis of early hominid masticatory morphology" (Kimbel et al, 1984:381).

CONCLUSIONS

While we see the need to develop this type of analysis, both by improvements in technique and by widening the scope of the morphological information, we believe, nonetheless, that some general conclusions and observations can be drawn from the results of the present study. We will present these in order, with those in which we have most confidence listed first.

1. There is strong evidence for incorporating the material assigned to H. habilis within the Homo clade; such a clade is defined by at least eight synapomorphic character states.

2. The distribution of character states emphasises the distinctive nature of the two 'robust' australopithecine taxa, A. robustus and A. boisei. Their association as a sister group is a good deal stronger than the association between either of them and any other taxon. This raises the question of whether the 'robust' clade should be given taxonomic recognition by reviving the genus Paranthropus.

3. A. africanus shares character states with both the Homo and the 'robust' australopithecine clade. The CI scores suggest that the position of A. africanus can be interchanged between these two clades (see Fig. 7 (ii) and (iii)) with no apparent detriment to the soundness of the cladograms. However, its removal from either of the two clades, and location as the sister taxon of all the other hominids (Fig. 7 (iv)) results in a lower CI score (0.71).

4. A. afarensis forms the sister group of the 'robust' australopithecine clade in three out of the four summary cladograms which are compatible with the regional cladograms. If this relationship is disrupted it results in a greater number of broken synapomorphies (Fig. 7(v) and (vi)).

5. The association of A. afarensis as the sister group of the Homo clade, or its location as the sister group of all other early hominid taxa results in cladograms with lower CI scores (0.71 and 0.70).

To return to the problem posed at the beginning of this paper, do we have the evidence necessary to determine whether Australopithecus is a clade or a grade? The results of this analysis are inconclusive. Three of the five regional cladograms are compatible with a clade, and two with a grade. The CI scores suggest that the distribution of the 39 character states used in this analysis can be interpreted as supporting two arrangements. One is that of an australopithecine clade (Fig. 7(i) and (ii)), and the other is compatible with a grade classification, making *Australopithecus* a paraphyletic group whose 'missing' member is *Homo*, which forms the sister group of *A. africanus*.

Both this more broadly based study and the narrower cladistic analysis of Corruccini & McHenry (1980), have shown that homoplasies are common enough to confound cladistic techniques which treat all characters equally. Nonetheless, the demonstration of the extent of the homoplasy reinforces the need for objective, and not subjective, character selection.

SUMMARY

The results of a cladistic analysis of the Hominidae are presented. The analysis adhered to a strict research protocol. Cranial material was assigned to four species of *Australopithecus* and three species of *Homo*; each species was assumed to be monophyletic. Quantitative data were obtained for a set of 39 size-corrected variables (characters) distributed across five functional regions of the cranium. Characters were subdivided into primitive and successively derived character states on the basis of inter-taxon variation; polarity was determined with respect to five extant and two extinct hominoid outgroup taxa. Regional and summary cladograms were assembled and compared using the Consistency Index as a criterion of parsimony.

The three species of *Homo* consistently form a monophyletic group (clade), as do the two robust species of *Australopithecus*. The evidence for the arrangement of the other two taxa assigned to *Australopithecus* is equivocal. These two 'gracile' species, together with the two 'robust' species of *Australopithecus* may make up an australopithecine clade. Alternatively, the 'gracile' species *Australopithecus africanus* may be the sister group of *Homo*, leaving *A. afarensis* as the sister taxon of the 'robust' australopithecines. The proposal that *A. afarensis* is the sister group of all other hominids is a less parsimonious arrangement according to these data.

The discovery of new fossils, while to be welcomed, does not necessarily clarify the relationships among those taxa already known from the fossil record. Attention must also be paid to improving and augmenting the ways in which such material is analysed.

ACKNOWLEDGEMENTS

We are grateful to the Director and Trustees of the National Museums of Kenya and the Government of Tanzania for allowing BAW to make a detailed study of fossils in their care. We also thank the Director and Trustees of the Transvaal Museum and the Head of the Department of Anatomy, University of the Witwatersrand, for allowing us to examine fossils for which they are responsible. BAW is in receipt of a grant from the Natural Environment Research Council (NERC) and ATC is an NERC Research student.

We are grateful to Paula Sutton for typing the manuscript.

REFERENCES

Key references

Andrews, P. & Franzen J. (eds.) (1984). The early evolution of man. Cour. Forsch.-Inst. Senckenberg, No. 69, pp.1-273.

Delson, E. (ed.) (1985). Ancestors: the hard evidence. Alan R. Liss: New York.

Howell, F.C. (1978). Hominidae. V.J. Maglio & H.B.S. Cooke (eds.): Evolution of African Mammals. Cambridge: Harvard University Press, pp. 154-248.

Tattersall, I. & Eldredge, N. (1977). Fact, theory and fantasy in human paleontology. Amer. Scien., *65*:204-11.

Tobias, P.V. (1980). Australopithecus afarensis and A. africanus: critique and an alternative hypothesis. Pal. Afr., *23*:1-17.

Main references

Abbott, S.A. (1984). A comparative study of tooth root morphology in the great apes, modern man and early hominids. Ph.D. Thesis: University of London.

Albrecht, G.H. (1978) Some comments on the use of ratios. Syste. Zool., *27*:67-71.

Andrews, P.J. (1978). A revision of the Miocene Hominoidea of East Africa. Bull. Br. Mus. (Nat. Hist.) Geol., *30*:85-224.

P. Andrews & J. Franzen (eds.) (1984). The early evolution of man. Cour. Forsch.-Inst. Senckenberg. No.69, pp.1-273.

Atchley, W.R., Gaskins, G.T. & Anderson, D. (1976). Statistical properties of ratios. I. Empirical results. Syst. Zool., *25*:137-48.

Bonde, N. (1977). Cladistic classification as applied to vertebrates. *In* M.K. Hecht, P.C. Goody & B.M. Hecht (eds.) Major patterns in vertebrate evolution. Plenum: New York, pp. 741-804.

Broom, R. (1938). The Pleistocene anthropoid apes of South Africa. Nature, *142*:377-379.

Broom, R. (1949). Another new type of fossil Ape-Man. Nature, *163*:57.

Clarke, R.J. (1977). The cranium of the Swartkrans hominid SK 847 and its relevance to human origins. Ph.D. Thesis: University of the Witwatersrand.

Corruccini, R.S. & McHenry, H.M. (1980). Cladometric analysis of Pliocene hominids. J. Hum. Evol., *9*: 209-221.

Dean, M.C. & Wood, B.A. (1982). Basicranial anatomy of Plio-Pleistocene hominids from East and South Africa. Am. J. Phys. Anthropol., *59*: 157-74.

E. Delson (ed.) (1985). Ancestors: the hard evidence. Alan R. Liss: New York.

Estabrook, G.F., Strauch, J.G. & Fiala, K.L. (1977) An application of compatibility analysis to the Blackiths' data on orthopteroid insects. Syst. Zool., *26*: 269-76.

Howell, F.C. (1965) Comment on 'New discoveries in Tanganyika: their bearing on hominid evolution' (by P.V. Tobias). Curr. Anthrop.,

6: 399-401.

Howell, F.C. (1978). Hominidae. In V.J. Magiio and H.B.S. Cooke (eds.): Evolution of African Mammals. Cambridge: Harvard University Press, pp. 154-248.

Huxley, J.S. (1958). Evolutionary process and taxonomy with special reference to grades. Upps. Univ. Arssks., *6*: 21-38.

Johanson, D.C. & White, T.D. (1979). A systematic assessment of early African hominids. Science, *202*:321-330.

Johanson, D.C., White, T.D. & Coppens, Y. (1982). Dental remains from the Hadar Formation, Ethiopia: 1974-1977 collections. Am. J. Phys. Anthropol., *57*: 545-603.

Kay, R.F. (1982). Sivapithecus simonsi, a new species of Miocene Hominoid, with comments on the phylogenetic status of the Ramapithecinae. Int. J. Primatol., *3*:113-73.

Kennedy, G.E. (1983). A morphometric and taxonomic assessment of a hominine femur from the Lower Member, Koobi Fora, Lake Turkana. Am. J. Phys. Anthropol., *61*:429-34.

Kimbel, W.H., Johanson, D.C. & Coppens, Y. (1982). Pliocene hominid cranial remains from the Hadar Formation, Ethiopia. Am. J. Phys. Anthropol., *57*:453-499.

Kimbel, W.H., White, T.D. & Johanson, D.C. (1984). Cranial morphology of Australopithecus afarensis: a comparative study based on a composite reconstruction of the adult skull. Am. J. Phys. Anthropol., *64*:337-88.

Kluge, A.G. & Farris, J.S. (1969). Quantitative phyletics and the evolution of anurans. Syst. Zool., *18*:1-32.

Leakey, L.S.B., Tobias, P.V. & Napier, J.R. (1964). A new species of the genus Homo from Olduvai Gorge. Nature, *202*:7-9.

Leakey, M.D. (1978). Olduvai fossil hominids: their stratigraphic positions and associations. *In* C.J. Jolly (ed.) Early Hominids of Africa. Duckworth: London, pp. 3-16.

Olson, T.R. (1981). Basicranial morphology of the extant hominoids and Pliocene hominids: the new material from the Hadar Formation, Ethiopia, and its significance in early human evolution and taxonomy. *In* C.B. Stringer (ed.) Aspects of human evolution. Taylor & Francis: London, pp. 99-128.

Partridge, T.C. (1978). Re-appraisal of lithostratigraphy of Sterkfontein hominid site. Nature, *275*: 282-87.

Partridge, T.C. (1979). Re-appraisal of lithstratigraphy of Makapansgat Limeworks hominid site. Nature, *279*: 484-88.

Patterson, C. (1982). Morphological characters and homology. *In* K.A.Joysey & A. Friday (eds.) Problems of Phylogenetic Reconstruction. Academic Press: London, pp. 21-74.

Pilbeam, D.R., Rose, M.D., Badgley, C. & Lipschutz, B. (1980). Miocene hominoids from Pakistan. Postilla, *181*:1-94.

Smith, R.J. (1983). The mandibular corpus of female primates: taxonomic, dietary and allometric correlates of interspecific variations in size and shape. Am. J. Phys. Anthropol., *61*:315-30.

Swindler, D.R. (1976). Dentition of living primates. Academic Press: London.

Tattersall, I. & Eldredge, N. (1977). Fact, theory and fantasy in human paleontology. Amer. Scien., *65*:204-11.

Tobias, P.V. (1967). Olduvai Gorge, Volume 2. The cranium and maxillary dentition of Australopithecus (Zinjanthropus) boisei. Cambridge University Press: Cambridge.

Tobias, P.V. (1972). Progress and problems in the study of early man in sub-Saharan Africa. *In* R.H. Tuttle (ed.) The functional and evolutionary biology of primates. Aldine Atherton: Chicago, pp. 63-93.

Tobias, P.V. (1973). Implications of the new age estimates of the early South African hominids. Nature, *246*:79-83.

Tobias, P.V. (1980). "Australopithecus afarensis" and A. africanus: critique and an alternative hypothesis. Pal. Afr., *23*:1-17.

Trinkaus, E. (1983). The Shanidar neanderthals. Academic Press: New York.

Trinkaus, E. (1984). Does KNM-ER 1481A establish Homo erectus at 2.0 myr B.P? Am. J. Phys. Anthropol., *64*:137-39.

Washburn, S.L. & Patterson, B. (1951). Evolutionary importance of the South African man-apes. Nature, *167*:650-51.

Weidenreich, F. (1936). The mandibles of Sinanthropus pekinensis: a comparative study. Palaeont. Sin., Series D, 7:1-169.

Weidenreich, F. (1937). The dentition of Sinanthropus pekinensis: a comparative odontography of the hominids. Palaeont. Sinica. Series no. 101, D(10), 2 vols. pp. 1-180 and pp. 1-121.

Weidenreich, F. (1943). The skull of Sinanthropus pekinensis: a comparative study of a hominid skull. Palaeont. Sinica. Series D, *10*:1-484.

White, T.D. (1977). New fossil hominids from Laetoli, Tanzania. Am. J. Phys. Anthropol., *46*:197-230.

White, T.D. (1980). Additional fossil hominids from Laetoli, Tanzania: 1976-1979 specimens. Am. J. Phys. Anthropol., *53*:487-504.

White, T.D. & Johanson, D.C. (1982). Pliocene hominid mandibles from the Hadar Formation, Ethiopia: 1974-1977 collections. Am. J. Phys. Anthropol., *57*:501-44.

White, T.D., Johanson, D.C. & Kimbel, W.H. (1981). Australopithecus africanus: its phyletic position reconsidered. S. Afr. J. Sci., *77*:445-70.

Wilkinson, M.J. (1983). Geomorphic perspectives on the Sterkfontein australopithecine breccias. J. Arch. Sci., *10*:515-529.

Wood, B.A. (1975). The nature and basis of sexual dimorphism in the primate skeleton. J. Zool., Lond., *180*: 15-34.

Wood, B.A. (1978). Allometry and hominid studies. *In* W.W. Bishop (ed.) Geological background to fossil man. Scottish Academic Press: Edinburgh, 125-38.

Wood, B.A. (in preparation). Cranial remains from Koobi Fora, Kenya. Clarendon Press: Oxford.

Wood, B.A. & Chamberlain, A.T. (1985). A reappraisal of variation in hominid mandibular corpus dimensions. Am. J. Phys. Anthropol. *66*:399-407.

APPENDIX A: DEFINITION OF CHARACTERS

No	Character	Definition
V1	Postorbital constriction	Minimum frontal breadth (ft-ft)/biorbital breadth (ek-ek).100
V2	Parietal shape	Sagittal arc/coronal arc.100
V3	Parietal-occipital arc ratio	Parietal sagittal arc/occipital sagittal arc.100
V4	Occipital sagittal curvature	Occipital sagittal chord/occipital sagittal arc.100
V5	Relative vault height	Auricular height/biporionic breadth (po-po).100
V6	Relative vault length	Glabella-opisthocranion/bioporionic breadth (po-po).100
V7	Frontal angle	Angle between surface of frontal squame and the Frankfurt Horizontal
V8	Encephalisation Quotient	Observed cranial capacity/expected cranial capacity. (Expected cranial capacity is based on a regression of brain weight against body weight in anthropoids)
V9	Occipito-marginal sinus	Present or absent
V10	Relative biparietal width	Maximum biparietal breadth/biporionic breadth.100
V11	Relative posterior cranial	Posterior cranial length (po/op)/ glabella-opisthocranion.100
F1	Position of zygomatic	Location of middle of zygomatic process of the maxilla with respect to the upper tooth row
F2	Relative facial width	Width of mid-face (zyg-zyg)/width of upper face (fmt-fmt).100
F3	Facial shape	Facial height (na-alv) width of upper face (fmt-fmt).100
F4	Alveolar prognathism	Alveolar projection (por-alv)/upper facial projection (por-glab).100
F5	Nasal bone shape	Maximum breadth across nasals/maximum height (na-alv).100
F6	Subnasal projection	Horizontal projection (sn-alv)/vertical projection (sn-alv).100
F7	Malar projection	Projection from porion of zygomaxillare (por-zm)/alveolar projection (por-alv).100 100

P1	Incisor size heteromorphy	I^1(MD.BL)/I^2(MD.BL).100
P2	Premolar size heteromorphy	P^3(MD.BL)/P^4(MD.BL).100
P3	Premolar root number	Number of separate roots (Abbott, 1984)
P4	Relative length of molar row	Molar row length/maxillo-alveolar length (Flowers).100
P5	Maxillo-alveolar index	Maxillo-alveolar breadth (ectomol-ectomol)/maxillo-alveolar length (Flowers).100
P6	Relative palate depth	Palatal depth at M^1/M^2/internal palate breadth at M_1/M_2.100
P7	Alveolar tooth row divergence	Minimum distance between C^1 alveoli/ minimum distance between M2 alveoli. 100
B1	Petro-tympanic angle	Angle between the long axes of the petrous and tympanic parts of the temporal bone (i.e. between PA/CC and CC/TP (Dean & Wood, 1982)
B2	Location of basion	Location relative to bitympanic line (anterior or posterior)
B3	Relative length of tympanic	Length of long axis of tympanic (CC-TP)/bitympanic width (TP-TP).100
B4	Relative width of mandibular fossa	Maximum width of fossa (entoglenoid to lateral extent of articular surface)/ biporionic breadth.100
B5	Angulation of foramen magnum	Angle in sagittal plane between Frankfurt Horizontal and a line joining basion to opisthocranion
M1	Corpus robusticity	Corpus breadth at M_1/Corpus height at M_1.100
M2	Position of mental foramen	Location of the mental foramen with respect to the lower tooth row
M3	Relative height of mental foramen	Height of foramen from base of corpus/ height of corpus at mental foramen.100
M4	Alveolar tooth row divergence	Minimum distance between C_1 alveoli/ minimum distance between M_2 alveoli.100
M5	Relative height of foramen spinosum	Height of foramen from base (gn-fs)/ height of symphysis (gn-id).100
M6	Relative length of canine socket	MD length of canine alveolus/summed alveolar tooth row length.100
M7	Premolar root number	Number of separate roots (Abbott, 1984)
M8	P_3 crown shape	MD/BL.100
M9	Molar size heteromorphy	M_2(MD.BL)/M_3(MD.BL).100

APPENDIX B: SUMMARY OF THE METRICAL DATA AVAILABLE FOR EACH TAXON

		Dryo	Hylo	Pong	Gori	Pan_t	Pan_p	Rama	afar	afri	robu	bois	habi	erec	nean	sapi
V1	$\overline{X}$	-	82	81	66	81	-	-	79	82	80	69	83	77	-	99
	Max	-	84	89	72	89	-	-	-	88	85	73	90	-	-	104
	Min	-	75	73	59	75	-	-	-	78	76	61	80	-	-	94
	N	-	10	10	10	10	-	-	1	3	2	3	4	1	-	10
V2	$\overline{X}$	-	53	105	107	93	-	-	$\gg 100$	117	-	123	109	101	-	118
	Max	-	83	123	125	118	-	-	-	127	-	141	126	108	-	134
	Min	-	35	88	81	73	-	-	-	110	-	109	88	91	-	102
	N	-	10	10	10	10	-	-	1	3	-	3	5	8	-	10
V3	$\overline{X}$	-	73	99	99	105	-	-	-	120	-	110	83	86	98	109
	Max	-	97	120	110	136	-	-	-	123	-	118	101	94	-	135
	Min	-	58	84	80	81	-	-	-	115	-	99	64	78	-	83
	N	-	10	10	5	10	-	-	-	3	-	3	3	8	1	10
V4	$\overline{X}$	-	86	87	83	85	-	-	85	83	-	79	81	74	80	82
	Max	-	92	93	89	93	-	-	87	89	-	86	86	79	-	89
	Min	-	62	84	78	79	-	-	84	79	-	71	77	67	-	74
	N	-	10	10	10	10	-	-	2	4	-	3	4	8	7	10
V5	$\overline{X}$	-	62	57	56	57	-	-	<60	70	-	62	72	76	-	91
	Max	-	67	62	65	62	-	-	-	74	-	75	74	85	-	98
	Min	-	58	50	50	53	-	-	-	67	-	56	70	60	-	79
	N	-	10	10	10	10	-	-	1	3	-	4	3	7	-	10
V6	$\overline{X}$	-	126	108	124	120	-	-	-	137	-	129	139	156	-	162
	Max	-	137	114	132	127	-	-	-	146	-	129	149	160	-	171
	Min	-	119	102	111	113	-	-	-	129	-	129	131	153	-	151
	N	-	10	10	10	10	-	-	-	2	-	2	3	6	-	10
V7	$\overline{X}$	-	40	40	25	35	-	-	(30)	45	30	29	54	39	-	85
	Max	-	45	55	30	40	-	-	-	45	-	34	60	43	-	90
	Min	-	30	30	20	25	-	-	-	45	-	20	47	35	-	70
	N	-	10	10	10	10	-	-	1	2	1	3	3	4	-	10
V8	$\overline{X}$	1.31	1.13	1.24	0.92	1.39	1.27	-	1.59	1.90	1.86	1.71	2.22	2.88	-	3.84
	Max	-	-	-	-	-	-	-	-	-	-	-	-	-	-	-
	Min	-	-	-	-	-	-	-	-	-	-	-	-	-	-	-
	N	-	-	-	-	-	-	-	-	-	-	-	-	-	-	-

		Dryo	Hylo	Pong	Gori	Pan_t	Pan_p	Rama	afar	afri	robu	bois	habi	erec	nean	sapi
V9	$\overline{X}$	x	-	x	x	x	-	-	(✓)	x	✓	(✓)	x	x	-	(x)
	Max	-	-	-	-	-	-	-	-	-	-	-	-	-	-	-
	Min	-	-	-	-	-	-	-	-	-	-	-	-	-	-	-
	N	-	-	-	-	-	-	-	-	-	-	-	-	-	-	-
V10	$\overline{X}$	-	94	84	76	88	-	-	80	104	-	85	96	111	-	121
	Max	-	100	90	86	92	-	-	-	110	-	92	98	117	-	128
	Min	-	88	72	67	81	-	-	-	100	-	82	94	108	-	112
	N	-	10	10	10	10	-	-	1	3	-	4	2	8	-	10
V11	$\overline{X}$	-	34	30	37	33	-	-	-	31	-	28	40	39	-	44
	Max	-	38	34	43	43	-	-	-	33	-	29	42	41	-	48
	Min	-	28	27	30	29	-	-	-	29	-	27	39	37	-	39
	N	-	10	10	10	10	-	-	-	2	-	2	2	2	-	10
F1	$\overline{X}$	M^1/M^2	M^2	M^1/M^2	M^1/M^2	M^2	-	M^1/M^2	P^4/M^1	P^4	P^3/P^4	P^4	M^1	-	-	M^2
	Max	-	M^2/M^3	M^2	M^2	M^2/M^3	-	M^2	M^1	M^1	P^4	P^4	M^1	-	-	M^2
	Min	-	M^2	M^1	M^1	M^1/M^2	-	M^1	P^4/M^1	P^3/P^4	P^3	P^3/P^4	P^4	-	-	M^1/M^2
	N	1	10	10	10	10	-	3	5	12	11	3	4	-	-	10
F2	$\overline{X}$	-	114	146	130	121	-	-	-	135	135	154	117	121	-	120
	Max	-	119	154	140	135	-	-	-	137	136	159	-	122	-	129
	Min	-	106	141	121	114	-	-	-	134	135	146	-	120	-	114
	N	-	10	10	10	10	-	-	-	2	2	3	1	2	-	10
F3	$\overline{X}$	-	52	96	88	87	-	-	-	79	75	87	72	65	-	64
	Max	-	59	115	101	100	-	-	-	81	-	97	79	66	-	70
	Min	-	43	83	82	71	-	-	-	77	-	77	66	64	-	57
	N	-	10	10	10	10	-	-	-	2	1	2	2	2	-	10
F4	$\overline{X}$	-	123	156	141	140	-	-	-	131	126	122	124	108	104	103
	Max	-	130	164	146	145	-	-	-	134	-	-	128	-	-	110
	Min	-	115	149	134	130	-	-	-	128	-	-	121	-	-	98
	N	-	10	10	10	10	-	-	-	2	1	1	3	1	5	10
F5	$\overline{X}$	-	104	30	51	45	-	35	-	46	40	54	53	52	81	78
	Max	-	114	50	63	65	-	-	-	56	44	67	64	-	-	100
	Min	-	77	11	38	34	-	-	-	38	36	43	46	-	-	65
	N	-	10	10	10	10	-	1	-	4	2	3	3	1	4	10

		Dryo	Hylo	Pong	Gori	Pan_t	Pan_p	Rama	afar	afri	robu	bois	habi	erec	nean	sapi
F6	$\bar{X}$	-	107	158	112	123	-	-	-	105	122	104	81	36	-	64
	Max	-	150	208	131	171	-	-	-	117	156	120	105	-	-	83
	Min	-	75	129	94	91	-	-	-	81	92	96	57	-	-	50
	N	-	10	10	10	10	-	-	-	4	7	3	3	1	-	10
F7	$\bar{X}$	-	51	56	57	55	-	-	-	62	71	73	58	61	56	60
	Max	-	56	58	60	59	-	-	-	65	72	-	60	-	-	66
	Min	-	47	49	52	49	-	-	-	59	71	-	56	-	-	56
	N	-	10	10	10	10	-	-	-	2	2	1	2	1	5	10
P1	$\bar{X}$	1.43	1.40	2.55	1.52	1.49	1.41	2.91	1.65	1.75	1.65	1.55	1.62	1.14	1.14	1.35
	Max	1.51	-	-	-	-	-	3.00	1.79	1.82	-	-	1.67	-	-	-
	Min	1.37	-	-	-	-	-	2.83	1.54	1.69	-	-	1.57	-	-	-
	N	3	-	-	-	-	-	2	4	2	1	1	2	1	1	-
P2	$\bar{X}$	124	109	101	114	114	122	108	101	92	84	85	98	110	98	105
	Max	136	-	-	-	-	-	122	105	100	91	90	105	-	110	-
	Min	116	-	-	-	-	-	89	97	86	82	77	94	-	83	-
	N	3	-	-	-	-	-	4	5	5	7	4	8	1	4	-
P3	$\bar{X}$	-	3	3	3	3/2	-	3/2	2	2	3	3	2	1	2	1
	Max	-	-	-	-	-	-	-	-	-	-	-	-	-	-	-
	Min	-	-	-	-	-	-	-	-	-	-	-	-	-	-	-
	N	-	-	-	-	-	-	-	-	-	-	-	-	-	-	-
P4	$\bar{X}$	-	43	42	44	40	-	-	54	49	55	55	59	-	-	49
	Max	-	47	45	51	43	-	-	-	51	62	57	60	-	-	54
	Min	-	39	38	38	37	-	-	-	48	49	52	59	-	-	44
	N	-	10	10	10	10	-	-	1	2	6	3	2	-	-	8
P5	$\bar{X}$	-	85	80	72	81	-	67	90	92	97	100	108	111	-	123
	Max	-	90	89	78	86	-	-	-	100	108	105	110	-	-	132
	Min	-	81	72	61	74	-	-	-	86	86	95	107	-	-	110
	N	-	10	10	10	10	-	1	1	3	7	3	2	1	-	10
P6	$\bar{X}$	-	32	37	42	32	-	-	29	48	37	55	45	77	-	29
	Max	-	37	45	49	46	-	-	34	57	52	59	56	-	-	38
	Min	-	29	28	34	22	-	-	25	31	28	53	36	-	-	18
	N	-	10	10	10	10	-	-	2	5	9	3	4	1	-	10

		Dryo	Hylo	Pong	Gori	Pan_t	Pan_p	Rama	afar	afri	robu	bois	habi	erec	nean	sapi
P7	$\bar{X}$	81	89	101	103	104	-	-	83	83	78	71	74	-	70	60
	Max	-	100	113	122	112	-	-	87	98	91	77	81	-	71	68
	Min	-	80	88	90	92	-	-	80	66	66	65	69	-	70	50
	N	1	10	10	10	10	-	-	2	7	9	2	4	-	2	10
B1	$\bar{X}$	-	64	57	67	63	-	-	40	54	22	30	36	46	-	29
	Max	-	74	-	-	-	-	-	45	58	28	34	41	-	-	-
	Min	-	58	-	-	-	-	-	35	51	17	27	30	-	-	-
	N	-	40	30	30	30	-	-	3	4	2	4	3	1	-	30
B2	$\bar{X}$	-	-6	-4	-7	-3	-	-	+	+2	+7	+5	+3	-	-	-1
	Max	-	-	-	-	-	-	-	-	+3	+8	+7	+7	-	-	-
	Min	-	-	-	-	-	-	-	-	-1	+6	+4	-1	-	-	-
	N	-	40	30	30	30	-	-	1	4	2	4	3	-	-	30
B3	$\bar{X}$	-	30	30	32	.31	-	-	-	24	27	28	23	-	-	22
	Max	-	-	-	-	-	-	-	-	27	30	33	29	-	-	-
	Min	-	-	-	-	-	-	-	-	21	25	25	18	-	-	-
	N	-	40	30	30	30	-	-	-	4	2	4	3	-	-	30
B4	$\bar{X}$	-	20	27	29	24	-	-	27	29	-	29	24	19	-	20
	Max	-	23	30	32	26	-	-	-	32	-	31	26	20	-	24
	Min	-	16	23	27	21	-	-	-	26	-	27	19	19	-	18
	N	-	10	10	10	10	-	-	1	4	-	4	3	2	-	10
B5	$\bar{X}$	-	-29	-	-	-22	-18	-	-6	-20	-	-8	+9	+6	-	+8
	Max	-	-	-	-	-	-	-	-	-	-	-7	+13	+9	-	-
	Min	-	-	-	-	-	-	-	-	-	-	-10	+5	+3	-	-
	N	-	40	-	-	46	36	-	1	1	-	2	2	2	-	27
M1	$\bar{X}$	48	50	49	54	53	48	49	59	60	66	69	62	55	50	46
	Max	53	67	61	65	71	61	55	76	65	75	80	68	64	-	59
	Min	44	37	43	45	44	38	37	48	56	55	57	59	46	-	33
	N	7	10	20	37	35	32	10	11	4	5	24	7	7	19	75

		Dryo	Hylo	Pong	Gori	Pan_t	Pan_p	Rama	afar	afri	robu	bois	habi	erec	nean	sapi
M2	$\bar{X}$	-	P_3	P_4	P_4	P_4	-	P_4	P_4	P_4/M_1	P_4	P_4	P_4	P_4	M_1	P_4
	Max	-	P_3/P_4	P_4/M_1	P_4/M_1	P_4	-	M_1	P_4/M_1	-	P_4	P_4	P_4/M_1	M_1	M_1	P_4/M_1
	Min	-	P_3/C	P_3/P_4	P_3	P_3/P_4	-	P_3/P_4	P_3	-	P_3/P_4	P_3	P_3/P_4	P_3/P_4	P_4/M_1	P_3/P_4
	N	-	10	10	10	10	-	4	16	1	4	20	7	8	3	10
M3	$\bar{X}$	-	35	47	49	45	-	35	44	45	-	53	50	47	-	46
	Max	-	50	58	55	54	-	41	54	-	-	65	55	51	-	58
	Min	-	29	41	45	38	-	31	38	-	-	39	45	43	-	41
	N	-	10	10	10	10	-	4	10	1	-	13	7	3	-	10
M4	$\bar{X}$	59	53	74	73	73	-	-	55	46	57	45	53	50	52	42
	Max	75	59	82	85	86	-	-	59	-	-	47	54	-	53	51
	Min	44	48	62	65	62	-	-	47	-	-	43	53	-	52	35
	N	2	10	10	10	10	-	-	3	1	1	3	2	1	3	10
M5	$\bar{X}$	-	25	25	28	25	-	-	38	42	42	46	-	39	-	49
	Max	-	30	33	32	31	-	-	-	46	55	48	-	-	-	60
	Min	-	19	20	25	16	-	-	-	39	31	42	-	-	-	41
	N	-	10	10	10	10	-	-	1	2	4	4	-	1	-	10
M6	$\bar{X}$	-	14	13	13	15	-	-	10	10	7	7	8	11	-	10
	Max	-	16	16	15	19	-	-	11	11	-	9	9	12	-	11
	Min	-	12	9	10	13	-	-	10	9	-	6	6	9	-	7
	N	-	10	10	10	10	-	-	5	4	1	6	3	2	-	8
M7	$\bar{X}$	2	-	2	2	2	-	-	2	2	2	2	2/1	1	-	1
	Max	-	-	-	-	-	-	-	-	-	-	-	-	-	-	-
	Min	-	-	-	-	-	-	-	-	-	-	-	-	-	-	-
	N	-	-	-	-	-	-	-	-	-	-	-	-	-	-	-
M8	$\bar{X}$	154	149	145	137	117	113	93	88	81	82	80	99	87	80	91
	Max	160	-	-	-	-	-	102	119	88	92	85	109	97	92	-
	Min	142	-	-	-	-	-	85	71	75	79	74	93	75	76	-
	N	6	-	-	-	-	-	4	21	3	12	5	8	13	6	-
M9	$\bar{X}$	83	106	108	108	110	102	97	94	102	93	97	96	115	99	100
	Max	86	-	-	-	-	-	102	108	113	106	114	102	136	107	-
	Min	78	-	-	-	-	-	92	82	85	84	84	90	102	95	-
	N	3	-	-	-	-	-	3	9	3	9	7	2	3	5	-

14 HOMO AND PARANTHROPUS: SIMILARITIES IN THE CRANIAL BASE AND DEVELOPING DENTITION

M.C. Dean,
Department of Anatomy and Embryology,
University College London,
London, WC1E 6BT, U.K.

INTRODUCTION

Until recently the major cranial characters of reference for identifying morphologically distinct populations of early fossil hominids have been chosen from the neurocranium, face, dentition and the masticatory system. Apart from frequent reference to the relative position of the foramen magnum, it is only in the last few years that the cranial base has begun to figure more prominently in the literature dealing with hominid phylogeny (Lieberman & Crelin, 1971; Laitman, 1977; Laitman et al. 1978, 1979; Leitman & Heimbuch, 1982; Olson, 1978, 1981; Dean & Wood, 1981a, 1982; Falk & Conroy, 1983; Kimbel, 1984 and Kimbel et al, 1984). Despite being established historically as one of the more conservative regions of the skull (Huxley, 1863, 1867), evidence about the relationships of the basicranium with other regions of the skull in primates has not been consistent. Neither have earlier comparative studies of the cranial base in the sagittal plane revealed any trends that have proved useful in distinguishing between groups of primates or fossil hominids. Moreover, certain regions of the cranial base of hominoids (such as the mastoid and digastric regions) appear not to share obvious or straightforward bony homologies, but rather appear to be more complicated than had previously been recognised (Krantz, 1963; Dean, 1985a). This makes data from these regions difficult to interpret and incorporate into an evolutionary model of hominid evolution. Nonetheless, recent comparative studies of the cranial base seen in norma basilaris, of early fossil hominids, the great apes and modern *Homo sapiens* (Dean & Wood, 1981a, 1982), have provided evidence that suggests whereas *Australopithecus africanus* and the great apes appear in some respects to share primitive characters, *Paranthropus*, early *Homo* and *Homo sapiens* by comparison apparently share a suite of derived characters. Because a similar pattern of apparent relatedness has also been noted previously for the eruption pattern of the permanent dentition in hominoids (Broom & Robinson, 1951, 1952) (in which *Australopithecus africanus* and the great apes appear similar but *Paranthropus* and *Homo* stand apart as distinct from these and yet similar to each other) it seemed important to re-examine these two lines of evidence as they conflict with the consensus view of early hominid relationships. Whereas the claim that similarities between *Homo* and *Paranthropus* can be attributed to parallel evolution is justified when there is apparently good reason (DuBrul, 1977), it becomes less likely when a number of similarities occur independently in several anatomical regions. For this reason the major aim of this paper is to reassess the comparative evidence from the cranial base and the developing dentition with due regard to their combined potential importance for studies of hominid phylogeny.

THE CRANIAL BASE

Kimbel, White & Johanson (1984) have recently summarized several "Homo-like" features of the cranial base that have been reported as characteristic of Paranthropus. These include a strongly flexed cranial base in the sagittal plane and coronally orientated petrous temporal bones (DuBrul, 1977; Tobias, 1967; Dean & Wood, 1981a, 1982); an anteriorly-positioned foramen magnum and a broad cranial base in the region of the carotid canals and styloid region (Dean & Wood, 1981a, 1982); a steep articular eminence of the glenoid fossa, a posteriorly projecting entoglenoid process (Du Brul, 1977) and vertically inclined tympanic plates (Tobias, 1967). Kimbel et al attribute all of these similarities to parallel evolution occurring independently in Homo and Paranthropus as a result of marked cranial base flexion in the mid-saggital plane caused by an "upper facial orthognathism" (Kimbel et al, 1984:377).

Despite the claim of Huxley (1863, 1867) that the degree of spheno-ethmoidal flexion in the mid-sagittal plane is related to facial prognathism, this has been shown not to be a consistent relationship in subsequent comparative studies of primates. Cameron (1930) noted a marked degree of cranial base flexion in Tarsius (151°) relative to other primates and Ashton (1957) and Scott (1958) made similar observations about Papio (148°). Both these values are nearer to the values for modern Homo sapiens (133°, Björk, 1955) than any other primates, yet Tarsius and Papio represent extremes of facial prognathism.

Subsequent studies have shown that the spheno-ethmoidal angle in Pan paniscus (140°) is closer to that of modern man than other primates (Cramer, 1977; Fenart & Deblock, 1973). Data for cranial base flexion in modern human races (Björk, 1972) also suggest that the cranial base and face are to some extent independent of each other. European and Australian skulls (the most flexed of the modern human races) show different degrees of facial prognathis. These racial differences in facial prognathism have been attributed mostly to the size of the jaws, but also to the length of the neck and the resulting altered position of the tongue and oropharyngeal structures (Proffitt, McGlone & Barrett, 1975). Laitman (1977) and Laitman et al (1978, 1979; Laitman & Heimbuch, 1982) have, however, pointed to a close relationship between cranial flexion and the position in the neck of both the pharynx and larynx. Laitman (1977) hypothesised that the descent of the tongue and larynx in man may have drawn on the attachment of the pharyngeal constrictor muscles at the pharyngeal tubercle and resulted in flexion of the cranial base. This suggests that facial (or mandibular) height together with length of the neck may be more closely related to cranial base flexion than facial prognathism.

Rak (1983) has carefully distinguished between the retruded orthognathic face of Paranthropus, where the palate remains long, and the retruded face of Homo, where palatal length is considerably reduced. It is noteworthy that in Paranthropus as well as in Australopithecus africanus, the sphenoid bone is long anteroposteriorly to accommodate the increased length of the jaws retruded under the neurocranium, whereas in Homo the sphenoid is shortened (Dean & Wood, 1981a, 1982; Wood & Dean, 1985). A relationship between the face and the length of the anterior part of the cranial base has been noted by Björk (1950) who, in a study of prognathism, concluded that differences in the size of the jaws account for major differences in prognathism between the crania belonging to different modern human racial types and among mammals. However, within racial groups, small increases in spheno-ethmoidal flexion, or shortening of the cranial base, result in minimal increases in mandibular prognathism.

What does emerge from comparative studies of the cranial base in the sagittal plane, is the much clearer influence of the brain upon the morphology of the cranial base. Expansion and 'swinging down' of the nuchal and occipital region have often been linked with cranial base flexion (Aeby, 1867; Keith, 1910; Bolk, 1910; Weidenreich, 1941; Moss, 1958, 1963, 1975; Biegert, 1963 and Björk, 1972). DuBrul and Laskin (1961) also provide additional evidence which demonstrates that restricted growth at the spheno-occipital synchondrosis in the rat results in an apparent neural expansion that brings about cranial base flexion and inwardly-rotated petrous temporal bones. In addition Anderson & Popovich (1983) have recently presented good data to show that lower cranial height (representing cranial height in the occipital region) is 50% greater in modern human crania which have the most 'closed' or flexed cranial base angles.

Holloway (1972) has noted that an endocast of Paranthropus (SK 1585) shows an "advanced cerebellar morphology" relative to Australopithecus africanus. This point was also noted by Tobias (1967) for OH 5, and it is possible that this may prove to be so in other crania attributed to Paranthropus. This morphology is characterised by "a more human-like shape, size and disposition of the cerebellar lobes, reflected by the lateral (or neo-cerebellar) lobe enlargement" in Paranthropus (Holloway, 1972:185). If these observations about the cerebellum in Paranthropus prove to be consistent they by no means offer a solution about whether it is likely that the basicranial similarities between Homo and Paranthropus occur in parallel, as parallelisms are probably as likely to occur in the brain as anywhere else. Nevertheless, in the light of previous studies relating to the cranial base, they strongly suggest that the disposition of the cerebellum is a more likely cause of similarity in cranial base morphology than 'facial orthognathism' (Kimbel et al, 1984). Falk & Conroy (1983) and Kimbel (1984) have also noted the presence of occipital marginal venous sinuses and the absence of large transverse sinuses in the posterior cranial fossa of Paranthropus. It may be that pronounced lateral expansion of the cerebellum during fetal development favours the development of occipital marginal venous sinuses along the path of 'least resistance'.

Another, and perhaps more convincing, argument that similar characters in the cranial base of Homo and Paranthropus result from parallel evolution has been put forward by DuBrul (1977). DuBrul has proposed that the degree of adaptation to bipedalism, and the need for efficient poise of the head on the vertebral column, requires that either a large neurocranium, (in modern Homo), or a massive face, (in Paranthropus), bring about a 'buckling' of the cranial base at sella. This view, that cranial base flexion is to some extent posture-related, is supported both by evidence from pathological conditions (Björk & Kuroda, 1968) and evidence from several workers linking the forward position of the foramen magnum with a low spheno-ethmoidal angle (DuBrul, 1950; Le Gros Clark, 1971). However, Moore & Spence (1969) have argued that changes in posture may only bring about flexion of the face, leaving the cranial base unaffected. Differences in cranial base flexion resulting from adaptation to posture are perhaps credible between small specimens of Australopithecus africanus and large specimens of Paranthropus boisei. However, it is less likely that specimens of early Homo (eg KNM-ER 1813, OH 24) which are not greatly dissimilar to Australopithecus africanus in overall cranial proportions (and indeed which have even been attributed to them) should exhibit markedly contrasting cranial base patterns (Dean & Wood, 1981a, 1982) which result solely from the degree of adaptation to bipedalism. Erecting one 'postural' hypothesis to explain cranial base flexion in Paranthropus and another 'neural expansion' hypothesis, to explain cranial base flexion in early Homo, is not consistent.

Among the features of the cranial base that Paranthropus shares with Homo, the coronally rotated petrous temporal bones emerge as the most distinct (Dean, 1982; Dean & Wood, 1981a, 1982). Data for the great apes and Australopithecus africanus indicate that there has been no shift in the petrous angulation which remains high and in the sagittal orientation in these taxa as it does for all other primates represented in Table 1.

Another notable feature of the cranial base of Australopithecus africanus and the great apes that sets them apart from Homo and Paranthropus is the prominent eustachian processes (associated with the muscles of the palate) and the elongated, scalloped markings of the longus capitis muscles on the basioccipital (Dean, 1985a). Given the lack of good evidence for clear differences in cranial base flexion in fossil hominids (Table 2) it seems unreasonable to attribute these differences in the markings for the attachments of the superior constrictor, levator and tensor palati and longus capitis muscles simply to the degree of flexion of the cranial base. Nonetheless whatever their aetiology, they represent further evidence for a suite of apparently primitive cranial base characters that set Australopithecus africanus apart from other early hominids.

Data for the angle the petrous temporal bone makes with the coronal plane passing through the carotid canals in several extant primate groups are given in Table 1. These data indicate that the sagittal orientation of this part of the temporal bone is independent of known variations in the size and disposition of the facial skeleton of the primates represented and that it is apparently independent of the degree of cranial base flexion in this group of primates. The degree of sagittal flexion of the cranial base is particularly difficult to measure in fossil hominid specimens, but estimates can be made from more intact specimens and from sagittal sections of endocasts that preserve the landmarks, foramen caecum (or some part of the anterior end of the cranial base), sella and basion

(Table 2). These values for cranial base flexion reveal a great deal of variation among hominids, the lowest value being for Sts 19 (120°) and the highest (160°) for Solo skull VI (Weidenreich, 1951). Laitman & Heimbuch (1982) have provided data for exocranial base flexion in nine early fossil hominid specimens and after a craniometric analysis of the basicrania of these individuals concluded that they showed marked similarities to those of extant great apes. However, it appears that petrous angulation is independent of cranial base flexion where estimates of both can be made on the same specimen, (see Table 1).

Data for extant primates have been provided by Cramer (1977) and Fenart & Deblock (1973) who have noted a more flexed cranial base in the sagittal plane for *Pan paniscus* than for the other African great apes, and Ashton (1957) has provided data for *Papio*, *Macaca*, *Pan*, *Pongo* and *Gorilla* indicating some differences in the values for the spheno-ethmoidal angle in this group (Table 1). The fact that petrous angulation also appears independent of body size as judged by the contrasting sample of primates represented in Table 1 further strengthens the case for the significance of the distinctions observed in the cranial base of *Australopithecus africanus* and the great apes on the one hand and *Homo* and *Paranthropus*

Table 1: Cranial base data for extant primates

	n	mean value for Petrous Angulation	n	mean value for Cranial Base Angle
Tarsius	-	-	?	151
Macaca	16	63.3	12	161
Papio	16	64.5	19	149
Theropithecus	6	63.3	-	-
Hylobates	30	66.2	30	170
Pan paniscus	30	68.4	58	140
Pan troglodytes	30	68.9	28	157
Gorilla	30	71.9	29	168
Pongo	30	68.1	28	165
Homo sapiens	30	45.7	243	133

Data for Cranial Base or Spheno-Ethmoidal angle taken from: Cameron 1930; Ashton, 1957; Cramer, 1977; Bjork, 1955.

Data for Petrous Angulation from: Dean & Wood, 1981; Dean, in preparation and Luboga, in preparation.

Table 2: Cranial Base Data for Hominid Fossils

Fossil Groups	Petrous Angulation	Cranial Base Angle
Australopithecus africanus		
Sts 5	65	145
Sts 19	59	120
Sts 25	72	-
MLD 37/38	60	145
Sts 60	-	135
Paranthropus boisei/robustus		
KNM-ER 406	44	-
OH 5	45	146
SK 47	45	-
SK 1585	-	141
Homo erectus		
KNM-ER 3733	48	165
KNM-ER 3883	55	150
OH 9	50	-
Solo Skull XI	50	160
Individual Fossils		
KNM-ER 407	49	-
" " 732	45	-
" " 1813	46	135
" " 1805	46	158
OH 24	54	148
SK 847	50	-
TM 1517	42	-

Data for petrous angulation taken from Dean & Wood (1981). Values (to within 5°) for cranial base angles estimated from midsagittal sections of endocasts and from original fossil hominid specimens.

on the other (Dean & Wood, 1981a, 1982). The apparent independence of petrous angulation and cranial base flexion in the sagittal plane in this sample of primates and hominids may result from differences in remodelling and mechanisms of growth that are known to occur between primates in the region of the sella turcica and the spheno-occipital synchondrosis (Latham, 1972; Sirianni & Van Ness, 1978; Gould, 1977). Although on the basis of the data and discussion presented here it is impossible to exclude the fact that similarities in the cranial base of *Homo* and *Paranthropus* may have occurred as a result of parallel evolution, it seems less likely that these similarities are the result of adaptations to posture or to changes occurring in the facial skeleton, and more likely that they result from changes occurring in the brain. It would, then, be wise not to dismiss apparently shared derived characters in the cranial base of *Homo* and *Paranthropus* as parallelisms until there is good corroborative evidence for this.

THE DEVELOPING DENTITION

The taxonomic significance of eruption patterns in primates has always, quite rightly, been viewed with caution, and eruption sequences have by and large been regarded as a less reliable indicator of affinity, than has the sequence of calcification of tooth germs. This is largely due to differences that exist between alveolar and gingival eruption sequences and the difficulty of assessing either in fossil material or dried skulls.

Broom & Robinson (1951, 1952) claimed that the eruption pattern of the permanent teeth in *Paranthropus* was almost identical to that of modern man, and different to that of *Australopithecus*. However, Garn & Lewis (1963) dismiss this claim as a "guess at the likely order of eruption of still unerupted teeth" but more recently discovered early fossil hominid specimens, of equivalent dental developmental age, tend to support the earlier observations of Broom & Robinson (Dean, 1985b).

Arguments about gingival eruption sequences can be misleading. For example data from Hurme (1949) indicates that M_2 and P_4 erupt within 0.71 years of each other in *Homo sapiens* and data from Willoughby (1978) and Nissen & Riesen (1964) indicate that in *Gorilla* and *Pan* M_2 and P_4 erupt on average 0.56 years and 0.6 years apart respectively (see also Dean, 1985b). Seen in the context of the total growth period these differences are small and it is well known that in *Pan paniscus* the more usual sequence is actually P_3,P_4,M_2 although the M_2,P_3,P_4 sequence of the other great apes can occur (Remane, 1960; Fenart & Deblock, 1973). Clements & Zuckerman (1953) have reported the P_3,P_4,M_2 sequence in *Pan troglodytes*, so underlining both the close concurrence and the variability of premolar and second permanent molar eruption within the growth period of *Homo sapiens* and the great apes. However, what stands out in the documentation of the eruption sequence given by Broom & Robinson (1951, 1952) for *Paranthropus* is not a shift in events occurring together in the growth period, but a major reorganisation of the eruption sequence of the dentition. Thus, in *Paranthropus* the I_1, M_1, I_2, ($\bar{C}$ and P_3), M_2, P_4, M_3 eruption sequence differs from the sequence seen in *Australopithecus*, the great apes and the majority of old world monkeys ($M_1, I_1, I_2, M_2, P_3, P_4, M_3, \bar{C}$, (Schultz, 1935)) in the earlier eruption of the permanent incisors and canines.

The M_1I_1 sequence of eruption in Homo sapiens is the most usual one, with these events occurring, on average, 0.33 years apart (Hurme, 1949). However, this is commonly reversed to an I_1, M_1 sequence with both events effectively occurring together in the growth period. In the great apes, however, M_1 eruption occurs first at 3.5 years in Gorilla and at 3.26 years in Pan, but I_1 eruption occurs at 5.75 years in Gorilla and 5.74 years in Pan, i.e. some 2.25 and 2.49 years later than M_1 eruption in Gorilla and Pan respectively (Willoughby, 1978; Nissen & Riesen, 1964). It would be exceptional to expect I_1 eruption to advance by nearly 2½ years within the great ape growth period and be coincident with M_1 eruption. It appears then that whereas certain differences in eruption pattern may simply reflect slight variations within the available growth period, others result from a much larger alteration in the timing of events during development. There is good evidence to support the proposals of Broom & Robinson (1951, 1952) from comparisons of the now increased sample of early fossil hominid mandibles of comparable developmental age representing individuals attributed to Australopithecus africanus (Taung, Sts 24), Australopithecus afarensis (LH 2), Paranthropus (SK 61, SK 62, SK 63 and TM 1536) and early Homo (KNM ER 820). Of these eight specimens those attributed to Paranthropus and early Homo show more advanced incisor eruption and calcification status with respect to the first permanent molar than do those attributed to Australopithecus africanus or afarensis. In these taxa, the first permanent molars are in occlusion, but the deciduous incisors are still present and the developing permanent incisor crowns are still low in the alveolus with little or no root formed (LH 2, Sts 24, Taung).

Evidence from published radiographs (Skinner & Sperber, 1981) as well as measurements made on the original fossil hominid specimens (Dean, 1985b), indicates that in all these specimens M_1 calcification has advanced to the stage of root bifurcation formation, or beyond, and in two specimens attributed to Paranthropus (SK 61 and SK 62) incisor eruption appears ahead of M_1 eruption. It seems likely that advanced incisor eruption represents a major change in the eruption pattern of both Homo sapiens and Paranthropus which sets them apart from Australopithecus and the great apes and other primates (Remane, 1960).

Several factors are known to affect the eruption of teeth and before any significance can be attached to this eruption pattern the likely underlying factors should be explored. Within primates there is a trend to delay the eruption of the molars. The two extremes are the insectivores, which erupt all their permanent molar teeth before any deciduous teeth are shed, and the modern human races where exfoliation of the deciduous incisors may occur before the eruption of the permanent first molars (Schultz, 1935; Clements & Zuckerman, 1953). Swindler & Gavan (1962) have shown that when certain calcification stages for first and second permanent molars in Homo sapiens and Macaca mulatta are compared they were found to occur at essentially equivalent times within the growth period. Likewise, first permanent molar eruption in Homo sapiens and the great apes occurs at roughly one third of the age of dental maturity (Moore & Lavelle, 1974; Dean & Wood, 1981b). These observations suggest that it is unlikely to be a change in first permanent molar eruption time that underlies the shift in incisor/molar eruption sequence in Homo or in Paranthropus. Fanning (1962) has demonstrated that early loss of deciduous teeth can under some circumstances promote premature eruption of successional teeth and it is possible that a prolonged growth period in modern man may lead to 'early' loss because of increased wear and tear on deciduous incisors.

The most likely cause of early eruption of permanent anterior teeth in Paranthropus is the differential size of the large first permanent molars and the small canines and incisors (Robinson, 1956; Wood & Stack, 1979). However, this does not explain the eruption pattern in early Homo, where the molars are smaller and the incisors larger than those of Paranthropus. The small size of the deciduous incisors in Paranthropus (Dean, 1985b), and the very large molars, suggest that the larger permanent incisors would be more compatible with the molars and so may underlie their earlier eruption. The fact that the large great ape male canines and small great ape female canines do not apparently erupt at different times, or in a different sequence, and the fact that according to data given by Willoughby (1978), lower canines in Gorilla erupt 2.2 years earlier than upper canines (unlike those of Pan that erupt together (Nissen & Riesen, 1964)) point to a more complicated eruptive control mechanism in hominoids than one that is dictated by size alone. Mechanisms exist whereby teeth of similar size, for example deciduous great ape canines and permanent human canines, can grow quickly in 3 years, or slowly in 12 years (Dean & Wood, 1981b), presumably to meet the requirements and constraints of the available growth period. Schumaker & El Hadary (1962) have also shown that the eruptive movements of teeth are to some extent independent of the rates of dental development and it remains possible that rates of eruption and the control of timing of events within the growth period are regulated by more general hormonal control mechanisms (Swindler, Olshan & Sirianni, 1982) rather than by local dento-alveolar factors. Another influence upon early incisor eruption in Paranthropus may be the torsional forces transmitted across the symphyseal region during mastication. It may in some way be disadvantageous for developing incisor tooth germs to occupy space in this region for a long period of time. It is not impossible that the shared incisor/first permanent molar eruption sequence in Homo and Paranthropus is occurring in parallel as a result of similar or different developmental mechanisms underlying tooth formation but these are not easily explained in the absence of histological evidence, and it remains equally likely that the timing of dental eruption is a shared-derived character of Homo and Paranthropus.

It can be concluded from the foregoing account of the cranial base and developing dentition that there are a group of characters that set Homo and Paranthropus apart from Australopithecus africanus and the great apes. The evidence linking the form of the face with cranial base flexion is not conclusive and neither, it appears, are the differences in cranial base flexion between Australopithecus, Homo and Paranthropus as clear cut as is often implied. It would then be wrong to attribute all the similarities of the cranial base in *Homo* and *Paranthropus* to a flexed cranial base, and equally wrong to imply they could not have been acquired for reasons other than parallel evolution.

SUMMARY

The cranial base and the developing dentition represent only narrow lines of evidence which, when considered in isolation from other cranial and postcranial evidence, provide no more than a hypothetical model of phylogenetic affinity among hominids. Nevertheless, the bulk of the available information about hominid comparative anatomy should be consistent with a hypothetical hominid phylogeny, and it is useful to compare schemes based on small numbers of characters with those which are apparently more broadly based. In this study, the eruption pattern of the incisors and first permanent molars, the sagittal orientation of the petrous temporal bones and the pronounced muscle markings for the longus capitus and levator and tensor palati muscles set the great apes, a range of other extant primates and *Australopithecus* apart from *Homo* and *Paranthropus*. *Homo* and *Paranthropus* share an apparently derived incisor eruption pattern, coronally rotated petrous temporal bones and have both lost the pronounced muscle markings in the upper pharyngeal region. The two cladistic schemes presented in Fig. 1 are compatible with these data. One view of relationships between early hominids holds that *Australopithecus africanus* and *Paranthropus* share derived characters (Johanson & White, 1979; White *et al.*, 1981; Rak, 1983; Kimbel *et al.*, 1984), and that any characters shared between *Paranthropus* and *Homo*, and which differ from the condition present in *Australopithecus africanus*, must have been acquired in parallel (Fig. 1A). Data presented in this paper do not support the proposal that either an increase in cranial base flexion in the sagittal plane, or the retrusion of the facial skeleton, are linked with coronally rotated petrous temporal bones in extant primates or fossil hominids. However, a more likely mechanism for the changes in the cranial base, that is the expansion and downward disposition of the cerebellum in *Paranthropus* and *Homo*, still does not exclude the possibility of parallel evolution.

A second and more parsimonious explanation for the distribution of basicranial characters that also fits in with the distribution of characters related to the developing dentition in hominoids, is one that requires no parallel evolution in the basicranium or developing dentition. In this model (Fig. 1B) *Homo* and *Paranthropus* share derived characters in the basicranium and developing dentition with respect to *Australopithecus africanus* and the great apes. This model (Fig. 1B) requires, however, that any characters (for example from the face (Rak, 1983)) that *Australopithecus africanus* and *Paranthropus* share must have occurred as a result of parallel evolution and in addition implies that any such characters cannot be truly homologous in a phylogenetic sense even if they appear to be functionally homologous (i.e. they would be analogous characters).

One implication of this latter model is that *Homo* and *Paranthropus* share a common ancestor to the exclusion of *Australopithecus*. This ancestor should exhibit the same derived characters that Homo and Paranthropus share together, yet possess none of the derived characters unique to either. The simplest hypothesis from the data presented here would be an ancestral form that has acquired the derived characters of Homo and Paranthropus in the cranial base and developing dentition, but retained primitive characters for the hominid clade in all other respects.

Another important implication of this second model (Fig. 1B) is that it

Figure 1. Two cladograms illustrating either (A) that characters shared between Homo and Paranthropus may have occurred in parallel, or (B) that these characters may have been inherited from a common ancestor distinct from Australopithecus. Solid circles beneath a genus indicate autapomorphies; solid circles at nodes indicate the presence of apomorphies.

A.

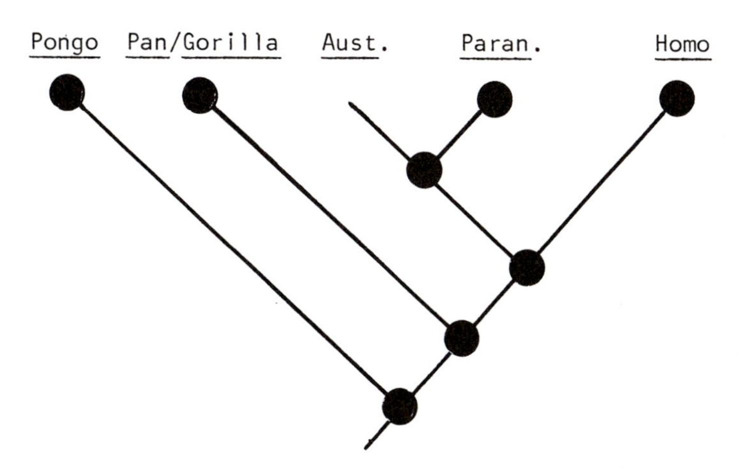

B.

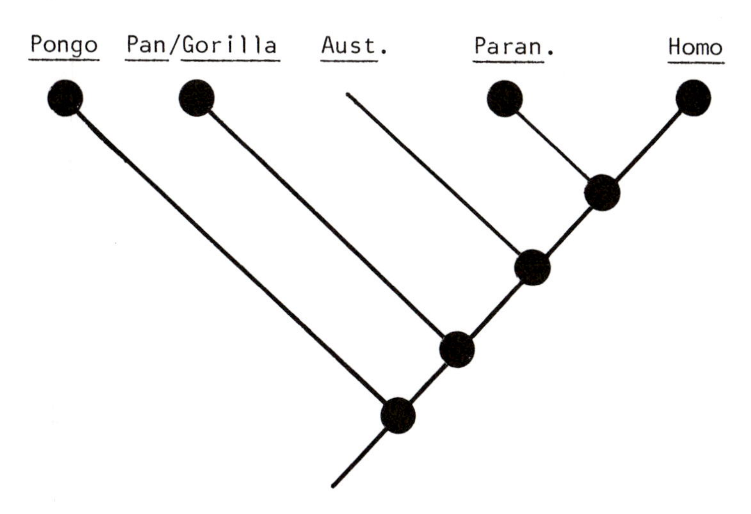

points to a clear distinction between Australopithecus africanus and Paranthropus. Other distinctions between the two (based upon cranial and dental criterion) have been contentious because of the possible effects of scaling and body size (Pilbeam & Gould, 1974, but see Wood & Stack, 1979). However the basicranial and dental developmental characters in this analysis do not appear to demonstrate any variation that can be linked with body size. This, then, must be evidence for recognising Australopithecus and Paranthropus as separate genera.

Possibly the most useful outcome of recent basicranial studies in norma basilaris has been to provide a sound basis for discriminating between hominid specimens whose attribution has been uncertain. Whereas petrous angulation on its own is of little use in distinguishing between Paranthropus and Homo (Dean & Wood, 1981a, 1982) it clearly suggests that specimens previously of uncertain affinity such as OH 24, KNM-ER 407, 732 and 1813, which have been variously attributed to Australopithecus, Paranthropus and Homo (Tobias, 1980, 1981; Holloway, 1976, 1978; Leakey, 1970, 1974; Howell, 1978; Olson, 1978; Leakey & Walker, 1980; Walker, 1981) are unlikely, on the basis of basicranial anatomy, to belong with Australopithecus (Wood & Dean, 1985). Should any of these specimens of 'uncertain affinity' be shown to share the derived characters common to Homo and Paranthropus, but none of the derived characters unique to either, they would then be a good hypothetical common ancestor for both in the scheme shown in Fig. 1B.

Future research may well shed light upon the cranial base morphology and the dental eruption sequence of Australopithecus afarensis and such data can then be incorporated into studies of hominid systematics. It would also be useful to accumulate detailed comparative morphological and ontogenetic evidence for any derived features shared by Australopithecus and Homo. These data would then set debates about the presence of parallelisms between Homo and Paranthropus, and Paranthropus and Australopithecus, in the proper perspective.

ACKNOWLEDGEMENTS

I am grateful to the Governments of Kenya and Tanzania for allowing me access to fossil material in their care, and also to C.K. Brain, R.E.F. Leakey, M.D. Leakey, P.V. Tobias and E.S. Vrba. I would like to thank Bernard Wood, Lawrence Martin and Peter Andrews for their helpful comments on this manuscript and Paula Sutton for typing it. I am also grateful to Dr. Sam Luboga for permission to quote material in preparation.

REFERENCES

Key references

Anderson, D. & Popovich, F. (1983) Relation of cranial base flexure to cranial form and mandibular position. Am. J. Phys. Anthropol., *61*: 181-87.

Björk, A. (1950). Some biological aspects of prognathism and occlusion of the teeth. Acta. Odontol. Scandinav., *9*:1-40.

Dean, M.C. (1985b). The eruption pattern of the permanent incisors and first permanent molars in Australopithecus (Paranthropus)

robustus. Am. J. Phys. Anthropol. (in press).
Dean, M.C. & Wood, B.A. (1982). Basicranial anatomy of Plio-Pleistocene hominids from East and South Africa. Am. J. Phys. Anthropol., *59*:157-74.
Dean, M.C. & Wood, B.A. (1981). Developing pongid dentition and its use in ageing individual crania in comparative cross-sectional growth studies. Folia. Primatol., *36*:111-27.

Literature cited

Aeby, C. (1867). Die schädelformen der menschen und affen. Leipzig.
Anderson, D. & Popovich, F. (1983). Relation of cranial base flexure to cranial form and mandibular position. Am. J. Phys. Anthropol. *61*: 181-87.
Ashton, E.H. (1957). Age changes in the basicranial axis of the Anthropoidea. Proc. Zool. Soc., Lond., *129*:61-74.
Biegert, J. (1963). The evaluation of characteristics of the skull, hands and feet for primate taxonomy. *In*: classification and human evolution. S.L. Washburn, ed. Aldine: Chicago, pp. 116-45.
Björk, A. (1950). Some biological aspects of prognathism and occlusion of the teeth. Acta. Odontol. Scandinav., *9*:1-40.
Björk, A. (1955). Cranial base development. Am. J. Orthod. *41*:198-225.
Björk, A. (1972). The role of genetic and local environmental factors in normal and abnormal morphogenesis. Acta Morphol. Neerl.-Scand. *10*:49-58.
Björk, A. & Kuroda, T. (1968). Congential bilateral hypoplasia of the mandibular condyles associated with congenital bilateral palpebral ptosis. A radiographic analysis of the craniofacial growth by the implant method in one case. Am. J. Orthodont. *54*: 584-600.
Bolk, L. (1910). On the slope of the foramen magnum in primates. Verk. Akad. Wet. Amst. *12*:525-34.
Broom, R. & Robinson, J.T. (1951). Eruption of the permanent teeth in South African Fossil Ape-Men. Nature *167*:443.
Broom, R. & Robinson, J.T. (1952). The Swartkrans Ape-Men Paranthropus crassidens. Transv. Mus. Mem., *6*: 1-123.
Cameron, J. (1930). Craniometric memoirs No. II. The human and comparative anatomy of Cameron's craniofacial axis. J. Anat., *64*: 324-36.
Clark, W.E. Le Gros (1971). The antecedents of man (3rd ed.), pp. 393. Edinburgh University Press.
Clements, E.M.B. & Zuckerman, S. (1953). The order of eruption of the permanent teeth in the Hominoidea. Am. J. Phys. Anthropol. *11*: 313-32.
Cramer, D.L. (1977). Craniofacial morphology of Pan paniscus; A morphometric and evolutionary appraisal. Contributions to primatology 10. S. Karger, Basel.
Dean, M.C. (1982). Ph.D. Thesis University of London. The comparative anatomy of the hominoid cranial base.
Dean, M.C. (1985a). The comparative myology of the hominoid cranial base. Folia primatol. I and II, *43*:234-48 and *44*:40-51.

Dean, M.C. (1985b) The eruption pattern of the permanent incisors and first permanent molars in Paranthropus robustus. Am. J. Phys. Anthropol. (in press).

Dean, M.C. & Wood, B.A. (1981a). Metrical analysis of the basicranium of extant hominoids and Australopithecus. Am. J. Phys. Anth Anthropol. *54*:63-71.

Dean, M.C. & Wood, B.A. (1981b). Developing pongid dentition and its use for ageing individual crania in comparative cross-sectional growth studies. Folia Primatol. *36*:111-27.

Dean, M.C. & Wood, B.A. (1982). Basicranial anatomy of Plio-Pleistocene hominids from East and South Africa. Am. J. Phys. Anthropol. *59*: 157-74.

DuBrul, E.L. (1950). Posture, locomotion and the skull in Lagomorpha. Am. J. Anat. *87*:277-313.

DuBrul, E.L. & Laskin, D.M. (1961). Preadaptive potentialities of the mammalian skull: An experiment in growth and form. Am. J. Anat. *109*: 117-32.

DuBrul, E.L. (1977). Early hominid feeding mechanisms. Am. J. Phys. Anthropol. *47*: 305-20.

Falk, D. & Conroy, E.G. (1983). The cranial venous sinus system in Australopithecus afarensis. Nature. *306*:779-81.

Fenart, R. & Deblock, R. (1973). Pan paniscus et Pan troglodytes - craniometrie - Etude comparative et ontogénique selon les méthodes classiques et vestibulaire. Tome 1. Musée Royale de l-Afrique Centrale - Tervuren, Belgique Annales - Serie IN - 8° Sciences Zoologiques - No. 204.

Fanning, E.A. (1962). Effect of extraction of deciduous molars on the formation and eruption of their successors. Angle Orthod. *32*: 44-53.

Garn, S.M. & Lewis, A.B. (1963). Phylogenetic and intra-specific variations in tooth sequence polymorphism. *In* Dental Anthropology Ed. Brothwell, D.R. (53-77).

Gould, S.J. (1977). Ontogeny and phylogeny. The Belknap press of Harvard University Press, Cambridge, Massachusetts. London, England.

Holloway, R.L. (1972). New Australopithecine endocast, SK 1585 from Swartkrans, South Africa. Am. J. Phys. Anthropol. *37*:173-86.

Holloway, R.L. (1976). Some problems of hominid brain endocast reconstruction, allometry and neural organization. *In:* P.V. Tobias & Y. Coppens (eds.) Les Plus Anciens Hominides, Paris: CNRS, 69-119.

Holloway, R.L. (1978). Problems of brain endocast interpretation and African hominid evolution. *In*: C.J. Jolly (ed.) Early hominids of Africa, Duckworth, London, 379-401.

Howell, F.C. (1978). Hominidae. *In*: V.J. Maglio & H.B.S. Cooke (eds.) Evolution of African Mammals, Cambridge: Harvard University Press, 154-248.

Hurme, V.O. (1949). Ranges of normalcy in the eruption of permanent teeth. J. Dent. Child. *16*: 11-15.

Huxley, T.H. (1863). Evidence as to man's place in nature. London: Williams & Norgate, 159 pp.

Huxley, T.H. (1867). On two widely contrasted forms of the human cranium. J. Anat. Physiol., London, *1*: 60.77.

Johanson, D.C. & White, T.D. (1979). A systematic assessment of early African hominids. Science, *202*: 321-330.

Keith, A. (1910). Description of a new craniometer and of certain age changes in the anthropoid skull. J. Anat. Physiol., London, *44*: 251-70.

Kimbel, W.H. (1984). Variation in the pattern of cranial venous sinuses and hominid phylogeny. Am. J. Phys. Anthropol. *63*: 243-63.

Kinbel, W.H., White, T.D. & Johanson, D.G. (1984). Cranial morphology of Australopithecus afarensis: A comparative study based on a composite reconstruction of the adult skull. Am. J. Phys. Anthropol. *64*: 337-88.

Krantz, G.S. (1963). The functional significance of the mastoid processes in man. Am. J. Phys. Anthropol., *21*: 591-93.

Laitman, J.T. (1977). The ontogenetic and phylogenetic development of the upper respiratory system and basicranium in man. PhD. dissertation, Yale University.

Laitman, J.T., Heimbuch, R.C. & Crelin, E.S. (1978). Developmental change in a basicranial line and its relationship to the upper respiratory system in living primates. Am. J. Anat. *152*: 467-83.

Laitman, J.T., Heimbuch, R.C. & Crelin, E.S. (1979). The basicranium of fossil hominids as an indicator of their upper respiratory systems. Am. J. Phys. Anthropol., *51*: 15-34.

Laitman, J.T. & Heimbuch, R.C. (1982). The Basicranium of Plio-Pleistocene hominids as an indicator of their upper respiratory systems. Am. J. Phys. Anthropol. *59*: 323-43.

Latham, R.A. (1972). The sella point and postnatal growth of the cranial base. Am. J. Orthod., *61*: 156-62.

Leakey, R.E.F. (1970). Fauna and artefacts from a new Plio-Pleistocene locality near Lake Rudolf in Kenya. Nature, *226*: 223-24.

Leakey, R.E.F. (1974). Further evidence of Lower Pleistocene hominids from East Rudolf, North Kenya, 1973. Nature, *248*:653-56.

Leakey, R.E.F. & Walker, A. (1980). On the status of Australopithecus afarensis. Science, *207*: 1103.

Leiberman, P. & Crelin, E.S. (1971). On the speech of Neanderthal man. Linguistic Inquiry, *11*: 203-22.

Moore, W.J. & Spence, T.F. (1969). Age changes in the cranial base of t the rabbit (Oryctolagus cuniculus). Anat. Rec. *165*:355-61.

Moore, W.J. & Lavelle, C.B.L. (1974). Growth of the facial skeleton in the hominoidea. Academic Press. London. New York.

Moss, M.L. (1958). The pathogenesis of artificial cranial deformation. Am. J. Phys. Anthropol., *16*: 269-85.

Moss, M.L. (1963). Morphological variations of the crista galli and medial orbital margin. Am. J. Phys. Anthropol., *21*: 259-64.

Moss, M.L. (1975). The effect of rombencephalic hypoplasia on posterior cranial base elongation in rodents. Arch. Oral. Biol., *20*: 489-92.

Nissen, H.W. & Riesen, A.H. (1964). The eruption of the permanent dentition in the chimpanzee. Am. J. Phys. Anthropol. *22*: 285-94.

Olson, T.R. (1978). Hominid phylogenetics and the existence of Homo in Member 1 of the Swartkrans Formation, South Africa. J. Hum.

Evol., 7: 159-78.

Olson, T.R. (1981). Basicranial morphology of the extant hominoids and Pliocene hominids: the new material from the Hadar Formation, Ethiopia, and its significance in early human evolution and taxonomy. *In*: C.B. Stringer (ed.): Aspects of human evolution. London: Taylor & Francis, 99-128.

Pilbeam, D. & Gould, S.J. (1974). Size and scaling in human evolution. Science *186*: 892-901.

Proffitt, W.R., McGlone, R.E. & Barrett, M.J. (1975). Lip and tongue pressures related to dental arch and oral cavity size in Australian aborigines. J. Dent. Res., *54*: 1161-72.

Rak, Y. (1983). The Australopithecine face. Academic Press. London. New York.

Remane, A. (1960). Zähne & Gebiss *In*: Primatologa, pp. 637-846. Ed. Hofer, H., Schultz, A.H. & Stark, D. Vol. III/2. S. Karger, Basel.

Robinson, J.T. (1956). The dentition of the Australopithecinae. Trans. Mus. Mem. *9*: 1-179.

Schultz, A.H. (1935). Eruption and decay of the permanent teeth in primates. Am. J. Phys. Anthropol. *19*: 489-581.

Schumaker, B.D. & El Hadary, M.S. (1961). Roentgenographic study of eruption. J. Am. Dent. Ass. *61*: 535-41.

Scott, J.H. (1958). The cranial base. Am. J. Phys. Anthropol. *16*: 319-48.

Sirianni, J.E. & Van Ness, A.C. (1978). Postnatal growth of the cranial base in Macaca nemestrina. Am. J. Phys. Anthropol. *49*:329-40.

Skinner, M.F. & Sperber, G.H. (1982). Atlas of radiographs of Early Man. Alan R. Liss, Inc., New York.

Swindler, D.R., Olshan, A.F. & Sirianni, J.E. (1982). Sex differences in permanent mandibular tooth development in Macaca nemestrina. Human Biology, *54*: 45-52.

Swindler, D.R. & Gavan, J.A. (1962). Calcification of the mandibular molars in Rhesus monkeys. Arch. Oral. Biol. *7*: 727-34.

Tobias, P.V. (1967). Olduvai Gorge, Vol. 2., The cranium and maxillary dentition of Australopithecus (Zinjanthropus) boisei. Cambridge University Press, Cambridge.

Tobias, P.V. (1980). The natural history of the helicoidal occlusal plane and its evolution in early Homo. Am. J. Phys. Anthropol., *53*: 173-87.

Tobias, P.V. (1981). A survey and synthesis of the African hominids of the late Tertiary and early Quaternary periods. *In*: L.K. Königgson (ed.) Current Argument on Early Man. Oxford: Pergamon Press, 86-113.

Walker, A. (1981). The Koobi Fora hominids and their bearing on the origins of the genus Homo. *In*: B.A. Sigmon & J.S. Cybulski (eds.) 'Homo erectus: papers in honour of Davidson Black' pp. 193-251, Toronto: University Press.

Weidenreich, F. (1941). The brain and its role in the phylogenetic transformation of the human skull. Trans. Am. Phil. Soc., *31*: 321-442.

Weidenreich, F. (1951). Morphology of solo man, 43: Part 3. Anthropological papers of the American Museum of Natural History New

York: 205-90.

White, T.D., Johanson, D.C. & Kimbel, W.H. (1981). *Australopithecus africanus*: its phyletic position reconsidered. S. Afr. J. Sci., 77: 445-70.

Willoughby, D.P. (1978). All about Gorillas. A.S. Barnes & Co., New Jersey.

Wood, B.A. & Stack, C.G. (1979). Does Allometry explain the difference between 'robust' and 'gracile' Australopithecines? Am. J. Phys. Anthropol.

Wood, B.A. & Dean, M.C. (1985). Patterns of basicranial anatomy in hominid evolution: an exercise in systematic and phylogenetic analysis. Table Ronde: C.N.R.S. Morphigenese du crane de l'Homme: 1-15.

15 THE CREDIBILITY OF HOMO HABILIS

C.B. Stringer,
Dept. of Palaeontology,
British Museum (Natural History),
Cromwell Road,
London SW7 5BD, England.

INTRODUCTION

The creation of new hominid species was a regular occurrence in the earlier part of this century (see e.g. Campbell, 1965), but in the last 20 years such events have become rare, and the subject of heated debate in the case of A. afarensis and H. habilis. At a time when major studies of the material referred to H. habilis from Olduvai and Koobi Fora are in preparation, or in press, it seems a hard task to present a significant review about this species which will not shortly be considered insignificant or redundant, particularly when the author has had virtually no contact with the original specimens concerned. But rather than attempt a dry review of the subject based entirely on the original work of others, I wanted to try to introduce some new data and anlyses. Through the co-operation of colleagues I have managed to rework some of their original data, and I have measured available casts of relevant crania wherever possible. This latter activity gave me my first shock when I realised that some casts were more than 5% smaller than the actual fossils for which I had reliable data, so I decided to rely on cranial angles instead of raw data in comparisons of casts and originals.

This paper will first briefly review the history of the species H. habilis from its inception in 1964, then turn to aspects considered important by other workers viz endocranial size and dental characters, then present new analyses of cranial and facial variation in H. habilis specimens, and finally conclude with comments on the credibility of the species. The material allocated to H. habilis from Olduvai must form an essential part of any discussion of the species, while the material from Koobi Fora has become important to the problem of H. habilis because of its excellent preservation and remarkable morphological variation. However for this exploratory paper I do not intend to discuss in any detail the postcranial material allocated to H. habilis from these sites, nor will I discuss claimed H. habilis material from South Africa (Tobias, 1983) or Ethiopia (Howell, 1978), except where I have been able to study at least a cast of the relevant material (SK 847).

HOMO HABILIS AT OLDUVAI

From 1959, a series of fossil hominid remains was recovered from Bed I and lower Bed II at Olduvai Gorge, Tanzania, which clearly differed from those of the Australopithecus (Zinjanthropus) boisei material identified at the site. In 1964, Leakey, Tobias & Napier named a new species of the genus Homo for these fossils, Homo habilis ("handy man").

Table 1. Homo habilis material (from Leakey, Tobias & Napier, 1964)

	Bed I	Bed II
TYPE	O.H.7: mandible, upper M*, parietals, hand bones of juvenile	
PARATYPES	O.H.4: mandible frag. with M and tooth frags.	O.H.13: mandible, maxillae, cranial parts of adolescent?
	O.H.6: P_3, M^1 or M^2, cranial frags (and tibia and fibula?)**	
	O.H.8: hand bones, foot bones, tooth, clavicle of adult***	
REFERRED		O.H.14: cranial frags of juvenile.
		O.H.16: partial skull and dentition of adult

* later renumbered O.H.45

** tibia and fibula later renumbered O.H.35

*** tooth later renumbered O.H.46, clavicle O.H.48, hand bones not hominid?

The holotype, paratype and referred specimens are listed in Table 1, and the main distinctive features of the species were given as a mean brain size intermediate between that of Australopithecus and H. erectus; maxillae and mandibles in the size range of H. erectus and H. sapiens rather than of Australopithecus; relatively large anterior dentition; a tendency to buccolingual narrowing of the dentition, especially in the lower premolars and molars; external sagittal curvature of the occipital slight as in H. sapiens, but not like that of Australopithecus or H. erectus; various features of the clavicle, hand bones and foot bones resembling H. sapiens sapiens, with some distinctive features such as greater robusticity.

Subsequent research has led to the renumbering of certain of the specimens as probably representing different individuals (Table 1). The tibia and fibula (now numbered O.H.35) were found at the site of O.H.5 ("Zinjanthropus") but Davis (1964) refrained from referring them to the same species and Leakey et al (1964) tentatively associated these bones with those of O.H.6 (H. habilis - Table 1). Despite the fact that the remains were found over 200 metres apart, Susman & Stern (1982) suggested that O.H.7 and O.H.8 and O.H.35 might all represent one subadult individual (age c. 13-14 years).

The erection of this new species for the Olduvai material was not well received on the grounds that while some of the Bed I specimens were not clearly differentiated from *A. africanus*, the Bed II specimens closely resembled known specimens of *H. erectus* (in 1964 the time span of Bed I and lower Bed II, now known to be c. 0.3 myr, was unknown but thought to be as much as 1 myr by some workers). Following direct comparison of the Olduvai and Indonesian fossils (Tobias & von Koenigswald, 1964) Tobias modified his view of the Bed II material, suggesting that some of it might represent the same grade of hominid as some of the "*Pithecanthropus*" material from Java (i.e. *H. erectus*), while O.H.16 was perhaps an australopithecine rather than representing *H. habilis*, to which it had been referred (Table 1). However by 1967, Tobias had apparently reconsidered O.H.13 as an *H. habilis* specimen (Tobias, 1967), and by 1978 O.H.16 was once again considered to represent this species (Tobias, 1978).

A useful summary of early opinions about the status of *H. habilis* can be gathered from the comments following an article by Tobias on *H. habilis* (Tobias 1965). Howell tended to agree with Robinson (1966) that the Bed I material represented an advanced australopithecine, while the Bed II material represented a descendant early form of *H. erectus*. In fact Howell saw enough cranial and dental resemblances between the Bed II and Indonesian fossils to suggest that *H. erectus modjokertensis* or *H. modjokertensis* might eventually prove to be the appropriate name for these fossils (see also Boaz & Howell, 1977; Howell, 1978). Howell disagreed with the view that O.H. 16 might represent an australopithecine, and both he and Napier anticipated Susman & Stern (1982) by noting the excellent match of the O.H.8 foot with the O.H.35 tibia and fibula. However they both cautioned that the lack of comparable australopithecine material precluded the use of any of the Olduvai postcranial specimens in taxonomic arguments about the distinctiveness of H. habilis. Robinson (1966) repeated his view that the Bed I and Bed II materials were more like, respectively, *Australopithecus* and *H. erectus* than they were like each other. This view was echoed by Simons, Pilbeam & Ettel (1969), Brace *et al* (1972) and others.

A number of more recent discoveries from Olduvai have been assigned to *H. habilis* (Tobias, 1978), the most complete being the cranium of O.H.24 from Bed I (Leakey *et al*, 1971). This crushed and distorted specimen was skilfully reconstructed by Clarke to produce a cranium which was considered to be similar to O.H.13 from Bed II, with an initial estimate of endocranial volume of c. 560 ml, later increased to c. 600 ml (Tobias, 1972; Holloway, 1983b). Following suggestions (Anon, 1971) that the specimen did not represent the genus *Homo* as defined by Leakey *et al* (1964) because of its low endocranial capacity and "dished" face, Tobias (1972) defended its inclusion in the genus, while refraining from actually assigning it to *H. habilis*. Tobias' caution in 1972 no doubt reflected the poor preservation of the specimen and the doubts of Leakey *et al* (1971) that, while it was possible that O.H.13 and 24 represented females, and O.H.7 and O.H.16 males of *H. habilis*, occipital differences between the two presumed sexes perhaps indicated further taxonomic variation. However Tobias (1978) later accepted O.H.24 as *H. habilis*, while others remained doubtful (Howell, 1978; Walker, 1981).

CLASSIFICATION OF THE KOOBI FORA HOMINIDS

Work at Koobi Fora, Kenya, from 1968 has provided a large number of specimens of relevance to the status of Homo habilis (Leakey et al, 1978). Initial descriptions of the material, where a basic classification of the first specimens into two genera, viz. Australopithecus and Homo was proposed (Leakey, 1972), were criticised by Robinson (1972) on the grounds that there was nomenclatural confusion between Australopithecus and Paranthropus. He proposed that the South African and East African forms should instead both be divided into two genera, Paranthropus and Homo, and that the gracile forms of both areas probably represented H. africanus. Further discoveries greatly enlarged and complicated the picture with the appearance of large and small-brained "gracile" hominids with apparent affinities variously to H. erectus, H. habilis and A. africanus (Walker, 1976, 1981). Walker (1976) discussed the classification of the large brained fossil KNM-ER 1470 and tested the specimen using indices and other characters which distinguish Australopithecus from Homo. Overall the specimen appeared more like Australopithecus than Homo using these criteria, and while he did not actually classify the specimen as an australopithecine, he certainly urged caution about classifying it as anything else. In a later paper (Walker, 1981) he appeared to prefer classifying specimens such as KNM-ER 1470 with O.H.7 and 16 as Australopithecus habilis, while assigning the small-brained gracile cranium KNM-ER 1813 to a late surviving form of A. africanus along with O.H.13 and 24. The model favoured by Walker appeared to be that A. africanus gave rise to A. habilis through body size increase, and this species was then probably ancestral to H. erectus (represented by specimens such as KNM-ER 3733 and SK 847). Wood provided a longer review of the East African material (Wood, 1978) and also suggested the possibility of a third hominid lineage in East Africa, apart from H. habilis/H. erectus and A. boisei. Although differing from Walker in preferring to classify specimens such as KNM-ER 1470 as Homo, he also regarded the habilis-erectus groups as representing a Homo lineage and noted the distinctiveness of KNM-ER 1813 from this lineage. But rather than regarding this latter specimen as a small-toothed australopithecine, he regarded it as probably distinct, without naming it.

Many other workers have been prepared to place all the non-erectus gracile hominids of Koobi-Fora in one taxon, usually H. habilis, and view variation between specimens such as KNM-ER 1470 and KNM-ER 1813 as related to sexual dimorphism and anagenetic change (e.g. Howell, 1978; Wolpoff,1980; White et al, 1981; Cronin et al, 1981; Tobias, 1983). Opinions vary about the classification of individual specimens such as KNM-ER 1805, regarded as H. habilis by some workers (and possibly abnormal, White et al, 1981) or as H. erectus by others (Howell, 1978; Wolpoff, 1980). Recent work on the cranial base by Dean & Wood (1982) has demonstrated that specimens such as KNM-ER 1470, 1813, and O.H.24 share features in common with KNM-ER 3733 which favour their inclusion in Homo, rather than A. africanus. KNM-ER 1805 is also unlike A. africanus but the evidence of the basicranium is compatible with its inclusion in either Homo or A. boisei.

H. ergaster

Differences between the Olduvai H. habilis specimens and

gracile hominids from Koobi Fora led Groves & Mazák (1975) to reaffirm the specific identity of the Olduvai material, but also to recognise a new species of Homo, H. ergaster. The holotype of this new species was the mandible KNM-ER 992. They accepted that H. habilis could be clearly distinguished from A. africanus (termed by them H. africanus) in features such as lower premolar shape and endocranial capacity, while H. ergaster was probably distinct from both in having smaller molars and premolars, with reduced numbers of premolar roots. H. ergaster was also characterised by a rather thick and massive mandible, and a large endocranial capacity compared to A. africanus. Specimens in the hypodigm included KNM-ER numbers 730, 820 and, less certainly, 1805, while the resemblances between H. ergaster and the "Telanthropus" mandible (SK 15) from Swartkrans were noted. The authors also speculated that earlier specimens from Koobi Fora, such as KNM-ER 1470, may have been ancestral to H. ergaster.

The main problem with this new species was that it was not properly differentiated from specimens assigned to H. erectus and this appears difficult to achieve for the holotype KNM-ER 992. Partly because of this, the new species name has been ignored, or considered premature (Wood, 1978). However the naming of H. ergaster was significant because if a new species were to be created for any of the gracile hominid material from East Africa, and the taxon included the ergaster holotype, this name would be available for the new species. The relevance of this to the classification of specimens such as KNM-ER 1813, which has been linked to the KNM-ER 992 mandible (Leakey et al, 1978), will be disucssed later.

ENDOCRANIAL VOLUME

Since the days of Keith's "cerebral Rubicon", brain size (as reflected by endocranial volume) has loomed large in studies of human evolution. Although absolute brain size is obviously significant in differentiating modern humans from apes, the relationship between brain size and body size becomes more important in assessing fossil hominids (Holloway, 1983b). For this paper I do not intend to use absolute endocranial volume as an important taxonomic character for the recognition of the genus Homo, as I feel that body size variation is still too poorly known in early Homo individuals to allow an appropriate adjustment for relative brain size. Instead I will examine variation in endocranial volume to identify whether one or more groups may be represented in a widely-defined Homo habilis taxon, encompassing specimens such as KNM-ER 1805, 1813 and 1470, as well as O.H.7 and 13.

Table 2 presents data on variation of endocranial volume in living and fossil hominoid samples, determined in most cases by the use of whole endocranial casts. In common with previous estimates, it is evident from these data that a coefficient of variation of about 10% is a reasonable figure for a hominoid species, with the largest variation found in Gorilla (figures as high as 13% are sometimes reported for this ape - Tobias, 1971; Wood, 1976). Turning now to a broadly defined sample of H. habilis crania, Fig. 1 presents determined and estimated values for the sample taken from Holloway (1983b) and my own estimates for incomplete specimens. The values quoted vary greatly in reliability, but the minimum and maximum values are from whole endocast determinations, rated as the most reliable by

Holloway (1983b). Of the other data, the figure for O.H.16 is merely the mean of the lowest (633 ml) and highest (700 ml) values mentioned by Tobias (1971) and Holloway (1973, 1978) respectively, while a relatively large value for O.H.7 is used here (Holloway, 1978, 1983a,b; Vaišnys et al, 1984) rather than Wolpoff's (1981) lower estimates. Values for KNM-ER 1590 and KNM-ER 3732 are my own guesses based on comparisons with other specimens of known endocranial size such as KNM-ER 1813, 1805, 3733, 3883 and O.H.24. These last figures cannot be considered as anything more than approximations but are conservative given the known volume of KNM-ER 1470 and the similarities in size of the comparable parts of the other Koobi Fora specimens. It is certainly possible that the immature KNM-ER 1590 already possessed an endocranial volume greater than that of KNM-ER 1470 (Walker, 1981; Wood, in press).

The coefficients of variation shown in Fig. 1 are uncorrected for the small sample sizes and may therefore be underestimates. Even so the whole sample has a value of 12.4%, larger than that of any hominoid sample in Table 2, suggesting that more than one taxon is being sampled. If exception is taken to the estimated values used, calculation will demonstrate that removing the most suspect values does little, or nothing, to lower the coefficient of variation of the remaining sample, since the extreme specimens KNM-ER 1470 and 1813 remain. While two subsets can easily be created which each have the expected variation of 10% or less, this cannot be achieved while KNM-ER 1470 and 1813 are grouped together, and as Wood (in press) has also indicated, on this basis alone it is unlikely that these two fossils are conspecific unless a measure of special pleading is allowed.

DENTAL DATA

Initial discussions of the species H. habilis laid a strong emphasis on dental metrics (Tobias, 1965, 1966; Robinson, 1966; Brace et al, 1972), while recent work has identified similarities in dental dimensions (plesiomorphies?) between A. afarensis and early Homo samples, while apparently differentiating both groups from the more derived A.

Table 2. Coefficient of variation for living and fossil hominoid endocranial values. Data from Holloway (pers. comm., 1982) and Trinkaus (1983)

TAXON	N	VOLUME (ml)	S.D.	C.V.
G. gorilla	39	522	57	10.9
P. troglodytes	33	395	35	8.9
P. paniscus	40	343	33	9.5
H. sapiens (mod.)	13	1475	149	10.1
H. erectus (Solo)	5	1151	99	8.6
Neanderthal	6	1510	150	9.9

africanus, A. robustus and A. boisei (Johanson & White, 1979; White et al, 1981). Looking at the early Homo samples as a whole it is evident that variation in this group is generally comparable with that found in australopithecine samples (Tobias, 1978; White et al, 1981).

Martin (1983) suggested criteria for establishing the existence of more than one species in a dental sample using characteristics of the range, mean and standard deviation of tooth measurements, compared with one-species samples of living hominoids. The methods would identify an unusually variable sample as probably representing at least two species, although some samples actually consisting of two species might not always be variable enough for this to show from the criteria used. Applying these criteria to the posterior dentitions of the early Homo group of White et al, (1981), only the M^1 (m-d length) and M_3 (b-l breadth) values immediately suggest the presence of more than one species in the sample, although the M^2 (m-d and b-l), M^3 (b-l) and P_3 (b-l) dimensions are also quite variable. Since the White et al (1981) samples appear to contain some individuals otherwise assignable to African H. erectus, the size variation in the sample is perhaps not surprising and indicates that this approach is not

Figure 1. Coefficients of variation for endocranial volumes of early Homo crania. C.V. values are uncorrected for small sample sizes, so are probably underestimates. Most values were obtained from Holloway (pers. comm., 1982, 1983b) but see text for discussion of my estimates for O.H. 16, KNM-ER 1590 and 3732.

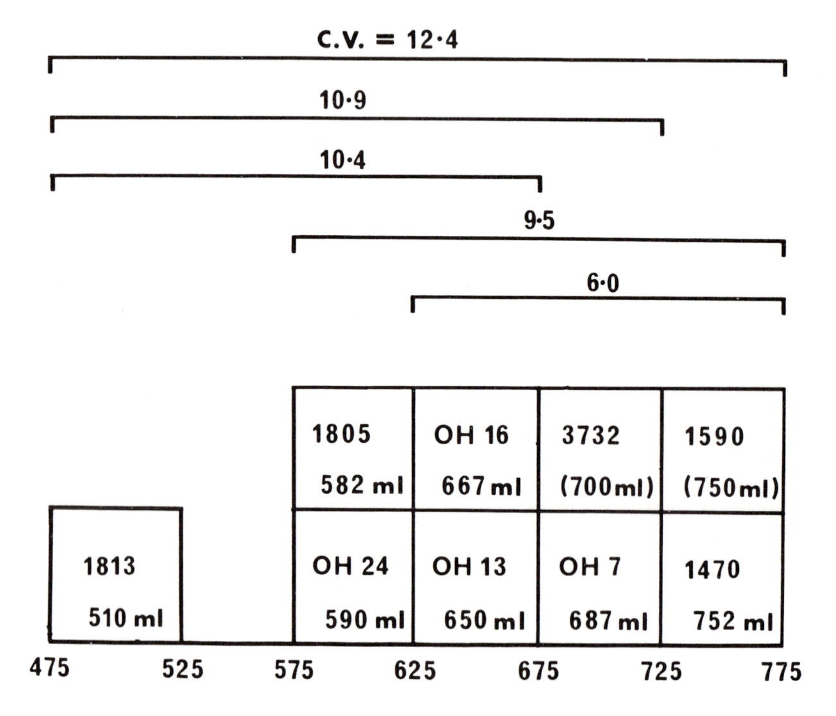

likely to produce significant data to separate possible subgroups in the early Homo sample - if two groups do exist, they are similar in most dental dimensions.

An alternative approach is to examine dental proportions, since these have been used to distinguish H. habilis from australopithecines (Tobias, 1965, 1978) and to distinguish specimens within the early Homo group (Groves & Mazák, 1975). In Table 3 I have presented data on premolars and first molars of living and fossil hominoids. Looked at in terms of primitive and derived characters, it is probable that broad premolars and M^1 (i.e. a small L/B index) are plesiomorphous for hominoids, as is a relatively small P_4/large M_1 (low index value). For M_1, which is distinctively narrow in many of the early Homo specimens (especially O.H.7), the polarity of this character is unclear, and a fairly high value of the index (c.110) may have been the plesiomorphous condition, based on living and fossil hominoid values. If this is so, it is the A. afarensis specimens which seem especially derived in their broad M_1, while the Homo specimens are closer to the assumed plesiomorphous condition.

For the fossil groups considered, the early Homo sample does appear derived in the narrow P^4 and M^1, but relatively primitive in P_4 and P_4/M_1 indices. Of the individual specimens listed, KNM-ER 992 and the Olduvai hominids are most contrasted in P_4 and P_4/M_1 indices, while the rest of the individual specimens are rather similar, save for the (plesiomorphous?) broad P^4 in KNM-ER 1805 and the (apomorphous?) large P_4 in KNM-ER 1802. The similarity of KNM-ER 992 and 1802 in the P_4/M_1 index may look odd to anyone who is familiar with these specimens, placing them both as more derived in the direction of A. boisei than is A. africanus, but examination of the tooth areas suggests that for the former specimen it is the very small M_1 size which is the significant factor, whereas for KNM-ER 1802 the size, shape and crown form of the P_4 does indeed resemble that of some South African australopithecines.

Overall the data of Table 3 confirm the narrowness of P^4 and M^1 as probable synapomorphies for the early Homo specimens, while suggesting that the P_4, M_1 and P_4/M_1 indices overall remain plesiomorphous. It is somewhat ironic that the holotype mandible of H. habilis (O.H.7) should be extreme for the group in its narrow P_4 and M_1, partly explaining Tobias' emphasis on this character (Tobias, 1965, 1966) and Groves & Mazák's (1975) differentiation of H. habilis and H. ergaster. The tooth shape data, then, confirms the presence of relatively narrow premolars and molars in the Homo sample, but does not suggest any clear division of the material beyond the differences already mentioned.

Dental morphology

In a study of the morphology of mandibular molars, Wood & Abbott (1983), and Wood, Abbott & Graham (1983) examined variation in crown areas, cusp areas, crown shape, crown profile and cusp numbers and patterns. They establish criteria which characterised the four main taxonomic groups studied, South African gracile hominids (SAFGRA), South African robust hominids (SAFROB), East African gracile hominids (EAFHOM) and East African robust hominids (EAFROB). They then applied these criteria

Table 3: Indices and areas of premolars and molars in living and fossil hominoids. Data from Weidenreich (1937), Tobias & von Koenigswald (1964), Leakey *et al.* (1971), Groves & Mazák (1975), Day *et al.* (1976), Wolpoff (1979), White *et al.* (1981), Kay (1982), Martin (1983), Wood (pers. comm.), Andrews (pers. comm.). Highest and lowest fossil values are underlined. *Homo* group data modified from White *et al* (1981).

	P_4 L/B	M_1 L/B	P^4 L/B	M^1 L/B	P_4 AREA	M_1 AREA	P_4/M_1 AREA
Miocene Siwalik	87.4	110.4	65.8	87.9	78.9	124.0	0.636
P. pygmaeus	91.2	107.6	74.8	90.8	118.6	152.3	0.778
G. gorilla	86.8	115.9	72.5	94.7	144.5	202.0	0.715
P. troglodytes	86.5	109.3	68.6	88.4	68.5	102.8	0.667
P. paniscus	90.9	111.4	69.7	90.0	53.9	86.2	0.625
A. afarensis	88.4	101.7	74.3	92.5	106.0	163.7	0.647
A. africanus	88.1	105.9	72.7	92.4	119.0	176.2	0.675
A. robustus	88.3	108.3	71.5	90.6	147.4	199.3	0.740
A. boisei	94.7	106.8	74.4	92.5	193.1	255.9	0.755
KNM-ER 992	76.6	110.6			94.4	129.1	0.731
1802	94.2	110.6			136.8	194.1	0.705
O.H. 7	100.0	116.3			114.5	175.9	0.651
O.H. 13	90.8	109.5	75.7	99.2	87.2	147.3	0.592
O.H. 16	88.3	115.7	75.0	99.6	108.8	179.3	0.607
O.H. 24			72.8	96.9			
O.H. 45				103.4			
KNM-ER 1590			74.7	96.4			
1805			67.8	95.1			
1813			79.6	97.6			
Homo group	89.6	114.0	77.0	99.6	103.2	159.2	0.648
Zhoukoudian	91.8	107.2	69.8	89.4	88.2	148.1	0.596
Krapina	84.4	108.7	74.3	98.4	77.5	143.1	0.542

in order to classify a number of "unknown" specimens in the two geographical areas into one of the two local groups (e.g. in South Africa, the Taung M_1 was classified as SAFROB using relative cusp area and fissure pattern, while crown area and additional cusps of the tooth could not clearly allocate it to either the SAFROB and SAFGRA groups. Similarly KNM-ER 1802 was predominantly classified as EAFHOM rather than EAFROB using these criteria).

In order to examine variation in these characters in the early *Homo* group I was allowed access to the original data used in the study and as an experiment I extended the classification criteria used for two members of the EAFHOM group, and KNM-ER 1802, to include SAFGRA as an additional category. The revised classification of these three lower dentitions from the EAFHOM sample is shown in Table 4, excluding the criteria of crown profile, which I did not examine.

It is evident from this table that in tooth area all the specimens considered more closely resemble the South African australopithecine sample than the East African *Homo* sample to which two of them have been allocated (and in fact, for M_2, KNM-ER 1802 and O.H. 16 resemble South African *robust* australopithecines even more closely!). There is less differentiation for the cusp and fissure patterns, while for cusp areas the classification of M_1 and M_2 in KNM-ER 1802 and O.H. 7 is divided between the australopithecine and *Homo* groups. Thus morphological characters suggest variation in the direction of *A. africanus* in some of the larger early *Homo* specimens, but an overlap with *A. africanus* in both morphology and size of teeth was

Table 4: Classification of lower molars of KNM-ER 1802, O.H.7 and O.H. 16 using criteria of Wood & Abbott (1983) and Wood, Abbott & Graham (1983). GRA = South African gracile hominid group, HOM = East African gracile hominid group, NA = not available, - = no discrimination.

		AREA	CUSPS	CUSP AREA	FISSURE
1802	M_1R	GRA	HOM?	HOM	HOM
	M_1L	GRA	HOM?	HOM	HOM
	M_2R	GRA	-	GRA	-
	M_2L	GRA	-	GRA	-
O.H. 7	M_1R	GRA	-	GRA	NA
	M_1L	GRA	-	GRA	NA
	M_2L	GRA	-	HOM	HOM
O.H.16	M_1R	GRA	-	-	GRA
	M_2R	GRA	-	HOM	HOM
	M_3R	GRA	-	HOM	NA
	M_3L	GRA	-	HOM	NA

not excluded by the original describers of *H. habilis*. What is needed next is for this useful work on molar morphology to be extended to *A. afarensis* and modern hominoid samples in order to test resemblances to the other groups analysed and to begin to establish probable plesiomorphous and apomorphous conditions.

MIDSAGITTAL CRANIAL ANGLES

Cranial angles measured between various points in mid-sagittal section have a long history of use (Howells, 1973; Krukoff, 1978) and abuse (Baker, 1974) in anthropology. Here I intend to use them, as I have previously (Stringer, 1978; Stringer & Trinkaus, 1981) to examine variation in crania with the aim of identifying polarities in values of cranial angles between different hominoid species. The values of the cranial angles used here are dependent entirely on the relative position of four osteometric points on the cranium (prosthion, nasion, bregma and basion, Howells, 1973; Stringer, 1978), and as such may be affected by various ontogenetic, functional and allometric considerations. Nevertheless I hope to demonstrate below that a consistent pattern of cranial angles can be recognised in most adult living hominoids regardless of overall size, allowing recognition of plesiomorphous and apomorphous hominoid conditions in cranial angle values, against which the fossil crania of the early *Homo* group can be assessed.

Three midsagittal cranial angles have been calculated for the facial triangle (between nasion, basion and prosthion) consisting of nasion angle (NAA), basion angle (BAA) and prosthion angle (PRA) and for the anterior cranial triangle (between nasion, bregma and basion) consisting of nasion-bregma angle (NBA), bregma angle (BRA) and basion-bregma angle (BBA). Samples of male and female ape crania of *G. gorilla*, *P. pygmaeus*, *P. troglodytes* and *Symphalangus syndactylus* were measured to provide a comparative base, and the calculated cranial angles were compared with those of modern *H. sapiens* (data from Howells, 1973) as in Table 5.

From Table 5 it is evident that two main patterns of cranial angles can be discerned in living hominoids. One is the pattern found in ape crania, characterised by high NAA, low PRA, low NBA, high BRA and low BBA values. The other pattern, found in *H. sapiens* crania, is the opposite of the condition found in the apes in each case, and can therefore be assumed to be the derived condition as against the shared primitive condition of the ape crania. For the purposes of this paper the data for *Pan* will be used to represent the (plesiomorphous) ape condition. Variation in BAA values will not be discussed here since clear polarities cannot be identified.

The mean values for the *Pan* and *Homo* sample are shown diagrammatically in Fig. 2, with the figures scaled to the same basion-nasion length to eliminate size differences in this visual comparison. The contrast between the two is immediately apparent, with *Pan* showing a relatively large and prognathic facial triangle (high NAA, low PRA), and a small and low anterior cranial triangle (low NBA and BBA, high BRA). Values for the fossil sample were compared with those obtained for *Pan* and *H. sapiens*. Original Middle and Upper Pleistocene crania were measured for the analysis, but the Plio-Pleistocene crania could only be studied as casts, and in the case of specimens such as KNM-ER 1470 and O.H. 24 casts of reconstructions were used, which must be taken into account in any conclusions reached. It was also necessary to reconstruct the position of some of the osteometric points in some specimens, and where the osteometric point nasion lay near glabella, the measurement was actually taken 5 mm below glabella, to provide a more comparable data point.

Figure 3 shows the cranial angles calculated for Pan, modern H. sapiens, a Neanderthal sample (Stringer & Trinkaus, 1981) and the Plio-Pleistocene specimens. Where polarity could be clearly determined on the basis of the modern hominoid data (Table 5), an arrow has been added to indicate the derived condition, which is in all cases found in H. sapiens and to a lesser extent, the Neanderthals. Crania KNM-ER 3733 and 3883, assigned here to H. cf. erectus, are generally further from modern human values than the Neanderthals, lying in an approximately intermediate position between H. sapiens and Pan for NAA, PRA, BRA and NBA values. These crania will be used here as a guide to the probable plesiomorphous Homo condition in these angles, against which the other Plio-Pleistocene specimens can be tested.

Figure 3 shows that all three australopithecine crania (Sts 5 - A. africanus; SK 48 - A. robustus (Paranthropus); KNM-ER 406 - A. boisei (Paranthropus)) are close to the mean values for Pan in NAA, PRA and BBA angles and can be considered to be primitive in these values. But where SK 48 and KNM-ER 406 are also primitive in BRA and NBA values, Sts 5

Figure 2. Cranial angles of sample of Pan troglodytes (n = 16) and modern H. sapiens (n = 1652, Howells, 1973), scaled to same basion-nasion length.

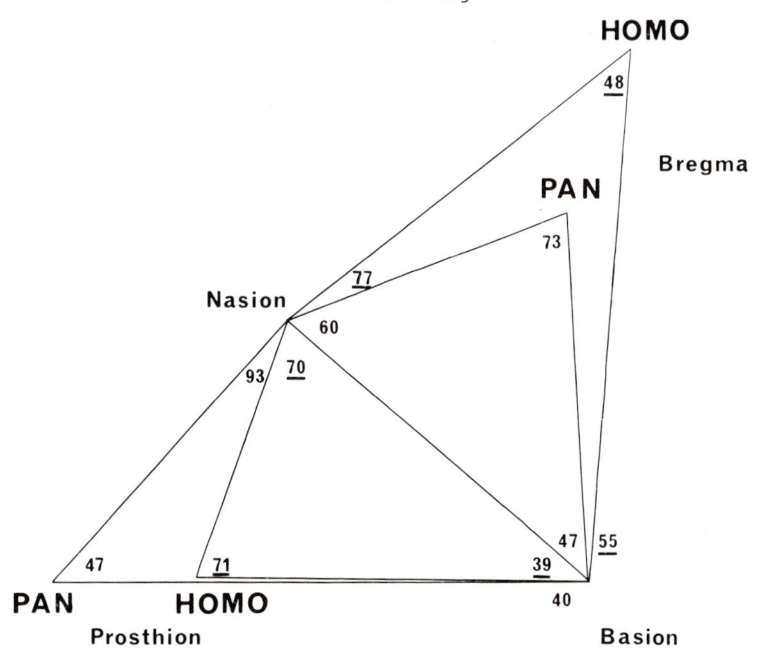

Table 5: Data on cranial angles of modern hominoids, with standard deviations. Recent human data from Howells (1973), with overall standard deviations calculated from means of 34 modern subsets of the data.

Species		sample	NAA		BAA		PRA		NBA		BRA		BBA	
H. sapiens	♂	834	69.5	3.4	39.1	2.5	71.4	3.1	77.3	2.9	47.8	-	54.9	2.5
	♀	818	70.5	3.5	38.4	2.4	71.2	3.3	77.5	2.7	47.3	-	55.2	2.4
combined		1652	70.0	3.5	38.7	2.3	71.3	2.0	77.4	1.8	47.6	1.1	55.0	1.4
P. troglodytes	♂	8	94.3	3.5	40.0	1.6	45.7	2.5	60.1	1.5	73.5	1.9	46.4	1.6
	♀	8	91.9	5.5	39.8	3.0	48.3	3.6	60.1	1.4	73.1	2.0	46.8	1.6
combined		16	93.1	4.6	39.9	2.3	47.0	3.3	60.1	1.4	73.3	1.9	46.6	1.6
G. gorilla	♂	15	104.0	5.1	33.6	3.4	42.3	2.8	53.9	3.0	82.8	4.6	43.2	4.5
	♀	14	92.8	2.9	39.3	2.3	47.8	2.5	55.1	2.7	78.2	4.0	46.6	3.0
combined		29	98.6	6.6	36.3	4.0	44.9	3.8	54.5	2.9	80.5	5.0	44.9	4.2
P. pygmaeus	♂	16	110.3	5.3	33.3	3.8	36.3	2.7	65.6	4.4	70.6	4.1	43.6	2.8
	♀	12	102.7	5.3	33.6	4.4	43.6	1.9	68.4	3.5	68.7	1.9	42.7	3.6
combined		28	107.2	5.8	33.3	3.8	39.3	3.7	66.6	4.2	69.9	3.5	43.3	3.1
S. syndactylus	♂	4	108.2	5.4	24.0	1.8	47.6	3.9	50.0	1.9	82.6	3.5	47.4	3.1
	♀	5	105.5	5.6	24.6	2.2	49.8	4.0	51.1	3.3	77.3	5.6	51.5	3.8
combined		9	106.1	5.1	24.5	1.8	49.1	3.6	50.7	2.6	79.6	5.3	49.7	3.9

appears to share the distinctive derived *Homo* condition for these angles. Examining the data for the early *Homo* specimens, it is evident that KNM-ER 1813 appears most derived in NAA and PRA values, and O.H. 24 in BBA, BRA and NBA values, while KNM-ER 1470 tends to occupy an intermediate position between the other two crania.

Drawing out the values of the three early *Homo* crania together with KNM-ER 3733, scaled to the same basion-nasion length (Fig. 4), shows interesting similarities and contrasts. Now it can be seen that while KNM-ER 1470 and 1813 closely resemble KNM-ER 3733 in cranial angles, O.H.24 is distinctly more primitive (i.e. like *Pan* and *Australopithecus*) in the facial triangle, and quite aberrant (i.e. ultra derived even compared with *H. sapiens*) in the anterior cranial triangle. While the facial triangle data agree with the appearance of the specimen in indicating a high and primitive degree of total facial prognathism, the anterior cranial triangle data are peculiar, giving values for BBA and BRA which are more extreme than any modern human population. This seems an unlikely combination, and suggests that a large amount of distortion remains in the cranial vault of the O.H. 24 reconstruction, as was recognised by Leakey *et al* (1971) and Tobias (1972).

Figure 3. Cranial angles for *Pan* sample, modern *H. sapiens* and fossil hominids. Where polarity can be established, an arrow indicates the inferred derived condition. Neanderthal data from Stringer and Trinkaus (1981). NAA = nasion angle, PRA = prosthion angle, BAA = basion angle, BBA = basion-bregma angle, BRA = bregma angle, NBA = nasion bregma angle.

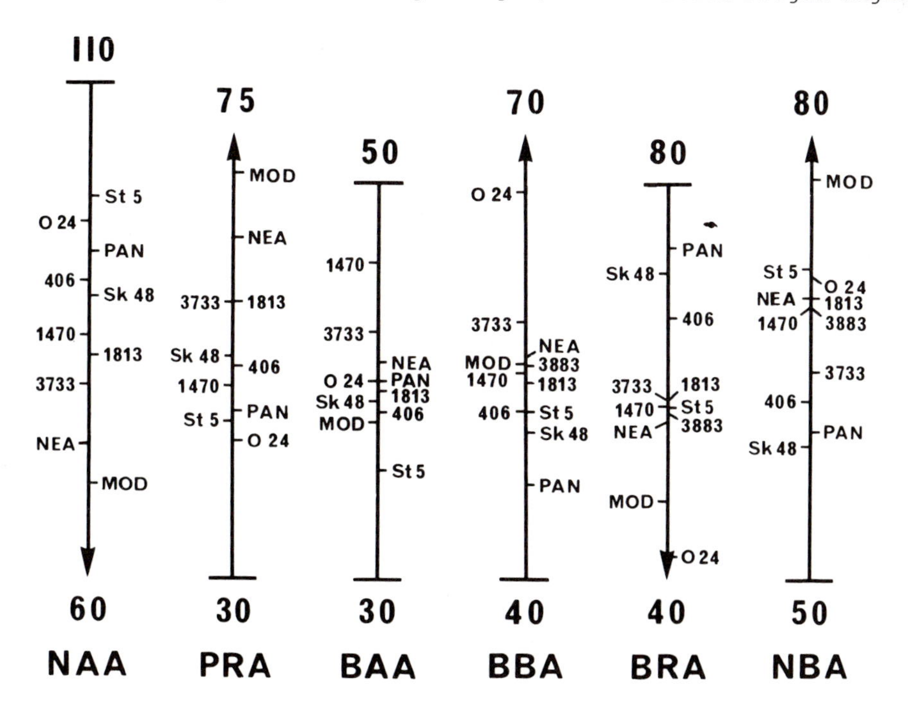

In order to display the overall results more clearly, I have plotted out values for three of the angles whose polarity is most clear in figures 5 and 6, adding data from the Middle Pleistocene crania from Broken Hill, Petralona and Bodo to provide further information on variation in the genus Homo. Figure 5 provides an overall measure of total facial prognathism and shows the plesiomorphous characteristics of the australopithecine crania, together with the KNM-ER 1470 and O.H. 24 reconstructions. KNM-ER 1813 is clearly much more Homo-like in these characters. Figure 6 shows a comparison of the two cranial angles at nasion, and here the spread of specimens reflects total facial prognathism (NAA values) and relative cranial height (NBA values). The ratio of the two angles is also a measure of the position of basion in relation to nasion. The robust australopithecine crania remain within the Pan range, while KNM-ER 3733 now appears more primitive than KNM-ER 1470 and 1813 by virtue of its relatively low cranial height. O.H. 24 and Sts 5 are distinct in their combination of (plesiomorphous) high NAA values and (derived) high NBA values. The significance of the cranial angle data will be discussed further in the concluding remarks.

FACIAL SHAPE

Variation in the facial form of the Plio-Pleistocene hominids has recently received much and deserved attention (Rak, 1983; Kimbel et al 1984). A series of facial, cranial and mandibular characteristics linking A. africanus, A. robustus and A. boisei has been identified and associated with trends in the expansion of the postcanine tooth row, an

Figure 4. Cranial angles of O.H. 24, KNM-ER 1470, 1813, and 3733, scaled to same basion-nasion dimension. Compare with figure 2.

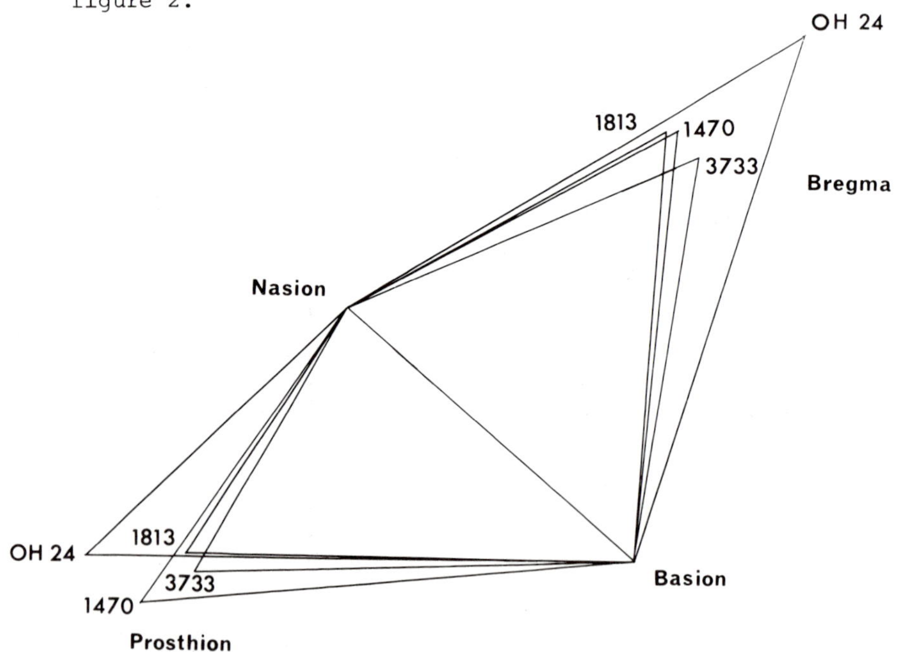

anteriorly positioned infraorbital region and origin of the masseter muscle, and the predominance of the anterior temporalis fibres. Primitive characters linking A. afarensis and early Homo have been noted, lending support to the phylogeny proposed by White et al (1981) which identifies a distinct australopithecine clade for A. africanus, A. robustus and A. boisei, from which A. afarensis is excluded on the basis of its primitive morphology. A. afarensis, not A. africanus, is identified as the probable last common ancestor of the Homo and australopithecine clades.

My own work on the Neanderthal face (e.g. Stringer & Trinkaus, 1981) has been much concerned with defining its unique characters, and in many ways the Neanderthal cranio-facial morphology is the exact antithesis of that described by Rak (1983) for the 'robust' australopithecines. In Neanderthals it is the anterior teeth which are relatively enlarged, while the premolars are small relative to molars (Table 3). The middle of the Neanderthal face is dominated by a large and projecting nasal opening, while the zygomaxillary region is swept back, associated with posterior placement of the masseter origin and the ascending ramus of the mandible. Since this distinctive Neanderthal morphology can be readily measured using simple transverse facial angles, I decided to calculate these same angles for the Plio-Pleistocene sample in order to investigate variation in these characters.

Figure 5. Scatter diagrams of values for prosthion angle (PRA) and nasion angle (NAA). Bars indicate 2 standard deviations for Pan sample (n = 16) and 34 subsets of modern H. sapiens (Howells, 1973).

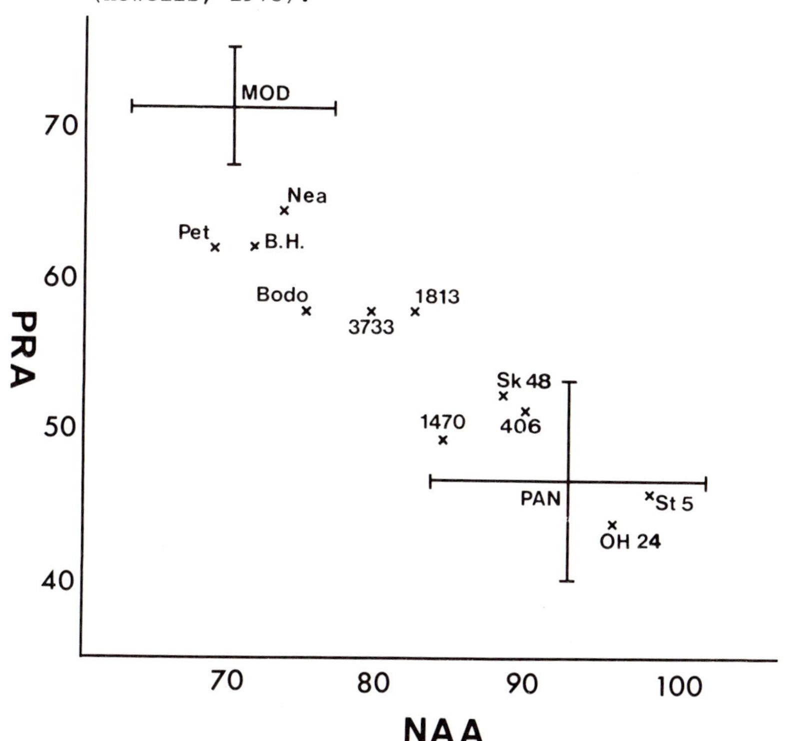

Two transverse facial angles can be used to measure the projection of the base of the nose (subspinale) in front of the zygomaxillare points (subspinale angle, SSA), and the projection of the origin of the nose (nasion) in front of the frontomalare points (nasiofrontal angle, NFA). As before, a sample of modern hominoid crania was measured in order to assess the probable primitive hominoid condition. A position of 'nasion' inferior to glabella was taken where necessary, and where subspinale could not be identified, the central point of a line projected between the lateral nasal borders was used instead (Howells, 1973). The values obtained for samples of modern apes were compared with those for modern H. sapiens (Howells, 1973), as shown in Table 6.

Two somewhat different patterns can be identified. One, found in the apes, is characterised by a low SSA value and relatively high NFA value, while the opposite condition is found in H. sapiens. Thus, while apes are generally characterized by a flattened upper face and projecting lower face, humans have a more projecting upper face and flatter lower face (Neanderthals are exceptional for the latter character). Data for Miocene fossil hominoids are not available, but examination of casts of Sivapithecus and "Ouranopithecus" specimens suggests that values for these specimens were probably similar to those of living apes, which show the assumed primitive pattern.

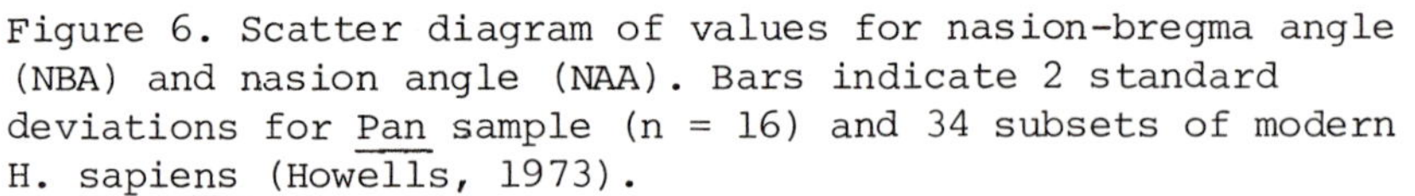
Figure 6. Scatter diagram of values for nasion-bregma angle (NBA) and nasion angle (NAA). Bars indicate 2 standard deviations for Pan sample (n = 16) and 34 subsets of modern H. sapiens (Howells, 1973).

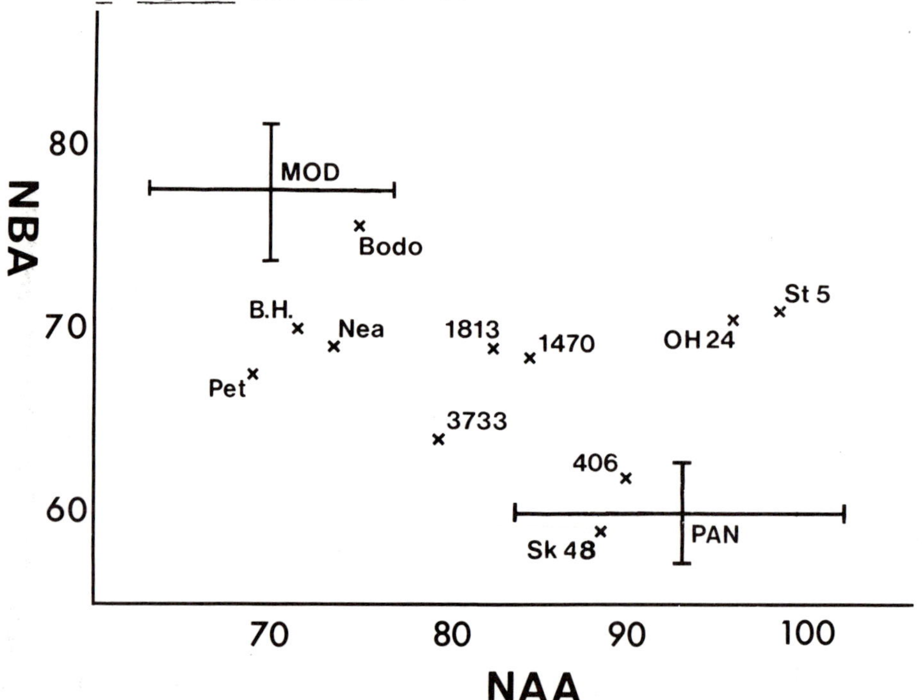

Data for fossil hominids were obtained using casts of Plio-Pleistocene crania and original specimens for Middle and Upper Pleistocene crania. In some cases (e.g. Sts 17 and 71, and SK 847) data had to be obtained by duplicating measurements from the midline, and once again it should be noted that casts of the existing reconstructions of specimens such as O.H. 5, O.H. 24 and KNM-ER 1470 were used. The peculiar and superiorly positioned zygomaxillary regions of some australopithecine specimens also led to measurement difficulties for SSA values.

The values obtained for the fossil hominid specimens are displayed in Fig. 7, together with the means of the *Pan* and *H. sapiens* samples. Two main 'trends' are apparent, and the *Pan* and *H. sapiens* data now appear quite similar to each other in comparison with some of the fossil specimens. A large number of fossil specimens retain the primitively flat upper face, while only KNM-ER 1813 and SK 847 closely resemble *H. sapiens* in showing a lower NFA value. In SSA values, all the living hominoids (including *H. sapiens*) are relatively similar when compared with the high values for this angle found in the 'robust' australopithecines

Table 6: Data on transverse facial angles of modern hominoids, with standard deviations. Recent data from Howells (1973) with overall standard deviations calculated from means of 34 modern subsets of the data

Species	Sample		Subspinale angle SSA		Nasiofrontal angle NFA	
H. sapiens	♂	834	128.8	5.0	142.1	4.3
	♀	818	128.7	5.1	142.8	4.3
combined		1652	128.7	4.4	142.4	2.5
P. troglodytes	♂	8	110.0	6.3	155.8	5.0
	♀	8	114.7	4.1	154.7	4.5
combined		16	112.3	6.8	155.2	4.6
G. gorilla	♂	15	104.9	4.8	159.4	6.5
	♀	14	105.3	3.1	157.0	4.3
combined		29	105.1	4.0	158.2	5.6
P. pygmaeus	♂	16	108.4	5.2	163.6	5.7
	♀	12	117.2	5.7	161.6	3.7
combined		28	112.3	7.0	162.7	4.9
S. syndactylus	♂	4	88.4	3.1	150.3	5.1
	♀	5	87.6	5.3	149.5	4.7
combined		9	88.0	4.2	149.9	4.6

while KNM-ER 1470, and to a lesser extent Sts 71 and O.H. 24, approach the highly derived condition found in the latter group. The KNM-ER 3733 and 3883 crania, usually assigned to H. erectus, seem more primitive than KNM-ER 1813 and SK 847 in NFA values, while the A. africanus specimens show variation in SSA, from values like H. sapiens to those approaching KNM-ER 1470 and the 'robust' australopithecines.

O.H.24 falls in a rather extreme position in Fig. 7, but taking the variation of the Pan and H. sapiens samples into account, it could belong to the same group as KNM-ER 1470 or Sts 5, but is much less likely to belong to a population also containing KNM-ER 1813 and SK 847. The supposedly 'dished' form of the O.H. 24 face has been discussed before by Tobias (1972), but he in fact did not consider the specimen to possess a transversely dished face in the manner of the 'robust' australopithecines. This is certainly true for the lower face, since the SSA value does not approach the 180° values found in 'robust' australopithecines, but the upper face appears extraordinarily flat even compared with australopithecine specimens. Once again some caution must be used in interpreting data such as these, where heavily reconstructed specimens are measured. The results of the analysis of facial angles will be discussed further in the next section.

THE CREDIBILITY OF HOMO HABILIS

The two main arguments against the existence of H. habilis have centred on the supposed lack of "morphological space" between A.

Figure 7. Scatter diagram of values for nasiofrontal angle (NFA) and subspinale angle (SSA). Bars indicate 2 standard deviations for Pan sample (n = 16) and 34 subsets of modern H. sapiens (Howells, 1973).

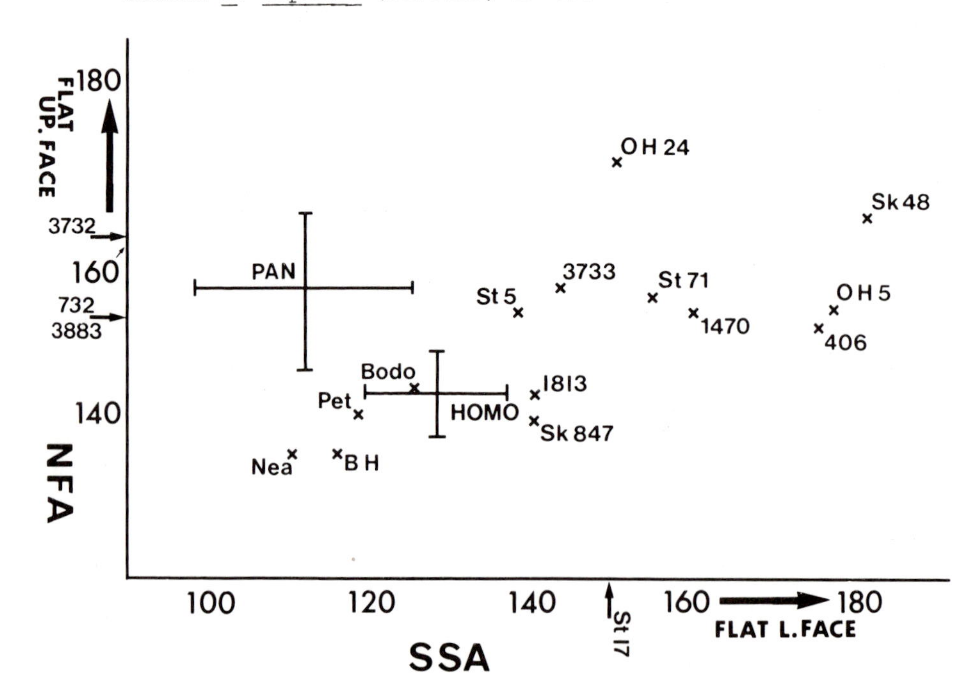

africanus and H. erectus for such a species, and the sheer variation in specimens assigned to the species. Regarding the first argument, which was based on an assumed evolutionary connection between A. africanus and H. erectus, it now appears invalid when viewed against the present uncertainty about the phylogenetic position of A. africanus. Regarding the second argument, I hope the preceding analyses of material from Olduvai and Koobi Fora have shown that while it is difficult to separate the early Homo specimens on dental grounds, the cranial, facial and endocranial data suggest a wide morphological and metrical variation. The problem remaining is whether this variation is sufficient to warrant the recognition of more than one species for the material generally assigned to H. habilis, and, if so, how the individual specimens should be allocated.

As we have seen, the dental distinctions drawn by Leakey et al (1964) do generally distinguish the early Homo sample by its relatively narrow teeth (Table 3), but it is the upper dentition which seems most derived for this character. For the lower dentition, many specimens seem merely primitive in P_4 and M_1 shape, although O.H. 7 does appear exceptional. However, it must be remembered that the early Homo sample of Table 3 is dominated by immature or young adult specimens with little interstitial wear, and hence relatively long teeth. It is fairly easy to match, or even exceed, most of the early Homo shape index values amongst South African australopithecine specimens when unworn specimens are compared, although this is more difficult for the premolar indices of O.H. 7 and KNM-ER 1813, where A. boisei or A. afarensis values are more likely to be comparable.

There is little basis for dividing up the early Homo sample using characters such as those presented in Tables 3 and 4, apart from separating off KNM-ER 992 for its broad and relatively large P_4. The low P_4 shape index for this fossil appears highly derived, most resembling specimens such as O.H. 22, Ternifine and late Pleistocene Homo, but the high P_4/M_1 index is atypical. It has been suggested by several workers that this mandible does not belong in the same group as O.H. 7, but while Groves & Mazák (1975) separated it from all the Olduvai specimens and made it the holotype of H. ergaster, Leakey et al (1978) grouped it with KNM-ER 1813, and by implication, with O.H. 13, O.H. 24 and A. africanus. I believe the cranial and dental resemblances between O.H. 13 and KNM-ER 1813 noted by Walker (1981) certainly indicate that these specimens should be classified together, in which case the O.H. 13 mandible becomes representative of the mandibular morphology expected for KNM-ER 1813. Unfortunately, as Groves & Mazák (1975) note, the O.H. 13 mandible is unlike KNM-ER 992 in several respects, making it more difficult to add this Koobi Fora mandible to the group. An alternative possibility is that KNM-ER 992 belongs with specimens such as KNM-ER 3733 and 3883 (H. cf. erectus), although the high P_4/M_1 index would again be unusual, as already indicated. Additionally, the ascending ramus of KNM-ER 992 looks rather low to match specimens such as KNM-ER 3733, but variation in samples such as Ternifine range from one specimen like the Koobi Fora mandible to another with an ascending ramus too large to match the KNM-ER 3733 cranium. Unfortunately I have to conclude this discussion on KNM-ER 992 by admitting that I cannot decide whether it belongs with the group containing KNM-ER 1813 and O.H. 13, or with H. cf. erectus. However, the characteristics of the recently discovered WT 15000 mandible of H. cf. erectus should be of value in resolving this question.

Regarding the mandible KNM-ER 1802, dental data presented here show it is essentially like other early Homo specimens in tooth shape indices examined, but resembles South African australopithecines in certain aspects of tooth size and morphology, as do specimens such as O.H. 7 and O.H. 16. In addition, according to data presented by Chamberlain & Wood (in press), the cross-sectional area of the mandible below M_1 is very large, falling above values for all A. afarensis mandibles except AL-333-32, all early Homo specimens (unless KNM-ER 1483 is included) and all four A. africanus mandibles measured. However, it falls within the ranges for A. robustus and A. boisei in this character. Additionally it lacks the supposed Homo-like form of the inner contour of the mandible discussed by Robinson (1966) and Tobias (1974), and instead resembles Australopithecus. The premolar crowns are large and rhomboidal, and as Abbott & Wood (in press) have discussed, they have "molarised" roots like those of australopithecines, particularly 'robust' specimens. Finally, even allowing for the subadult/young adult age of the individual, the anterior origin of the ascending ramus and development of an extramolar sulcus, apparent even at the M_2 position, are further indications of australopithecine affinities, using the criteria of White (1977). Putting these data together suggests that this specimen might have belonged to a population with a masticatory emphasis on vertical occlusal forces. This provides an interesting parallel with the anteriorly-placed zygomaxillary region (and inferred masseter origin) of KNM-ER 1470 (see Fig. 7), since these two specimens have generally been assumed to be closely related on other grounds (Leakey et al 1978; Wood, 1978).

The O.H. 7 holotype mandible of H. habilis can be distinguished more readily from australopithecines in some aspects of dental shape, but less so on size or morphology (Tables 3,4). Like KNM-ER 1802 it possesses a relatively narrow P_4, but the premolars of O.H. 7 are absolutely and relatively smaller (Table 3) and differ in morphology and root form. However the O.H. 7 mandible was probably also large in corpus area, and displays an anteriorly placed ascending ramus with a developed extramolar sulcus. However comparisons with KNM-ER 1802 must be tempered by the juvenile status of O.H. 7, which affects not only tooth wear but relative position of the ascending ramus origin. From a comparative perspective, the O.H. 7 ascending ramus certainly originates anteriorly compared with many later Homo specimens at an equivalent developmental stage, but is positioned somewhat more posteriorly than that of MLD 2, a juvenile A. africanus mandible. The parietal bones of O.H. 7 do indicate a large endocranial capacity, probably approaching 700 ml, and I agree with Holloway about similarities with the parietal arch of KNM-ER 1470. Thus despite dental differences between O.H. 7 and KNM-ER 1802 I am inclined to associate them both with KNM-ER 1470 as Homo habilis.

Regarding the crania studied here it must again be stressed that casts of reconstructed specimens were used, and for KNM-ER 1470 this may have particularly influenced measures of facial prognathism, while for O.H. 24

a number of other aspects of overall cranial form must be treated with great caution. Nevertheless the conclusions regarding KNM-ER 1813, and to a lesser extent, KNM-ER 1470 seem clear. They both show derived *Homo*-like features in the reduced NAA of the facial triangle, and in the three angles of the anterior cranial triangle (Figs. 3-6). KNM-ER 1470, however, shows a lower and more primitive PRA value which probably cannot be entirely attributed to reconstruction, and quite distinctive transverse facial angles compared to the *Homo*-like values shown by KNM-ER 1813, none of which appear to be due to reconstruction (Fig. 7). While differences between KNM-ER 1470 and 1813 in NFA values follow the pattern found in dimorphic modern hominoids (male values larger i.e. flatter upper face), the SSA values show the opposite pattern (SSA values in the great apes are generally smaller in males, probably related to the larger size of the anterior dentition). Only if it were argued that Plio-Pleistocene hominids had an atypical pattern of dimorphism in lower facial projection could the differences between KNM-ER 1470 and 1813 be readily attributable to sexual dimorphism, with the former specimen as a male individual. Alternatively, if we accept that variation in endocranial volume between these two specimens rules out the possibility that they are conspecific, the SSA value of KNM-ER 1470 appears derived in the direction of the australopithecine clade, combined with a primitively high NFA value, while KNM-ER 1813 instead appears relatively primitive in its lower SSA value, and derived in the low NFA value in the direction of *Homo*.

It is certainly possible that the O.H. 24 facial triangle values are affected by distortion, as those of the anterior cranial triangle certainly appear to be. But, as reconstructed, O.H. 24 is primitive in measures of facial prognathism (Figs. 3-6) and resembles Sts 5 in this respect, as does its combination of high NAA and NBA values (Fig. 6). In transverse facial angles (Fig. 7) it is characterised by a very large NFA value and a fairly high and derived SSA value similar to values for KNM-ER 3733 and *A. africanus* specimens. While variation between O.H. 24 and KNM-ER 1470 might be containable in a population range similar to the variation shown by living hominoids, the implied pattern of dimorphism, assuming O.H. 24 to be female and KNM-ER 1470 to be male, is again contrasted with that of the great apes (Table 6). Additionally, while on the basis of their cranial angles, it is possible to accept O.H. 24 and KNM-ER 1470, or KNM-ER 1470 and 1813, as belonging to one population, it is much more difficult to accept all three as conspecific. I am inclined, therefore, to initially group O.H. 24 with KNM-ER 1470 rather than with KNM-ER 1813, on the basis of their shared higher SSA values, and their more primitive PRA and NFA values. The alternative would be to classify O.H. 24 as distinct from either, and it is true that in its present state it is the most similar overall to *A. africanus* of the three crania studied here. However, the cranial shape and endocranial volume of the specimen seem to ally it primarily with KNM-ER 1470 rather than with *A. africanus*, so provisionally it can be assigned, like KNM-ER 1470, to *H. habilis*.

From the preceding remarks it is evident that the degree of distinction accorded to the differences between KNM-ER 1470 and KNM-ER 1813 is critical to the delineation of the species *H. habilis*. Figure 8 displays a division of the early *Homo* material into two groups, based on an initial separation of the two specimens on the grounds of endocranial volume. Specimens which can be grouped with KNM-ER 1470 on various grounds, including chronostratigraphic considerations where necessary, are placed in group 1, and those linked with KNM-ER 1813 in group 2. The marked overall similarities between KNM-ER 1813 and O.H. 13 in preserved parts make their classification together one of the most secure starting points for assessing other material. Cranial similarities between KNM-ER 1813 and O.H. 16, for example in frontal bone, supraorbital torus and occipital morphology, can further extend this grouping, giving an indication of the expected level of sexual dimorphism. The problematic KNM-ER 1805 cranium could be grouped here or with *H.* cf. *erectus* on chronostratigraphic grounds, but given the endocranial size and sagittal cresting of the specimen, it seems reasonable to classify it with O.H. 16, which, as reconstructed, has a somewhat larger endocranial volume but temporal lines which also converge markedly posterior to bregma. Apparent facial resemblances between KNM-ER 1805 and 1470, which might instead suggest its classification with the latter specimen must be viewed with caution until new reconstructions of the face of the specimen (Walker, 1981) can be studied.

Regarding body size estimates for the early *Homo* groups, it is evident from similarities in overall cranial dimensions that specimens such as KNM-ER 1470 were likely to be similar to smaller individuals of *H.* cf. *erectus* in body size, and a comparable conclusion can be reached from the size of the sub-KBS postcranial material assigned to early *Homo*. Thus, males of group 1 were probably similar to *H.* cf. *erectus* (and some modern humans) in body size, while specimens such as O.H. 24 (group 1) and KNM-ER 1813 (group 2) were probably smaller-bodied, in the size range of *A. africanus*.

Referring again to Fig. 8 it is evident that the most secure taxonomic assignments are the grouping of KNM-ER 1470 with 1590, 1802 and 3732, and the grouping of KNM-ER 1813 with O.H. 13 and 16. Both subsets can be distinguished from *A. africanus*, with the group 2 specimens in fact being more clearly delineated than the group 1 set which shows more retained primitive characters and a masticatory system apparently paralleling or sharing features with those of australopithecines. Nevertheless derived *Homo* features are also present in group 1 and with an extension of variation to include O.H. 7, the specific designation becomes *H. habilis*. The group 2 subset of KNM-ER 1813, O.H. 13 and O.H. 16 displays more *Homo*-like characters, particularly in an external cranial morphology recalling *H.* cf. *erectus* (Cronin *et al*, 1981; Hublin, 1983). However there is little evidence to support assignment of any of these specimens to the same taxon as Indonesian *H. erectus* (or *H. modjokertensis*, Howell, 1978). There are large differences in cranial robusticity and endocranial size between specimens such as KNM-ER 1813 and O.H. 13, and specimens such as Sangiran 4, and this contrast extends to differences in dental proportions, where the Sangiran material more closely resembles the Zhoukoudian sample and other later Pleistocene hominids.

Thus, *if* the difference in endocranial volume between KNM-ER 1470 and 1813 is a valid criterion for separating them taxonomically, as Wood (in press) indicated beside vertical lines, and new data added to group written above horizontal line (stratigraphic) or below (characters added). Probable primitive characters are bracketed. From this analysis, group 1 is the most plausible assemblage for the species H. habilis. Group 2 cannot certainly be identified with H. ergaster because the characters of KNM-ER 992 are as consistent with its classification as H. cf. erectus (Africa).

GROUP 1

1470 SUB-KBS	1590 3732	1802	Homo habilis O.H.7 (8 & 35?)	O.H.24	1481 3228
	Cranial size and shape similarities to 1470 (High NFA)	Large teeth Large jaws	Cranial size and shape similarities. Large and very narrow teeth. Large mandible? Anterior ascend. ramus? SUB-KBS & OLDUVAI I	Cranial shape similarities. Homo cranial base. High SSA. (High NFA) (Low PRA) High NBA? Abnormally low BRA Abnormally high BBA	Sub KBS Large Body size?
High SSA (High NFA) Low NAA (Low PRA) High BBA High NBA Low BRA Homo cranial base Homo-like brain ECV 750 ml	Large teeth Narrow M^1	Narrow teeth Large premolars Anterior ascend. ramus.	Premolars variable ECV 690-750 P-C resembles Homo?	NAA Variable Dimorphism in tooth size & ECV? ECV 590-750	Body size & post-crania in range of H. erectus?

GROUP 2

1813 POST-KBS?	O.H.13	O.H.16	1805	(Homo ergaster) 992	Postcrania not identified
	Cranial & dental similarities to 1813 POST-KBS? & OLDUVAI II	Similarities in cranial & dental shape to 1813 & O.H.13	Cranial & dental similarities to preceding specimens. Homo cranial base? Cranial and facial angles unknown POST-KBS & OLDUVAI II	Similar mandible size, but less robust. Overall tooth size small, but proportions somewhat distinct. Could also represent H. cf. erectus.	
(Low SSA) Low NFA Low NAA, High PRA High BBA, High NBA Low BRA Homo cranial base Small narrow teeth (ECV 510 ml)	Small premolars Gracile mandible (Posterior ascend. ramus) ECV 510-650	Large teeth & high temporal lines in ♂ ECV 510-670?	Cresting in some crania (brain may show primitive features)	Premolars variable in shape, and in proportion to M_1	Body size in range of A. africanus?

also believes, the group to which KNM-ER 1813 belongs is probably neither Australopithecus, H. habilis, H. erectus nor H. modjokertensis. Allocation of the KNM-ER 992 mandible becomes critical here, but as indicated previously, I can go no further than to suggest that the appropriate species name for KNM-ER 1813 may be H. ergaster, or a new presently unnamed species of the genus Homo. Regarding the relationships of these different groups, it is evident that both group 1 and 2 specimens share synapomorphies with members of the genus Homo, but on the characters assessed here, group 2 specimens appear closer to H. cf. erectus (Africa) and Homo sapiens. Group 1 specimens retain a more primitive cranial morphology in some respects, but also exhibit some characters found in australopithecines which may not be primitive for hominids. Such characters suggest that group 1 specimens were either derived from an ancestor sharing masticatory specialisations with australopithecines (excluding A. afarensis), or had developed such characters by parallel evolution. A gradualistic evolutionary explanation for the different characters shown by specimens such as KNM-ER 1470, O.H. 24 and KNM-ER 1813 is also possible (Cronin et al, 1981), where the apparently earlier specimens retain more characters from an A. africanus-like ancestor, and the supposedly later specimens (such as KNM-ER 1813) show more characters of H. cf. erectus, but the endocranial volume data contradict such a scheme, since the early segment of the sample has large values, approaching those of H. cf. erectus, while the late segment has small values, more like A. africanus. Even the argument that the early segment is dominated by male individuals, and the later segment by females seems insufficient to account for this situation.

SUMMARY AND CONCLUSIONS

Thus, the data presented here provide little support for the classification of KNM-ER 1813 as A. africanus, as long as the small endocranial volume of the specimen is not accorded undue significance. The same conclusion can also be reached using the criteria suggested by Walker (1976) for assessing the classification of KNM-ER 1470. It appears to represent a small-bodied sister species to H. cf. erectus and H. sapiens. As reconstructed, KNM-ER 1470 also shows Homo characters, but is both somewhat more primitive, and somewhat more like A. africanus and Paranthropus in maxillary shape. This is also true of O.H. 24, but this heavily reconstructed specimen is probably still too distorted to yield far-reaching conclusions. Combining KNM-ER 1470, 1802, and O.H. 24 as a group reveals the presence of more specific australopithecine characters than is sometimes recognised for H. habilis, where a large absolute endocranial volume and Homo-like postcranium have usually been emphasised. If, as this paper suggests, there were at least three Plio-Pleistocene species of 'early Homo', we must now consider the possibility that a Pliocene radiation of Australopithecus and Paranthropus was followed by a radiation of Homo-like forms. Given the existence of distinct species of australopithecine grade at opposite ends of the African continent, we must even consider the possibility that Homo-like forms might have evolved in parallel in different areas of the continent. In which case, of course, they should not all be called Homo.

ACKNOWLEDGEMENTS

I would like to thank B. Wood, R. Holloway, P. Andrews, M. Day, A. Chamberlain and G. Wilson for their help or advice; K. Hebb for assistance in collecting comparative data on modern hominoids, and R. Kruszynski and the B.M.(N.H.) photographic studio for preparation of figures.

REFERENCES

Key references

Howell, F.C. (1978). Hominidae. *In* Evolution of African Mammals, ed. V.J. Maglio & H.B.S. Cooke, pp. 154-248. Cambridge: Harvard University Press.

Kimbel, W.H., White, T.D. & Johanson, D.C. (1984). Cranial morphology of Australopithecus afarensis: a comparative study based on a composite reconstruction of the adult skull. Am. J. Phys. Anthrop. *64*, 337-88.

Leakey, L.S.B., Tobias, P.V. & Napier, J.R. (1964). A new species of genus Homo from Olduvai Gorge. Nature, *202*, 7-9.

Tobias, P.V. (1965). New discoveries in Tanganyika: their bearing on hominid evolution. Current Anthropology, *6*, 391-411 (including discussion).

Tobias, P.V. (1978). Position et rôle des australopithécinés dans la phylogenèse humaine, avec étude particulière de Homo habilis et des théories controversées avancées à propos des premiers hominidés fossiles de Hadar et de Laetoli. *In* Les origines humaines et les époques de l'intelligence, E. Boné et al., pp. 38-77. Paris: Masson.

Wood, B.A. (1978). Classification and phylogeny of East African hominids. *In* Recent advances in Primatology. Volume 3. Evolution, ed. D.J. Chivers & K.A. Joysey, pp. 351-72. London: Academic Press.

Main references

Abbott, S.A. & Wood, B.A. (in press). Mandibular premolar root form and evolution in the Hominoidea. J. Anat.

Anon (1971). Confusion over fossil man. Nature, *232*, 294-5.

Baker, J.R. (1974). Race. London: Oxford University Press.

Boaz, N.T. & Howell, F.C. (1977). A gracile hominid cranium from Upper Member G of the Shungura formation, Ethiopia. Am. J. Phys. Anthrop., *46*, 93-108.

Brace, C.L., Mahler, P.E. & Rosen, R.B. (1972). Tooth measurements and the rejection of the taxon "Homo habilis". Yrbk. Phys. Anthrop. *16*, 50-68.

Campbell, B.G. (1965). The nomenclature of the Hominidae. Occ. Pap. Roy. Anthrop. Inst. *22*.

Chamberlain, A.T. & Wood, B.A. (in press). A reappraisal of variation in hominid mandibular corpus dimensions. Am. J. Phys. Anthrop.

Cronin, J.E., Boaz, N.T., Stringer, C.B. & Rak, Y. (1981). Tempo and mode in hominid evolution. Nature, *292*, 113-22.

Davis, P.R. (1964). Hominid fossils from Bed I, Olduvai Gorge, Tanganyika. Nature, *201*, 967-70.

Day, M.H., Leakey, R.E.F., Walker, A.C. & Wood, B.A. (1976). New hominids

from East Turkana, Kenya. Am. J. Phys. Anthrop., *45*, no. 3, 369-436.
Dean, M.C. & Wood, B.A. (1982). Basicranial anatomy of Plio-Pleistocene hominids from East and South Africa. Am. J. Phys. Anthrop., *59*, 157-74.
Groves, C.P. & Mazák, V. (1975). An approach to the taxonomy of the Hominidae: gracile Villafranchian hominids of Africa. Casopis Mineral. Geol., *20*, no. 3, 225-47.
Holloway, R.L. (1973). New endocranial values for the East African early hominids. Nature, *243*, no. 5402, 97-9.
Holloway, R. (1978). Problems of brain endocast interpretation and African hominid evolution. *In* Early hominids of Africa, ed. C.J. Jolly, pp. 379-401. London: Duckworth.
Holloway, R.L. (1983a). The O.H. 7 (Olduvai Gorge, Tanzania) parietal fragments and their reconstruction: a reply to Wolpoff. Am. J. Phys. Anthrop., *60*, 505-16.
Holloway, R.L. (1983b). Human brain evolution: a search for units, models and synthesis. Canadian J. Anthrop. *3*, no. 2, 215-30.
Howells, W.W. (1973). Cranial variation in Man. Papers Peabody Museum, Harvard, *67*, 1-259.
Hublin, J.J. (1983). Les superstructures occipitals chez les prédécesseurs d'Homo erectus en Afrique: quelques remarques sur l'origine du torus occipital transverse. Bull. Mem. Soc. Anthrop. Paris, *10*, no. 13, 303-12.
Johanson, D.C. & White, T.D. (1979). A systematic assessment of early African hominids. Science, *203*, no. 4378, 321-30.
Kay, R.F. (1982). Sivapithecus simonsi, a new species of Miocene hominoid, with comments on the phylogenetic status of the Ramapithecinae. Int. J. Primatol., *3*, no. 2, 113-73.
Krukoff, S. (1978). Structures angulaires constantes au cours de l'evolution du crâne, chez l'homme actuel et fossile, et chez les singes supérieurs. *In* Les origines humaines et les epoques de l'intelligence. E. Boné et al., pp. 117-52, Paris, Masson.
Leakey, R.E.F. (1972). Further evidence of Lower Pleistocene hominids from East Rudolf, North Kenya, 1971. Nature,*237*, 264-9.
Leakey, R.E.F., Leakey, M.G. & Behrensmeyer, A.K. (1978). The hominid catalogue. *In* Koobi Fora research project. Volume 1: The fossil hominids and an introduction to their context 1968-1974, ed. M.G. Leakey & R.E.F. Leakey, pp. 86-90. Oxford: Clarendon Press.
Leakey, M.D., Clarke, R.J. & Leakey, L.S.B. (1971). New hominid skull from Bed I, Olduvai Gorge, Tanzania. Nature, *232*, no. 5309, 308-12.
Martin, L.B. (1983). The relationships of the later Miocene Hominoidea. Ph.D. thesis, University of London.
Rak, Y. (1983). The Australopithecine face. New York: Academic Press.
Robinson, J.T. (1966). Reply to Tobias. Nature, *209*, no. 5027, 957-60.
Robinson, J.T. (1972). The bearing of East Rudolf fossils on early hominid systematics. Nature, *240*, 239-40.
Simons, E.L., Pilbeam, D. & Ettel, P.C. (1969). Controversial taxonomy of fossil hominids. Science, *166*, 258-9.

Stringer, C.B. (1978). Some problems in Middle and Upper Pleistocene hominid relationships. *In* Recent advances in Primatology. Volume 3. Evolution, ed. D.J. Chivers & K.A. Joysey, pp. 395-418. London: Academic Press.

Stringer, C.B. & Trinkaus, E. (1981). The Shanidar Neanderthal crania. *In* Aspects of human evolution, ed. C.B. Stringer, pp. 129-65. London: Taylor & Francis.

Susman, R.L. & Stern, J.T. (1982). Functional morphology of Homo habilis. Science, *217*, 931-34.

Tobias, P.V. (1966). The distinctiveness of Homo habilis. Nature, *209*, no. 5027, 953-57.

Tobias, P.V. (1967). General questions arising from some Lower and Middle Pleistocene hominids of the Olduvai Gorge, Tanzania. S. Afr. J. Sci., *63*, no. 2, 41-48.

Tobias, P.V. (1971). The brain in hominid evolution. New York: Columbia University Press.

Tobias, P.V. (1972). "Dished faces", brain size and early hominids. Nature, *239*, no. 5373, 468-69.

Tobias, P.V. (1974). Does the form of the inner contour of the mandible distinguish between Australopithecus and Homo? *In* Perspectives in Palaeoanthropology: D. Sen Festschrift volume, ed. A.K. Ghosh, pp. 9-17. Calcutta: Mukhopadhyay.

Tobias, P.V. (1983). Hominid evolution in Africa. Canadian J. Anthrop., *3*, no. 2, 163-85.

Tobias, P.V. & von Koenigswald, G.H.R. (1964). A comparison between the Olduvai hominines and those of Java and some implications for hominid phylogeny. Nature, *204*, 515-18.

Trinkaus, E. (1983). The Shanidar Neandertals. New York: Academic Press.

Vaišnys, J.R., Lieberman, D. & Pilbeam, D. (1984). An alternative method of estimating the cranial capacity of Olduvai hominid 7. Am. J. Phys. Anthrop., *65*, no. 1, 71-81.

Walker, A. (1976). Remains attributable to Australopithecus in the East Rudolf succession. *In* Earliest man and environments in the Lake Rudolf basin, ed. Y. Coppens, F.C. Howell, G. Ll. Isaac & R.E.F. Leakey, pp. 484-89. Chicago: University of Chicago Press.

Walker, A.C. (1981). The Koobi Fora hominids and their bearing on the origins of the genus Homo. *In* Homo erectus - Papers in honor of Davison Black, ed. B.A. Sigmon & J.S. Cybulski, pp. 193-215. Toronto: University of Toronto Press.

Weidenreich, F. (1937). Dentition of Sinanthropus pekinensis. Palaeont. Sinica, n.s. D, no. 1, (whole volume).

White, T.D. (1977). The anterior mandibular corpus of early African hominidae: functional significance of shape and size. Ph.D. thesis, University of Michigan.

White, T.D., Johanson, D.C. & Kimbel, W.H. (1981). Australopithecus africanus: its phyletic position reconsidered. S. Afr. J. Sci., *77*, no. 10, 445-70.

Wolpoff, M.H. (1979). The Krapina dental remains. Am. J. Phys. Anthrop., *50* , no. 1, 67-113.

Wolpoff, M.H. (1980). Paleoanthropology. New York. Knopf.

Wolpoff, M.H. (1981). Cranial capacity estimates for Olduvai hominid 7. Am. J. Phys. Anthrop., *56*, 297-304.

Wood, B.A. (1976). The nature and basis of sexual dimorphism in the primate skeleton. J. Zool., *180*, 15-34.

Wood, B.A. (in press). Early Homo in Kenya, and its systematic relationships. *In* Ancestors: the hard evidence, ed. E. Delson. New York: Alan Liss.

Wood, B.A. & Abbott, S.A. (1983). Analysis of the dental morphology of Plio-Pleistocene hominids. I. Mandibular molars: crown area measurements and morphological traits. J. Anat., *137*, no. 1, 197-219.

Wood, B.A., Abbott, S.A. & Graham, S.H. (1983). Analysis of the dental morphology of Plio-Pleistocene hominids. II. Mandibular molars - study of cusp areas, fissure pattern and cross sectional shape of the crown. J. Anat., *137*, no. 2, 287-314.

16 THE NATURE, ORIGIN AND FATE OF HOMO ERECTUS

A. Bilsborough,
Department of Anthropology,
University of Durham,
43 Old Elvet, Durham, DH1 3HN, England.

B.A. Wood,
Department of Anatomy,
The University of Liverpool,
P.O. Box 147, Liverpool L69 3BX.

INTRODUCTION

The hominid fossil record consists of a series of individual specimens which exhibit different kinds and degrees of morphological variation. Much of the effort of palaeoanthropologists is directed towards investigating that variation and evaluating its significance, and this activity is a prerequisite for grouping specimens into phenetic groupings (or 'phena') which may subsequently be formally delineated as taxa. Likewise, assessments of the variability and polarity of morphological features lie at the heart of attempts to construct phylogenetic schemes which represent the possible evolutionary relationships between those phena.

Delineation of phena should not depend primarily upon accidents of discovery and preservation. However, the historical sequence in which fossil specimens are found undoubtedly does influence workers' perceptions of the resultant phenetic groupings, for the temptation is to 'build' phenetic groupings around the better preserved fossils. This, in turn, influences the form of any phyletic arrangement which is based upon the groupings. It is also the case that the resultant phenetic schemes and phylogenies are dependent on the theoretical 'models' which underlie their construction. In this paper we will take a particular case, Homo erectus, and investigate the influence of two different models on two problems associated with that taxon, namely its definition and phylogenetic context. More specifically, we shall compare and contrast the results of using both phenetic and cladistic models for the elucidation of these problems.

The phenetic analysis used in this study is based upon morphometric cranial data. Data such as these simply provide information about similarities, for phenetic data comprise a mixture of patristic, cladistic and functional information (Cain & Harrison, 1960; Martin, 1968). The second approach, cladistic analysis, which by attempting to identify inferred 'derived' (or 'apomorphous') character states, and by weighting these at the expense of inferred 'primitive' (or 'plesiomorphous') characters, should place greater emphasis on cladistic information. We have used these two different techniques (i) to define H. erectus, (ii) to explore its geographical distribution and variation through time and (iii) to investigate how it relates to fossil forms which precede and postdate it.

BACKGROUND AND HYPODIGM

Discoveries during the last quarter century have very considerably increased the 'classic' H. erectus hypodigm from Java (Dubois, 1891) and China (Black, 1934; Weidenreich 1936, 1937, 1943). These more recent discoveries include further Asian remains from the Sangiran dome area of Java (Jacob, 1975; Sartono, 1975, 1976; von Koenigswald, 1975) and Lantian in China (Woo, 1964, 1966) as well as material from the African and European continents. North African finds made three, or more, decades ago include Ternifine (Arambourg, 1955, 1963) and Sidi Abderrahman (Arambourg & Biberson, 1956), and more recently evidence has been recovered from Thomas (Ennouchi, 1969, 1972) and Salé (Jaeger, 1973, 1975). Remains from Southern and Eastern Africa include those from Swartkrans (SK15)(Broom & Robinson, 1949; Robinson, 1953), Olduvai (Leakey, 1961; Rightmire, 1979); Gombore II (Chavaillon & Coppens, 1975) and Koobi Foré (Leakey, 1973; Leakey & Walker, 1976; Walker & Leakey, 1978). European specimens referred to the taxon include the mandible from Heidelberg (or Mauer)(Wust, 1951) and remains from Montmaurin (Vallois, 1956), Petralona (Hemmer, 1972; Stringer et al., 1979); Arago (de Lumley & de Lumley, 1971); Vertesszöllös (Thoma, 1972) and Bilzingsleben (Vlcek, 1978). The fragmentary remains from Ubeidiyah, Israel (Tobias, 1966) have also been cited as evidence of H. erectus in the Levant.

Despite the impressive length of the list, many of the referred specimens are fragmentary and/or of uncertain status. In particular, a strong case can be made that many of the European finds claimed as H. erectus should more properly be assigned to H. sapiens and some workers (e.g. Howells, 1981; Stringer, 1981) have gone so far as to deny all claims for the existence of European H. erectus populations (see below). Thus, H. erectus sensu stricto is still best known from the collections from Sangiran and Trinil in Indonesia and Zhoukoudian (Beijing) in China (Stringer, 1984a; Wood, 1984) and we will use data from these as being characteristic of the taxon and these remains (or a subset of them) will function as the hypodigm for any subsequent comparisons.

The phenetic and cladistic analyses are based on the same samples, but inevitably detailed data are different in the two analyses. Both studies are restricted to cranial remains; the mandible will not be considered.

THE NATURE OF HOMO ERECTUS

Phenetic analysis

This analysis is based on a series of measurements which were designed to reflect regional phenetic variation in the cranium. The cranium was divided into eight regions on the basis of anatomical and functional considerations (Bilsborough, 1976, 1978). The regions are (1) the face, (2) maxilla and upper jaw, (3) mandible, (4) zygomatic arch and temporal area, (5) area of articulation with the mandible, (6) nuchal area, (7) basicranium and (8) cranial vault (Fig. 1). Nearly one hundred measurements were taken, but this study will concentrate on the results of measurements taken on the face (Region 1, n = 20), the maxilla and upper jaw (Region 2, n = 11) and the neurocranium or cranial vault

(Region 8, n = 14) i.e. a total of 45 measurements. The data were compared using simple univariate methods based on standard sample statistics of the raw data, and some indices derived from them. Subsequently samples and specimens were compared using multivariate methods, in particular Q-mode Canonical Variate analysis and Generalized Mahalanobis or D distances derived from a correlation matrix. Details of the programs and the methods may be found elsewhere (Bilsborough, 1984; Bilsborough & Wood, in preparation).

The *H. erectus* specimens included in the study have been divided into 'early' and 'late' sub-groups. The 'early' group comprises the East African remains attributed to, or compared with, *H. erectus*, and the 'late' group comprises the remains from Zhoukoudian (Lower Cave) and Indonesia. The *H. erectus* subgroups were compared with specimens representing the following early hominid taxa, *Australopithecus africanus*, *Australopithecus robustus*, *Australopithecus boisei* and *Homo habilis*, as well as with specimens whose taxonomic attribution is uncertain. The two final comparative series comprise a sample of Neanderthal crania and a series of modern *Homo sapiens* crania. Measurements were taken on the originals unless otherwise indicated, and details of the samples are given below:-

'Early' *Homo erectus*: KNM-ER 3733, KNM-ER 3883, OH9. (Walker & Leakey, 1978; Rightmire, 1979).
'Late' *Homo erectus*: Zhoukoudian (Pekin) crania II, (D), III (E), X (LI), XI (LII), XII (LIII) (casts). The Pekin remains are principally of neurocrania, and within the face as defined (Bilsborough & Wood, in preparation) only the supra- and periorbital regions are relatively intact. Measurements for the middle and lower face are therefore taken from

Figure 1. Diagrams of the skull divided into areas. The areas used in the present phenetic analysis are (1) the face, (2) the maxilla and upper jaw, and (8) the cranial vault, or neurocranium.

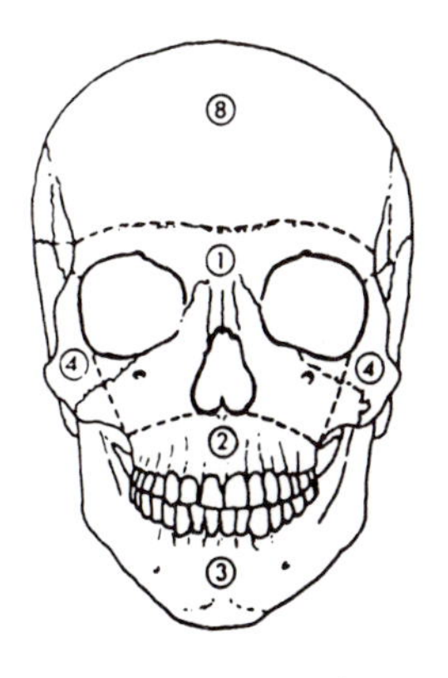

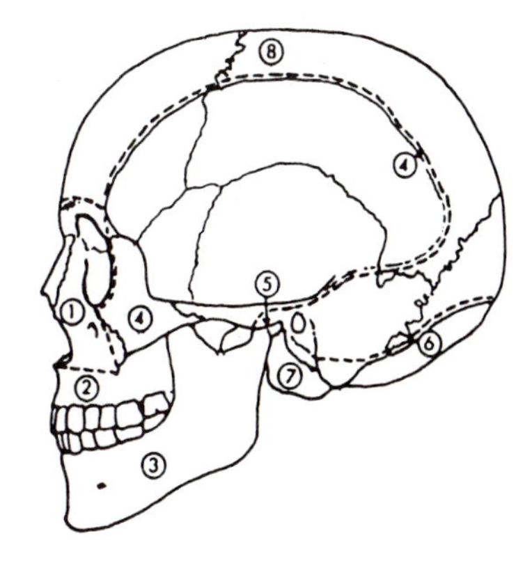

Weidenreich's reconstruction of the skull of a Pekin female which incorporates cranium XI (LII) and facial fragments I-III, together with an additional left maxilla (facial bone V)(Weidenreich, 1943).
Trinil 1 (cast).
Sangiran 2, 4, 10, 12 and 17.
Australopithecus africanus: Sts 5, Sts 17, Sts 52a, Sts 71, TM 1511. (Broom & Schepers, 1946; Broom et al, 1950; Robinson, 1954; Tobias, 1967; Tobias & Wells, 1967).
Australopithecus robustus: SK 12, SK 46, SK 48, SK 52, TM 1517. (Broom & Robinson, 1952; Robinson, 1954; Tobias, 1967; Tobias & Wells, 1967).
Australopithecus boisei: OH5, KNM-ER 406 and 733. (Tobias, 1967; Leakey & Leakey, 1978).
Homo habilis: OH13, OH16, OH24. (Leakey et al, 1964; Leakey et al, 1971).
Hominidae spec. indet: SK 847. (Clarke & Howell, 1972; Clarke, 1977).
KNM-ER 407, 732, 1470, 1805, 1813 and 3732. (Leakey & Leakey, 1978).
European Neanderthals (Homo sapiens neanderthalensis): Neanderthal, Spy I, Spy II, La Ferrassie I, La Quina, Mt. Circeo, La Chapelle-aux-Saints, Le Moustier (casts).
Modern Homo sapiens (H.s.sapiens): 50 crania in the collection of the Department of Physical Anthropology, Cambridge.

Neurocranium. The neurocranium of the H. erectus hypodigm varies in absolute size from being small to moderately capacious (775-1200 cm^3), with the maximum width towards the base, modest cranial height, a retreating postorbitally-constricted frontal, moderately-sized parietals which are flat sagittally, but arched coronally, and an occipital which is expanded in both sagittal and coronal planes and which shows marked sagittal angulation between the longer, nuchal portion of the occipital squama and the shorter upper portion. The available Indonesian crania are smaller than those from Zhoukoudian, but show similar proportions: frontal and parietal sagittal chord arc ratios are high (usually $\geq$ 90%) whereas that of the occipital is low (generally $\leq$ 80%) and the corresponding coronal chord arc ratios for the parietal (85-90%) and occipital (73-85%) indicate marked curvature of the bones in that plane. Of the 'early' H. erectus crania from East Africa, KNM-ER 3733, the most complete, is the smallest, and OH9, the largest. KNM-ER 3733 shows considerable sagittal convexity of the frontal squame, and the occipital is sagittally very long, being exceeded by only one Zhoukoudian specimen. In contrast, the parietals are small: sagittal chord and arc values are less than those of many pre-H. erectus hominids, and are even within the A. africanus range. Coronal development of the parietal is well below that of other H. erectus crania, and comparable to some H. habilis specimens. The biporionic diameter of KNM-ER 3733 is less than that of the Zhoukoudian crania and its total sagittal curvature, and coronal development of the mid-vault, are reminiscent of the Indonesian specimens rather than the Chinese crania.

The frontal of KNM-ER 3883 is comparable in size to that of KNM-ER 3733, but the parietals are bigger and essentially 'square'. However, temporal development is less than that of KNM-ER 3733, so that the total coronal dimensions of the mid-vault are virtually identical. The occipital is sagittally and coronally smaller than that of KNM-ER 3733; the chord in particular, is short, being comparable to some small H. habilis crania, but the arc is longer, and the degree of occipital curvature resembles that of many other H. erectus specimens; indeed, the angulation is greater than in KNM-ER 3733. Occipital breadth is less than that of other possible 'early' H. erectus specimens, but exceeds that of non-erectus hominids, while the total sagittal development of the neurocranium is greater than that of KNM-ER 3733 and the smaller Indonesian specimens, and approaches more closely to that of the smallest Zhoukoudian cranium.

The OH9 neurocranium is significantly bigger than that of the two Koobi Fora specimens, especially in the sagittal plane. This is largely a result of mid-vault expansion, with the sagittal dimensions of the parietals comparable to those of the largest Zhoukoudian crania; coronal development of the mid-vault is not quite as marked, but it is greater than in either of the Koobi Fora or the Indonesian specimens. The frontal, on the other hand, is sagittally no longer than in the Kenyan specimens, and is thus shorter than in any of the Zhoukoudian crania, but it is appreciably broader than in the Koobi Fora examples and indeed, all but the largest of the Zhoukoudian individuals. Sagittal and coronal dimensions of the occipital fall within the range of other African and Asian specimens, although coronally it is towards the upper limit of the Zhoukoudian sample.

Of the non-erectus specimens from East Africa, the Olduvai H. habilis sample (OH7, OH13, OH16, OH24) have small frontals, moderate parietals and unexpanded occipitals, which lie well within the A. africanus range and which are significantly below any of the H. erectus values. The gracile Koobi Fora cranium KNM-ER 1813, included by some within H. habilis, has a sagittally much longer occipital than any of the Olduvai specimens, and it falls within the H. erectus range. However, this is associated with a very short parietal, comparable to that of A. africanus crania, so that total sagittal development of the neurocranium is comparable to that in H. habilis. Moreover, KNM-ER 1813 lacks the biasterionic expansion of the occipital seen in H. erectus.

The more rugged KNM-ER 1805 has a markedly angulated occipital showing similarities with H. erectus, but the sagittal dimensions of the bone are less than in H. erectus specimens, and it lacks the coronal expansion of both parietal and occipital regions which characterise H. erectus. Despite some 'erectus-like' features, largely a consequence of the extensive nuchal area, phenetically KNM-ER 1805 is closer to the H. habilis sample.

The neurocranium of KNM-ER 1470 is relatively complete. It is substantially larger than that of any specimen within the H. habilis sample, showing considerable expansion of the frontal and occipital regions and also in the cranial base, as measured by biporionic diameter. Total mid-cranial expansion is considerable and is associated with a coronally-expanded occipital, so that estimated endocranial volume is comparable to that of the smallest Indonesia H. erectus cranium. However, KNM-ER 1470 fails to reveal a consistent pattern of detailed similarities with H. erectus; frontal size, especially breadth, is smaller and there are major contrasts in the rear of the neurocranium. Compared with KNM-ER 1470, H. erectus specimens have larger occipitals that are somewhat expanded in the sagittal plane, but to a much greater extent in the coronal one. These differences are emphasised if KNM-ER 1470 is compared with the Sangiran 2 cranium. While the occipital sagittal dimensions of KNM-ER 1470 are similar to those of the smaller Sangiran 2 cranium, the biasterionic diameters, both chord and arc, are considerably greater in this, the smallest of the 'later' H. erectus specimens.

In combination, the neurocranial dimensions which distinguish the early hominid taxa are those which reflect the coronal development of the mid-vault, and the sagittal and coronal development of the parietals and the frontal and occipital. They differentiate the 'early' H. erectus crania from Africa from the earlier, or synchronic, non-erectus specimens. The latter form a broad cluster which is relatively close to A. africanus, but not to the 'robust' australopithecines, but which is separate from the two H. erectus samples. The 'early' and 'late' H. erectus groups lie closer to the H. sapiens groups, so much so, that the 'late' specimens are only a little further from modern man than some late Pleistocene groups (i.e. the Neanderthals). The 'early' H. erectus sub-group is only slightly further away.

The 'early' H. erectus neurocranium is thus convincingly separate from that of early Homo, such as Olduvai H. habilis (and perhaps KNM-ER 1813, Wood (in press)) and even from larger, more rugged, specimens such as KNM-ER 1805, whose affinities both in combination, and singly, are with the early Homo specimens. Of the individual East African hominids, only KNM-ER 1470 approaches 'early' H. erectus, and even in this case there is still a significant gap, corresponding to the contrasts, particularly in the rear of the neurocranium, which have been noted above.

Face. The exceptionally good preservation of KNM-ER 3733 permits detailed analysis of its facial proportions, which are otherwise not well known in H. erectus. The only other broadly comparable specimens to preserve this region are Sangiran 17 and the much later Bodo d'Ar (Conroy et al, 1978) crania.

Morphometric data from KNM-ER 3733 indicate that its facial proportions differ strikingly from those of other late Pliocene/basal Pleistocene hominids (Bilsborough & Wood, in preparation). The face is absolutely longer and broader than in many early hominids, but there are also differences of shape and proportions compared with non-erectus faces. Whereas Australopithecus species and KNM-ER 1470 share a pattern of facial breadth in which the mid-face is wider than the upper face, in KNM-ER 3733 the proportions are reversed: bimaxillary diameter is only moderate, whereas supraorbital breadth is expanded. The degree of facial projection also differentiates KNM-ER 3733 from other early hominids; total prognathism is less than that of non-erectus forms, and there are contrasts in the relative contributions of different parts of the face to the total projection. The mid-face of KNM-ER 3733 is flatter with more salient nasal bones than those of other specimens, and the malar process is shallower, with its face sloping postero-inferiorly, rather than orientated vertically or sloping antero-inferiorly. There are, furthermore, differences in the degree of projection of the malar region relative to the remainder of the mid-face. Calibration of the upper border relative to the mid-face shows that the malar process as a whole, and not merely its

Figure 2. The order of groups along the first Canonical Axis (CVI) based on all twenty characters representing the face (left), and on the five facial characters with the greatest scaled loadings (right).

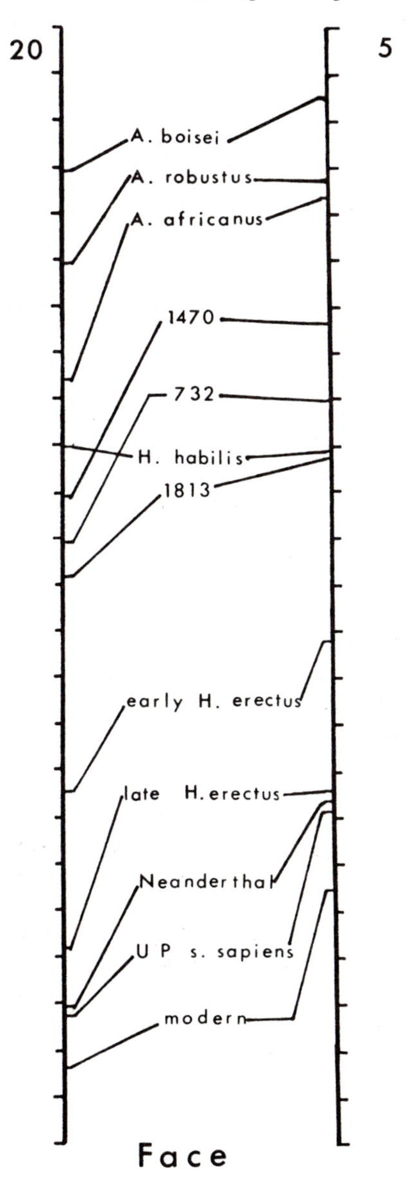

lower border, is retracted relative to the mid-face.

Combining the individual facial dimensions into a multivariate metric serves to emphasise the distinctiveness of the face of KNM-ER 3733. It is well separated from the other basal Pleistocene forms and is also some distance from 'late' *H. erectus* (Fig. 2). The morphological 'isolation' of the 'early' *H. erectus* face is underlined since the other earlier or synchronic hominids are all considerably further away from 'early' *erectus* than are the undoubtedly human and modern human groups of 1.0-1.5 my later. The features of the face of KNM-ER 3733 which are apparently further developed in 'late' *H. erectus* and in more recent hominids are the reduction of prognathism, the flatter mid-facial region, the more salient nasal bones, the shallow, inferiorly-retreating, malar process which is retracted, the lateral rather than the medial development of the supraorbital torus, and the relative breadth of the upper face.

Maxilla and upper jaw. The maxillary and upper jaw region does not so readily separate *H. erectus* from non-*erectus* forms as do the two preceding complexes. Here the major contrasts are generic: the australopithecine palate is long and relatively narrow, with large premolar and molar segments and a deep anterior alveolar margin, whereas the palate of *Homo* is shorter and broader, with reduced cheek teeth and an anterior dental arcade that is proportionately longer. This pattern is present in *H. habilis* and other gracile specimens such as KNM-ER 1813 which have a shallow anterior alveolar margin; other larger specimens such as SK 847 and KNM-ER 1470 have a deeper alveolar margin, but the anterior region of the palate is similarly broad.

Such palatal expansion and associated anterior tooth enlargement is continued in 'early' *H. erectus*. The incisor region of KNM-ER 3733 is especially broad, comparable with that of the absolutely larger Sangiran 4 maxilla, although the cheek teeth are larger and the posterior part of the palate markedly broader in the Indonesian specimen.

The 'late' *H. erectus* Zhoukoudian palate, on the other hand, shows a shallower nasoalveolar clivus and a reduced incisal region. The palate is narrower at the front; the premolar row is somewhat shortened compared with the earlier specimen, but the most marked feature is the greatly reduced molar segment.

Combining the individual dimensions into a multivariate analysis produces a tighter grouping than in the face, but one in which the relative positions are not dissimilar. The 'early' *H. erectus* specimens are again separated from the 'late' sample, which groups more closely with Upper Pleistocene *H. sapiens*. However, the 'early' *H. erectus* palate is at least as distant from other basal Pleistocene hominids as it is from late *H. erectus*, and it is evident that while total diversity in maxillary proportions within the genus *Homo* is less than in the preceding regions, 'early' *erectus*, and non-*erectus* forms of comparable date, are well separated relative to the overall spread.

Conclusions. In a morphometrically-based phenetic study, a taxon is characterized in either a uni-, bi- or multivariate sense i.e. it is given a numerical value with respect to other taxa. A phenetic definition is thus a metrical description of a taxon, and because it does not discriminate between the different origins of the sources of the size and shape components of the differences between it, and other taxa, it corresponds more closely to a 'grade' definition, than to a 'clade' one (Huxley, 1958). Size is clearly a strong influence in a phenetic definition, and Andrews (1984) and others have drawn attention to the influence, for example, of brain size on a 'grade' definition of H. erectus. However, while the phenetic descriptions of the three cranial regions of H. erectus presented in this study undoubtedly reflect size, they also reflect shape, and are thus more information-laden than a simple, crude, size discriminant.

These phenetic studies, therefore, suggest a distinct break between 'early' East African H. erectus and the earlier non-erectus remains. There are also phenetic differences between 'early' African and 'late' Asian H. erectus populations, but these are less in terms of multivariate distance than the differences between, for example, the former and H. habilis sensu stricto, or KNM-ER 1470.

Cladistic analysis

An alternative to a phenetically-based analysis is a cladistic analysis, which is based on the distribution of the 'states' of a series of selected characters. A cladistic analysis seeks to define a taxon in terms of its possession of unique character states, and it then investigates its relationships with other taxa on the basis of these uniquely shared characters. The distinction between phenetic and cladistic analysis is more than one of technique, for morphometric data have a place in cladistic as well as phenetic analysis; rather the difference is in the way such data are treated. Whereas in phenetic analyses metrical data are treated as continuous variables, in cladistic analysis it is the discontinuities that are emphasized, for it is these empirically-determined discontinuities which have provided the boundaries between character states (e.g. Corruccini & McHenry, 1980; Wood & Chamberlain, this volume).

A 'clade' (as opposed to a 'grade') definition and phyletic assessment of H. erectus has been part of several more broadly-based cladistic studies (Eldredge & Tattersall, 1975; Tattersall & Eldredge, 1977; Olson, 1978, 1981), but more recently particular attention has been paid to the cladistic analysis and assessment of H. erectus (Wood, 1984; Stringer, 1984a; Andrews, 1984). Wood (1984) and Stringer (1984a) are in the main consistent in their reference hypodigm for H. erectus sensu stricto. Both include the Trinil, Sangiran, Ngandong and the Zhoukoudian remains; Stringer (1984a) includes OH9 in his reference hypodigm, whereas Wood (1984) does not. In their papers, both Wood (1984) and Stringer (1984a) cull the literature for features which have been cited as 'defining' H. erectus. However, because the bulk of these features are taken from 'grade' orientated definitions, both authors conclude that the list of 'probable' unique, i.e. autapomorphic, features of H. erectus is likely to be much shorter than the list of 'possible' ones. Many features which are 'apparently' autapomorphic when *H. erectus* is compared with *H. sapiens* appear not to be so when *H. erectus* is compared with earlier *Homo* (e.g. *H. habilis*) and australopithecine taxa (Wood, 1984; Andrews, 1984). Stringer (1984a) concludes that of his characters, numbers 1 to 12 are probably autapomorphic. Wood (1984) suggests that autapomorphies are limited to the occipital, except for a few features such as an angular torus and the form of the frontal bones. Andrews (1984) draws on these contributions for his review of *H. erectus*, in which he enumerates seven features as being possibly characteristic, or uniquely derived, with respect to *H. erectus* (Table 1). Thus, the evidence from cladistic studies serves to both emphasize the distinction between *H. erectus* and *H. sapiens* and draw attention to the features which are shared between *H. erectus*, presumed earlier *Homo* populations (e.g. *H. habilis*) and synchronic individual specimens such as KNM-ER 1470 and 1813. Wood (1984) concludes that, despite its smaller brain, it is KNM-ER 1813 which shows more evidence that it may share synapomorphies with *H. erectus* (see below).

THE DISTRIBUTION OF HOMO ERECTUS

An important aspect of the distribution of *H. erectus* is the status of two of the crania in the 'early' *H. erectus* group, that is the crania KNM-ER 3733 and 3883. Rightmire's (1984) definition of *H. erectus* is essentially phenetic, but it is based mainly on meticulous and detailed morphological observation, and only partly on morphometric evidence. He concludes that the differences between KNM-ER 3733 and 3883 and Asian *H. erectus* are relatively minor and concludes that "the evidence suggesting overall similarity of cranial form is much more striking" (Rightmire, 1984:97). In contrast, while Wood (1984) and Stringer (1984a) similarly recognize the similarities between the two African crania and Asian *H. erectus*, their assessment is summarized by the following quotation, that the "Asian and African forms are similar in plesiomorphies rather than apomorphies" (Stringer 1984a:137). Wood (1984) concluded that if "the definition of *H. erectus* is widened to include the observed features of KNM-ER 3733 and 3883...the remaining differences between *H. habilis* and the newly defined *H. erectus* would, with the exception of size, be relatively trivial" (Wood, 1984:107). On this point, the differences between the two present authors (A.B. sides

Table 1: List of possible residual autapomorphic features of *Homo erectus* (after Andrews, 1984)

i. Frontal keel.

ii. Thick cranial vault bones.

iii. Parietal keel.

iv. Angular torus on the parietal.

v. Inion well separated from endinion.

vi. 'Developed' mastoid fissure.

vii. Recess between entoglenoid and tympanic plate.

with Rightmire, 1984) are a clear and simple reflection of the differences in the phenetic and cladistic models.

While these caveats apply to KNM-ER 3733 and 3883, the relationships of OH9 have been interpreted differently from author to author. Wood (1984) drew attention to two areas, frontal bone morphology and cranial thickness, in which the condition in OH9 was closer to that of Asia Homo erectus than is the case for KNM-ER 3733 and 3883. However, Stringer (1984a) and Andrews (1984) make no distinction between the three East African crania. Although Rightmire (1984) recognises differences between the African crania and Homo erectus from Asia, he makes no distinction between the more 'robust' OH9, and KNM-ER 3733 and 3883. While one of us (A.B.) is happy to include the two latter crania in H. erectus sensu stricto, B.W. would prefer to see the latter category restricted to specimens which share the characteristic morphology of the Asian forms.

THE ORIGINS OF HOMO ERECTUS

Just as the way of defining H. erectus was affected by the theoretical model which was used, the extent to which observations can be made about the origin, the morphological consistency through time, and the fate of H. erectus is also affected by the choice of model. In this section these problems will be discussed in the different ways that they are perceived by these using the cladistic and phenetic models. For the purposes of this discussion, we shall consider two of the crania belonging to the 'early' African H. erectus sample (KNM-ER 3733 and 3883) to have affinities with, but to be distinguishable from, the 'later' classic Asian H. erectus samples from Java and China.

Despite claims for an early appearance of H. erectus outside Africa and, by implication, the possibility of an extra-African origin for the taxon (see, for example, Stringer 1984b) much recent evidence suggests that the East African erectus specimens are the earliest known. Claims to the contrary appear to be based upon the widely quoted K-Ar date of 1.9 ± 0.4 my from the Puchangan beds at Modjokerto (Jacob & Curtis, 1971) and the persistent suggestion that non-erectus, possibly antecedent, hominids are present in those beds. However, the K-Ar date has a large associated error, and an isolated estimate is of little value; it does not, in any case, directly date any hominid specimens, but is derived from material some 50 m below the site of one fossil find, at a location removed from the majority of the specimens which were found on the main Sangiran dome. More recent studies (summarised by Pope, 1983 and Semah, 1984) suggest a relatively young age for the Javanese deposits with few, if any, of the hominid fossils being in excess of 1.0 my and thus considerably younger than the earliest African specimens.

Tobias and von Koenigswald (1964) drew attention to similarities between the Olduvai H. habilis material and some of the Sangiran specimens, and impressions of diversity within the Javanese material have been further fostered by the complex taxonomic and phylogenetic schemes proposed by some workers (e.g. Jacob, 1981; Sartono, 1982) including the suggestion that early (i.e. pre-erectus) Homo is present in Java. However, the resemblances between Javanese and East African specimens are in dental and gnathic features which exhibit considerable overlap between early hominid taxa. Cranial remains are well represented in the Javanese sample, but all show typically *erectus* features and there is no evidence of the lightly constructed neurocrania characteristic of the African early *Homo* specimens.

Neither chronological nor morphological evidence, therefore, demonstrates the existence of early, pre-*erectus* hominids in S.E. Asia. By contrast, the African evidence provides not only the oldest specimens of *erectus sensu lato*, but also other, earlier and morphologically distinctive hominids. In these circumstances we consider it likely that the origin of *erectus* was an African one, and that the most appropriate procedure is to search for possible ancestral morphologies among the array of pre-*erectus* fossils from that continent. In attempting to reconstruct phylogenetic history using phenetic evidence we have moved from the first to the second of the three levels of palaeontological analysis identified by Tattersall & Eldredge (1977). In doing so, we accept that chronological evidence now becomes relevant to such an exercise, and our approach is in broad agreement with the stratophenetic method (Gingerich & Schoeninger, 1978; Gingerich, 1979).

Both relative and absolute chronological placement, as well as the pattern of similarity in individual cranial characters, suggest that some of the earlier *H. habilis*/early *Homo* specimens (e.g. KNM-ER 1470, ER 1590, ER 3732 and H7) possibly document the initial phases of the development of an *erectus* grade. Morphometric comparison of the diversity with that evident in the later, better documented, phases of human evolution indicates that no spectacularly high rates of morphological change are required to derive *erectus* from such antecedent populations (Bilsborough, 1984). However, although definitive resolution is not possible, the available chronometric evidence also suggests that 'early' *H. erectus* was at least partly synchronic with some other early *Homo* specimens such as OH13 and OH16 (Olduvai Lower/Middle Bed II) and KNM-ER 1805 (Upper Member, Koobi Fora Formation)(the stratigraphic position of KNM-ER 1813 is at present unresolved, but it is likely to antedate the 'early' *H. erectus* crania from Koobi Fora).

In our view it is unlikely that such cranial diversity can be accommodated within a single species: no preceding or subsequent hominid taxon exhibits such a range, and to do so would imply a population range even greater than that of the available specimens (see also Wood, 1978; Howell, 1978; Walker & Leakey, 1978). The consequence of such an interpretation is that the appearance of the *erectus* morphotype does not reflect the simple anagenetic transformation of successive populations within a single lineage, but that later *H. habilis* and 'early' *H. erectus* represent not only different grades, but also distinct clades.

The basis of such a speciation event is unknown, but it presumably involves ecological contrasts between the two taxa. The morphometric distinctiveness of the 'early' *H. erectus* face from that of other Pliocene/basal Pleistocene hominids is impressive (Fig. 2), as is its apparent stability over much of the Lower/Middle Pleistocene - the very limited fossil evidence (i.e. KNM-ER 3733, Sangiran 17 and possibly Bodo) suggests that there was relatively little change through time. In this respect it is especially vexatious that facial remains are so rare; more material is needed to substantiate these contrasts but, if confirmed, they indicate structural and biomechanical differences, possibly associated with food processing, between basal erectus and other early hominid populations. The detailed differences in facial morphology therefore reinforce the conclusion that, quite apart from the 'robust' australopithecines, there are at least two distinct clades of early Pleistocene hominids (Bilsborough, 1983, 1984; Wood, 1978, in press).

Cladistic analysis provides a different approach to the problem of the origin of Homo erectus. Strict followers of the cladistic method would have to demur from discussing ancestor/descendant relationships, claiming that there is simply too little fossil evidence ever to know anything so detailed as the identity of an ancestor or descendant (Wiley, 1979). We do not take such a pessimistic view of phylogenetic hypotheses and, while recognizing that the exercise lacks the rigour which can be injected into the construction of cladograms, we are willing to speculate about the possible antecedants of H. erectus. The basic phenetic conclusions outlined above, i.e. that there are two phena of early Homo, the larger-brained H. habilis e.g. KNM-ER 1470, 1590 and OH7 and the smaller-brained Homo sp. indet. e.g. KNM-ER 1813, OH24, are the starting point of any cladistic assessment. In the former group, only the dominance of the occipital in the occipital-parietal arc ratio shows a resemblance to the possible derived condition seen in H. erectus (Wood, 1984). In the latter group, it is KNM-ER 1813 which shows most evidence of what might be termed incipient H. erectus autapomorphies. These include a weak occipital crest, a shallow post-toral sulcus on the frontal bone, a slight sagittal prominence on the frontal bone and an erectus-like occipital-parietal arc ratio (Wood, 1984). Thus, of the two 'non-erectus' phena within early Homo, it is, paradoxically, the smaller-brained of the two which shows more signs of providing the basis of the derived features which characterise Homo erectus (sensu stricto).

STABILITY VERSUS CHANGE IN HOMO ERECTUS

We have already emphasised the contrasts between two of the 'early' African H. erectus crania, and other specimens assigned to the taxon. Many other workers have analysed diversity within varying subsets of the total erectus sample, and interpreted its significance in terms of polytypism and/or phyletic change (e.g. Howell, 1978; Jelinek, 1978, 1980; Stringer, 1978, 1984a,b; Rightmire, 1979, 1984). In contrast to the above assessment, several workers have seen H. erectus as being relatively homogeneous, and cited it as a phase of stasis in Hominidae, compatible with a punctuated equilibria model of evolutionary change (Gould & Eldredge, 1977; Stanley, 1979, 1982; Eldredge & Tattersall, 1982).

The probable origin of the 'stasis' view is perhaps Leakey & Walker's (1976) comparison of ER 3733 with the much later Zhoukoudian material, but additional support for a punctuationist interpretation was provided especially by Rightmire's unsuccessful attempt to identify trends in H.

erectus (Rightmire, 1981). Rightmire's study is based upon variability in four cranial characters - endocranial capacity, biauricular diameter, M_1 breadth and mandibular corpus robusticity which were cited as providing 'indications of change or stasis in key regions of the skull and dentition' and as having diagnostic and/or biomechanical significance for H. erectus. The H. erectus sample included the 'early' African specimens, as well as some others not commonly included in the taxon. The metrical values for the characters did show fluctuations but they did not reveal any significant trends over time. This led Rightmire to conclude that 'a traditional gradualist view is not supported. Homo erectus was apparently a stable taxon, exhibiting little morphological change throughout most of its long history' (Rightmire, 1981:246).

This seems to us a sweeping conclusion to derive from the analysis of only four cranial characters. Other workers (e.g. Lestrel, 1976; Cronin et al, 1981 and Stringer, 1984b) detect a more definite trend towards endocranial enlargement within erectus; this is also likely to apply to Rightmire's data if small, late and taxonomically contentious specimens such as Salé (regarded by many as archaic H. sapiens) are excluded from the analysis. Rightmire's other variables appear to be examples of genuinely stable characters within H. erectus or ones which, by their definition, necessarily exhibit little or no significant change. Biauricular diameter is highly correlated with basicranial breadth and, indeed, is regarded by Rightmire as a measure of this. The basicranial region is relatively invariant in the Hominidae as a whole, probably because of functional constraints associated with truncal erectness etc. One of the distinguishing features of H. erectus is that the maximal breadth is at, or close to, the cranial base so that it is not unexpected that erectus specimens do not reveal a significant temporal trend in this feature; the characters' distribution has been largely predetermined by assignment of the specimens to H. erectus and such variability as occurs probably reflects fluctuations in overall cranial, or body, size.

First molar breadth is relatively stable over all Hominidae - certainly more so than other cheek tooth dimensions (Bilsborough, 1973, 1976) - and therefore does not provide evidence of exceptional stasis within H. erectus. In fact Rightmire's own data indicate some reduction in all but his earliest sample which does, however, include OH13, a specimen more usually assigned to H. habilis. Rightmire's analysis of mandibular robusticity, by utilising an index, obscures changes in the absolute dimensions upon which the index is based. In evaluating this and the lower molar data, the difficulty of identifying isolated mandibles as H. erectus needs to be borne in mind. In East Africa in particular, the problems of allocation are well recognised. If mandibles of the smaller, less robust H. habilis are erroneously assigned to H. erectus, the parameters of the latter taxon will be biased accordingly. Similar provisos apply to Rightmire's later erectus mandible sample, which includes specimens such as Montmaurin and Arago, considered by many to be H. sapiens (e.g. Stringer, 1981, 1984b). Given such uncertainties, assertions of stasis in H. erectus (e.g. Gould & Eldredge, 1977; Stanley, 1981) appear to be less than firmly grounded; while some regions (e.g.

the face, see above) and individual characters may show little temporal variation, there is no compelling evidence that H. erectus was exceptionally stable compared with other hominid taxa.

Bilsborough (1976, 1978, 1984) has elsewhere expressed the contrary view point, arguing for phyletic change as a prime determinant of the variability observed within H. erectus. While some of his earlier conclusions need to be reassessed in the light of the larger samples and revised chronology now current, analysis of multiple characters points to undoubted, temporally-directed, diversity in H. erectus, especially in neurocranial, palatal, dental and jaw dimensions. Stringer (1984a and b) similarly identifies chronological contrasts in H. erectus on the basis of different characters and, in part, different techniques. It is noteworthy that such diversity is evident even when the definition of H. erectus is tightly drawn by excluding the early East African specimens from the taxon.

THE FATE OF HOMO ERECTUS

One consequence of the diversity within H. erectus is the problem of adequately differentiating between specimens referred to H. erectus, and those of what has become known as 'archaic' H. sapiens. Recent discoveries, especially in Africa and Europe, have focussed attention on the morphological variability of Middle Pleistocene hominids, and taxonomically contentious specimens are known from all major continental areas. Such specimens include Mauer, Petralona, Arago, Vertesszollos and Bilzingsleben (Europe), Bodo I and II, Ndutu, Omo (Kibish) II and Salé (Africa) and Sambungmachan and Ngandong crania (S.E. Asia). While this disagreement may in part reflect workers' varying perceptions of subjective constructs such as taxonomic boundaries, it cannot be attributed solely to this cause since many of the specimens show combinations of features which are otherwise only found in specimens which can be confidently assigned to each of the two species. The two theoretical models call upon us to deal with this problem in contrasting ways.

For the cladist, there is wide agreement that modern H. sapiens displays many autapomorphies (Eldredge & Tattersall, 1975; Wood & Chamberlain, this volume) and despite their robustness and retention of primitive characters, crania are put in 'archaic' Homo sapiens if they display unequivocal evidence of one or more of these autapomorphic traits. It is noteworthy that as a result of recognising the many autapomorphies of H. sapiens, and the relatively few autapomorphies of H. erectus, Wood (1984) and Stringer (1984a) called into question the assumption that H. erectus is more closely related to H. sapiens than is any other hominid taxon. Indeed, in his review of these conclusions, Andrews (1984) comes down in favour of a cladogram which has H. sapiens and H. habilis as sister groups, with H. erectus as the sister group of this clade.

However, to a pheneticist the same distribution of features is interpreted differently. Whereas the phenetic analysis has tended to 'open out' a relatively large gap between the two groups of ('early' and 'late') H. erectus and the earlier hominines, the phenetic distance between H. erectus and H. sapiens is relatively small. Far from suggesting discontinuity, it is powerfully suggestive of continuity in the fossil record and some workers (e.g. Bilsborough, 1976, 1978; Jelinek, 1978, 1980) have stressed the similarities between specimens assigned to Homo erectus and 'archaic' Homo sapiens, and the arbitrary nature of the boundary. Some workers have attempted to straddle the divide between the two models and Stringer has offered both 'clade' and 'grade' approaches to the problem (Stringer, 1978, 1981, 1984b; Stringer et al, 1979).

SUMMARY AND CONCLUSIONS

By juxtaposing phenetic and cladistic analyses of the same hominid material we have aimed to draw attention to the contrasting principles that underlie the two approaches, and the ways in which they influence the results that are derived. The phenetic approach includes measures of both similarities and differences, but does not differentially weight characters entering the analyses, although prior weighting does occur in the special 'all or nothing' sense of character selection. The relevance of such selection, especially when performed by experienced workers, should not be underestimated, but the basis of selection for a particular investigation is rarely stated as an explicit protocol item.

In cladistic analysis character selection occurs both prior to, and within, the analysis, in the latter case according to clearly defined procedures. Derived features are weighted, emphasising contrasts between groups and thereby generating discontinuities. The technique allows investigation of the distribution of character states and permits the precise formulation of group relationships according to the distribution of those character states. Put in simple terms, the differences between the two techniques are these; the phenetic model incorporates similarities as well as differences between taxa, whereas the cladistic method seeks out, identifies and then emphasises, discontinuities.

Phenetic data suggest that the 'early' erectus remains from East Africa are at least specifically distinct from specimens usually included in H. habilis, H. ergaster (Groves & Mazak, 1975) or A. africanus. Phenetic differences also exist between the same specimens and the Asian hypodigm of H. erectus, but the authors differ in their interpretation of these (see below). Cladistically, it seems that in terms of individual characters H. erectus sensu stricto has to be defined on the basis of fewer features than is usually the case. The defining characters include details of the occipital bone, including the torus, the presence of an angular torus and perhaps a frontal sagittal keel; traditional H. erectus features such as post-orbital narrowing and thick cranial vault bones seem unlikely to be autapomorphies, and thus can only be included as part of a 'combination' definition (Wood, 1984). On the evidence of both traditional 'character-based' and combination definitions one of us (B.W.) concludes that KNM-ER 3733 and 3883, two of the three relatively complete crania in the 'early' H. erectus group, should not be assigned to H. erectus sensu stricto. That is not to say that H. erectus is not present in Africa (for A.B. and B.W. agree that OH9 is properly assigned to H. erectus), nor that the population from which the erectus clade originated may not have been African.

Even if the hypodigm of H. erectus is restricted to Asia, both authors see evidence of temporal variation within H. erectus and would reject the hypothesis that the hypodigm provides good evidence of stasis. However, whereas A.B. sees continuity, and thus affinity, between H. erectus and H. sapiens, B.W. follows Andrews (1984) in interpreting the cladistic evidence as suggesting that H. habilis has equal, if not greater, claims to be regarded as the sister group of H. sapiens.

We differ in our evaluation of the relative utility of phenetic and cladistic techniques for investigating hominid phylogeny, but we are agreed that both approaches have an established role in palaeoanthropology. Our differing viewpoints derive in large part from differing interests and a concern to ask different questions of the fossil record. Each technique yields information that the other cannot provide; both derive models that illuminate different aspects of early hominid diversity and neither model can adequately encompass all aspects of the hominid fossil record. However, such models, and the different nomenclatural devices that help us to convey information about them, should never be confused with the actuality of that record.

ACKNOWLEDGEMENTS

The authors are grateful to the Directors and Trustees of the Museums and Heads of Departments who have allowed them access to fossil material and especially to the Director and Trustees of the National Museums of Kenya.

Research incorporated in this paper was supported by the NERC and the Royal Society, the Boise Fund, Oxford, and King's College, Cambridge.

REFERENCES

Key references

Bilsborough, A. (1978) Some aspects of mosaic evolution in hominids. *In* 'Recent Advances in Primatology' *3*, (eds. D.J. Chivers & K.A. Joysey). Academic Press: London, pp. 335-50.

Howells, W.W. (1981). Homo erectus in human descent: ideas and problems. *In* 'Homo erectus: Papers in Honor of Davidson Black' (Eds. B.A. Sigmon & J.S. Cybulski). University of Toronto Press: Toronto, pp.63-85.

Rightmire, G.P. (1981). Patterns in the evolution of Homo erectus. Paleobiology, *7*: 241-46.

Rightmire, G.P. (1984). Comparisons of Homo erectus from Africa and Southeast Asia. Cour. Forsch. Inst. Senckenberg, *69*:83-98.

Wood, B.A. (1984). The origin of Homo erectus. Cour. Forsch. Inst. Senckenberg, *69*:99-111.

Main references

Andrews, P. (1984). An alternative interpretation of the characters used to define Homo erectus. Cour. Forsch. Inst. Senckenberg, *69*: 167-75.

Arambourg, C. (1955). A recent discovery in human paleontology: Atlanthropus of Ternifine (Algeria). Am. J. Phys. Anthropol., *13*, 191-201.

Arambourg, C. (1963) Le Gisement de Ternifine. Archives de L'Institut de Paléontologie Humaine. Mem., *32*, Part 2, 37-190.

Arambourg, C. & Biberson, P. (1956). The fossil human remains from the paleolithic site of Sidi Abderrahman (Morocco). Am. J. Phys. Anthropol., *14*:467-87.

Bilsborough, A. (1973) A multivariate study of evolutionary change in the hominid cranial vault, and some evolutionary rates. J. Hum. Evol., *2*:387-403.

Bilsborough, A. (1976) Patterns of evolution in Middle Pleistocene hominids. J. Hum. Evol., *5*:423-39.

Bilsborough, A. (1978) Some aspects of mosaic evolution in hominids. *In* 'Recent Advances in Primatology' *3*, (Eds. D.J. Chivers & K.A. Joysey). Academic Press: London, pp. 335-50.

Bilsborough, A. (1983) The pattern of evolution within the genus Homo. *In* 'Progress in Anatomy. Volume 3' (Eds. V. Navaratnam & R.J. Harrison). Cambridge University Press: Cambridge, pp.143-64.

Bilsborough, A. (1984) Multivariate analysis and cranial diversity in Plio-Pleistocene hominids. *In*'Multivariate Statistical Methods in Physical Anthropology' (Eds. G.N. van Vark & W.W. Howells). D. Reidel: Amsterdam, pp.351-75.

Bilsborough, A. & Wood, B.A. (in preparation) Cranial morphometry of early hominids. I. Facial region.

Black, D. (1934) On the discovery, morphology and environment of Sinanthropus pekinensis. Phil. Trans. R. Soc., *223*, Series B: 57-120.

Broom, R. & Robinson, J.T. (1949). A new type of fossil man. Nature, *164*:322-23.

Broom, R. & Robinson, J.T. (1952) Swartkrans Ape-Man, Paranthropus crassidens. Trans. Mus. Mem., *6*:1-123.

Broom, R., Robinson, J.T. & Schepers, G.W.H. (1950). Sterkfontein Ape Man, Plesianthropus. Trans. Mus. Mem., *4*:1-117.

Broom, R. & Schepers, G.W.H. (1946) The South African fossil ape-men, the Australopithecinae. Trans. Mus. Mem., *2*:1-272.

Cain, A.J. & Harrison, G.A. (1960). Phyletic weighting. Proc. Zool. Soc. Lond., *135*:1-31.

Chavaillon, J. & Coppens, Y. (1975) Découverte d'Hominidé dans un site Acheuléen de Melka-Kunturé (Ethiopie). Bull. Mém. Soc. Anthrop. Paris, *2*:125-28.

Clarke, R.J. (1977) The cranium of the Swartkrans hominid SK 847 and its relevance to human origins. Ph.D. Thesis: University of the Witwatersrand.

Clarke, R.J. & Howell, F.C. (1972) Affinities of the Swartkrans 847 hominid cranium. Am. J. Phys. Anthropol., *37*:319-36.

Conroy, G.C., Jolly, C.J., Cramer, D. & Kalb, J.E. (1978). Newly discovered fossil hominid skull from the Afar depression, Ethiopia. Nature, *276*:67-70.

Corruccini, R.S. & McHenry, H.M. (1980). Cladometric analysis of Pliocene hominids. J. Hum. Evol., *9*:209-221.

Cronin, J.E., Boaz, N.T., Stringer, C.B. & Rak, Y. (1981) Tempo and mode in hominid evolution. Nature, *292*,113-22.

Dubois, E. (1891) Palaeontologische onderzoekingen op Java. Versl. Mijnw. Batavia, *3*:12-14 and *4*:12-15.

Eldredge, N. & Tattersall, I. (1975) Evolutionary models, phylogenetic reconstruction and another look at hominid phylogeny. *In* 'Approaches to Primate Paleobiology' (Eds. F.S. Szalay) Karger: Basel, pp.218-42.

Eldredge, N. & Tattersall, I. (1982) The myths of human evolution. Columbia University Press: New York.

Ennouchi, E. (1969) Découverte d'un Pithecanthropien au Maroc. C.R. Acad. Sci, Paris, *269D*:763-65.

Ennouchi, E. (1972) Nouvelle découverte d'un Archanthropien au Maroc. C.R. Acad. Sci. Paris, *274D*:3088-90.

Gingerich, P.D. (1979) The stratophenetic approach to phylogeny reconstruction in vertebrate paleontology. *In* 'Phylogenetic Analysis and Paleontology' (Eds. J. Cracraft & N. Eldredge) Columbia University Press: New York, pp.41-77.

Gingerich, P.D. & Schoeninger, M.J. (1978) The fossil record and primate phylogeny. J. Hum. Evol., *6*:483-505.

Groves, C.P. & Mazak, V. (1975) An approach to the taxonomy of the Hominidae: Gracile Villafranchian hominids of Africa. Cas. Miner. Geol., *20*:225-47.

Gould, S.J. & Eldredge, N. (1977) Punctuated equilibria: the tempo and mode of evolution considered. Paleobiology, *3*:115-51.

Hemmer, H. (1972) Notes sur la position phylétique de l'homme de Petralona. Anthropologie, Paris, *76*:155-61.

Howell, F.C. (1978) Hominidae. *In* 'Evolution of African Mammals' (Eds. V.J. Maglio & H.B.S. Cooke) Harvard University Press: Cambridge, pp.154-248.

Howells, W.W. (1981) Homo erectus in human descent: ideas and problems. In 'Homo erectus: Papers in Honor of Davidson Black' (Eds. B.A. Sigmon & J.S. Cybulski) University of Toronto Press: Toronto, pp.63-85.

Huxley, J.S. (1958) Evolutionary processes and taxonomy with special reference to grades. Upps. Univ. Arssks., *6*:21-38.

Jacob, T. (1975) Morphology and paleoecology of early man in Java. *In* 'Paleoanthropology, Morphology and Paleoecology' (Ed. R. Tuttle) Mouton: The Hague, pp.311-325.

Jacob, T. (1981) Solo Man and Peking Man. *In* 'Homo erectus: Papers in Honour of Davidson Black' (Eds. B.A. Sigmon & J.S. Cybulski) University of Toronto Press: Toronto, pp.87-104.

Jacob, T. & Curtis, G.H. (1971) Preliminary potassium-argon dating of early man in Java. Contr. Univ. Calif. Archeol. Res. Facil., *12*:50.

Jaeger, J.J. (1973) Un pithecanthrope évolue. La Recherche., *39*:1006-7.

Jaeger, J.J. (1975) The mammalian faunas and hominid fossils of the Middle Pleistocene of the Maghreb. *In* 'After the Australopithecines. Stratigraphy, ecology and culture change in the Middle Pleistocene' (Eds. K.W. Butzer & G.Ll. Isaac). Mouton: The Hague, pp.399-418.

Jelinek, J. (1978) Homo erectus or Homo sapiens? *In* 'Recent Advances in Primatology, Volume 3' (Eds. D.J. Chivers & K.A. Joysey) Academic Press: London, pp.419-29.

Jelinek, J. (1980) European Homo erectus and the origin of Homo sapiens. *In* 'Current Argument on Early Man' (Ed. L-K. Konnigsson). Pergamon: Oxford, pp.137-44.

von Koenigswald, G.H.R. (1975) Early man in Java: catalogue and problems. *In* 'Paleoanthropology, Morphology and Paleoecology' (Ed. R. Tuttle). Mouton: The Hague, pp.303-9

Leakey, L.S.B. (1961) New finds at Olduvai Gorge. Nature, *189*, 649-50.

Leakey, L.S.B., Tobias, P.V. & Napier, J.R. (1964) A new species of the genus Homo from Olduvai Gorge. Nature, *202*:7-9.

Leakey, M.D., Clarke, R.J. & Leakey, L.S.B. (1971) New hominid skull from Bed I, Olduvai Gorge, Tanzania. Nature, *232*:308-12.

Leakey, M.G. & Leakey, R.E. (1978) The fossil hominids and an introduction to their context, 1968-1974. Clarendon Press: Oxford.

Leakey, R.E.F. (1973) Further evidence of Lower Pleistocene hominids from East Rudolf, North Kenya, 1972. Nature, *242*:170-3.

Leakey, R.E. & Walker, A.C. (1976) Australopithecus, Homo erectus and the single species hypothesis. Nature, *261*:572-74.

Lestrel, P.E. (1976) Hominid brain size versus time: revised regression estimates. J. Hum. Evol., *5*:207-12.

de Lumley, H. & de Lumley, M.A. (1971) Découvertes de restes humaines anté-néanderthaliens à la Caune de l'Arago (Tautavel, Pyrénées, Orientales). C.R. Acad. Sci. Paris, *272D*, 1739-42.

Martin, R.D. (1968) Towards a new definition of primates. Man, *3*:377-401.

Olson, T.R. (1978) Hominid phylogenetics and the existence of Homo in Member 1 of the Swartkrans Formation, South Africa. J. Hum. Evol., *7*:159-78.

Olson, T.R. (1981) Basicranial morphology of the extant hominoids and Pliocene hominids: the new material from the Hadar Formation, Ethiopia, and its significance in early human evolution and taxonomy. In 'Aspects of Human Evolution' (Ed. C.B. Stringer) Taylor & Francis: London, pp.99-128.

Pope, G.G. (1983) Evidence on the age of the Asian Hominidae. Proc. Nat. Acad. Sci., *80*:4988-92.

Rightmire, G.P. (1979) Cranial remains of Homo erectus from Beds II and IV, Olduvai Gorge, Tanzania. Am. J. Phys. Anthropol., *51*: 99-115.

Rightmire, G.P. (1981) Patterns in the evolution of Homo erectus. Paleobiology, *7*:241-46.

Rightmire, G.P. (1984) Comparisons of Homo erectus from Africa and Southeast Asia. Cour. Forsch. Inst. Senckenberg, *69*:83-98.

Robinson, J.T. (1953) Telanthropus and its phylogenetic significance. Am. J. Phys. Anthropol., *11*:445-501.

Robinson, J.T. (1954) The genera and species of the Australopithecinae. Am. J. Phys. Anthropol., *12*:181-200.

Sartono, S. (1975) Implications arising from Pithecanthropus VIII. *In* 'Paleoanthropology, Morphology and Paleoecology' (Ed. R.H. Tuttle) Mouton: The Hague and Paris, pp.327-60.

Sartono, S. (1976) The Javanese Pleistocene hominids: a re-appraisal. Proc. Union Int. des Sciences Préhist. et Protohist. Coll. VI. 'Les plus Anciens Hominidés' (Eds. P.V. Tobias & Y. Coppens). CNRS: Paris, pp.456-64.

Sartono, S. (1982) Characteristics and chronology of early men in Java. *In* Proc. 1er Congres Int. de Paléont. Hum., 2 CNRS: Paris, pp.491-533.

Semah, F. (1984) The Sangiran dome in the Javanese Plio-Pleistocene chronology. Cour. Forsch. Inst. Senckenberg., *69*:245-52.

Stanley, S.M. (1979) Macroevolution: pattern and process. W.H. Freeman: San Francisco.

Stanley, S.M. (1982) The new evolutionary timetable. Basic Books: New York.

Stringer, C.B. (1978) Some problems in Middle and Upper Pleistocene relationships. Recent Adv. Palaeont., *3*:395-418.

Stringer, C.B. (1981) The dating of European Middle Pleistocene hominids and the existence of Homo erectus in Europe. Anthropol., *19*: 3-14.

Stringer, C.B. (1984a) The definition of Homo erectus and the existence of the species in Africa and Europe. Cour. Forsch. Inst. Senckenberg, *69*:131-43.

Stringer, C.B. (1984b) Human evolution and biological adaptation in the Pleistocene. *In* 'Hominid Evolution and Community Ecology' (Ed. R. Foley). Academic Press: London, pp.55-83.

Stringer, C.B., Howell, F., Clark & Melentis, J.K. (1979) The significance of the fossil hominid skull from Petralona, Greece. J. Arch. Sci., *6*:235-53.

Tattersall, I. & Eldredge, N. (1977) Fact, theory and fantasy in human paleontology. Am. Scient., *65*:204-11.

Thoma, A. (1972) Cranial capacity, taxonomical and phylogenetic status of Vertesszöllös man. J. Hum. Evol., *1*:511-12.

Tobias, P.V. (1966) Fossil hominid remains from 'Ubeidiya', Israel. Nature, *211*:130-33.

Tobias, P.V. (1967) Olduvai Gorge, Volume 2. The cranium and maxillary dentition of Australopithecus (Zinjanthropus) boisei. Cambridge University Press: Cambridge.

Tobias, P.V. & von Koenigswald, G.H.R. (1964) A comparison between the Olduvai hominids and those of Java and some implications for hominid phylogeny. Nature, *204*:515-18.

Tobias, P.V. & Wells, L.H. (1967) South Africa. *In* 'Catalogue of Fossil Hominids' (Eds. K.P. Oakley & B.G. Campbell). BM(NH):London, pp.49-100.

Vallois, H.V. (1956) The pre-Mousterian mandible from Montmaurin. Am. J. Phys. Anthropol., *14*:319-23.

Vlcek, E. (1978) A new discovery of Homo erectus in Central Europe. J. Hum. Evol., *7*:239-51.

Walker, A. (1981) The Koobi Fora hominids and their bearing on the origins of the genus Homo. *In* 'Homo erectus: papers in honor of Davidson Black' (Eds. B.A. Sigmon & J.S. Cybulski). University of Toronto Press: Toronto, pp.193-215.

Walker, A. & Leakey, R.E.F. (1978) The hominids of East Turkana. Sci. Amer., 44-56.

Weidenreich, F. (1936) The mandible of Sinanthropus pekinensis: a comparative study. Palaeont. Sin., Series D, 7:1-169.

Weidenreich, F. (1937) The dentition of Sinanthropus pekinensis: a comparative odontography of the hominids. Palaeont. Sin., Series D(10), 2 vols., pp.1-180 and pp.1-121.

Weidenreich, F. (1943) The skull of Sinanthropus pekinensis: a comparative study of a hominid skull. Palaeont. Sin., Series D, *10*:1-484.

Wiley, E.O. (1979) Cladograms and phylogenetic trees. Syst. Zool., *28*, 88-92.

Woo, J.K. (1964) Mandible of the Sinanthropus - type discovered at Lantian, Shensi - Sinanthropus lantianensis. Scient. Sin., *13*:801-11.

Woo, J.K. (1966) The hominid skull of Lantian, Shensi. Vert. Pal., *10:* 1-22.

Wood, B.A. (1978) Classification and phylogeny of East African hominids. *In* 'Recent Advances in Primatology Vol.3' (Eds. D.J. Chivers & K.A. Joysey). Academic Press: London, pp.351-72.

Wood, B.A. (1984) The origin of Homo erectus. Cour. Forsch. Inst. Senckenberg, *69*:99-111.

Wood, B.A. (in press) Early Homo in Kenya, and its systematic relationships. *In* 'Ancestors: the Hard Evidence' (Ed. E.Delson). Alan R. Liss: New York.

Wust, K. (1951) Uber den Unterkiefer von Mauer (Heidelberg). Z. Morph. Anthrop., *42*:1-112.

17 THE ORIGIN OF HOMO SAPIENS: THE GENETIC EVIDENCE

J.S. Jones,
Department of Genetics & Biometry,
University College London,
London WC1

INTRODUCTION

Comparative anatomists have always been impressed by the close physical similarity of man to his living relatives; in Darwin's own phrase, by the fact that "Man still bears in his bodily frame the indelible stamp of his lowly origin". It is often assumed that physical similarity must necessarily reflect genetic relatedness, so that animals which show little divergence in body form underwent rather small genetic changes during their separation from a common ancestor. The number of genes involved in the origin and maintenance of man as a species distinct from his predecessors and from his living relatives might hence be rather small. This view is implicit in many early anthropological theories, such as the eighteenth century claim that African babies were born with tails because they were hybrids of man and chimp, and accords also with those models of human origins which emphasise the importance of social (rather than genetic) transformations in leading to the appearance of Homo sapiens. Information on the genetic differences between man and other living primates therefore has the potential to test competing theories of human evolution.

Comparative anatomy has been greatly extended by new techniques which allow us to dissect the genomes of related species at a level of precision approaching that of single nucleotide substitutions. Here I hope to show that although molecular biology has been extremely valuable in enabling us to compare the structure of various organisms (and has produced a useful measure of the *physical* similarity of man and his relatives) it has been less successful in explaining the nature and extent of the *genetic* events which accompany speciation in animals; and is hence of only limited value in understanding the evolutionary processes which led to the origin of man. The genetics of living primates can tell us something about human origins, but such information is indirect and can only supplement, and never replace, that which comes from palaeontology.

MOLECULAR ANATOMY OF MAN AND THE PRIMATES

It has long been known that man and primates share genes. Landsteiner found in 1925 that chimpanzees have O and A blood groups, and the Rhesus system depended for its detection on alleles shared by monkey and man. A celebrated early test of gene sharing was carried out in Edinburgh and London Zoos, when 20 of 27 chimpanzees tested with the chemical PTC (which tastes bitter to people who possess the appropriate sensitivity allele) found it unpalatable: one to an extent great enough to spit the test solution over the distinguished geneticist R.A. Fisher. The human and higher primate genomes contain about thirty thousand million nucleotides, which represent up to fifty thousand functional genes and many DNA sequences of unknown function (Shows et al, 1982). Molecular biology allows us to ask what proportion of these genes is common to man and his relatives and hence what structural changes accompanied the origin of man.

Cell surface polymorphisms

Blood groups are only one manifestation of the complex system of genetic variation in antigen structure present on the human cell surface. More than twenty are known in man, and some of these are present in bodily secretions of individuals possessing a 'secretor' allele. All non-human primates are homozygous for 'secretor', but they share other human blood-group polymorphisms (and have a number of species-specific blood groups). Although chimps segregate for A and O alleles in the ABO system, O is rare in other primates: gorillas, for example, are all blood group B. Groups similar to human MN and Rhesus occur in chimps and other primates (Socha and Ruffie, 1983).

A large part of human chromosome 6 codes for genes in the Major Histocompatibility Complex (MHC), a system of individually distinct antigenic cues coded for by what may prove to be more than a thousand tightly linked gene loci, many of which might possess up to 20 alleles (Thomson, 1983). The MHC has a number of functions associated with the immune system, and its general arrangement has been strongly conserved during primate evolution (Fig. 1). The ordering of MHC genes in man, chimpanzee and gorilla is remarkably similar. Tests of cross reaction for individual alleles show that there is also considerable antigen sharing between man and chimp, but rather fewer antigens in common between man and gorilla (Balner, 1981).

Variation in chromosome structure

The anatomy of primate chromosomes has been clarified by new staining techniques which can reveal up to 1000 chromosome bands per haploid set (Yunis and Prakash, 1982). These patterns show that, excluding 'constitutive heterochromatin' (which possesses few coding genes), man and chimp have 13 identical chromosome pairs, man and gorilla 9, and man and orang 8. The major structural difference between man and his relatives is that chromosome 2 (one of the largest human chromosomes) is split into two units in the other species. Many of the interspecific differences result from chromosome inversions: rearrangements of segments of the genetic material which lead to reversals in the ordering of chromosome bands. There are also some translocations, which involve the transfer of sequences of bands between chromosomes. If these structural changes are reversed, then there is complete homology of the 1000 bands in each of the four species, so that what structural change has taken place during their evolution results from the re-shuffling of chromosomal material rather than the gain or loss of large blocks of genes.

New information on the chromosomal location of particular DNA sequences supports this view (Page et al, 1984). One sequence of about 36,000 bases which is found on both the X and the Y chromosome in man is confined to the X in chimps, gorillas and orangs; translocation from X to Y has occurred since the divergence of the lineage leading to man.

The conservation of chromosome structure among the primates is emphasised by the discovery that about 40 gene loci map to homologous chromosomal locations in man and chimpanzee (Roderick et al, 1984).

Divergence in soluble enzymes

Gel electrophoresis, which detects differences in the charge and shape of protein molecules, can identify single mutations in genes which code for soluble enzymes. It has been widely used in studies of the

Figure 1. Two Primate Gene Clusters. Upper Group: the Beta Globin gene family. Lower Group: the Major Histocompatibility Complex. Boxes - expressed genes. Crosses - pseudogenes. The apparent absence of "C" genes in the chimp and rhesus MHC is probably due to incomplete information. After Jeffreys (1982) and Balner (1981).

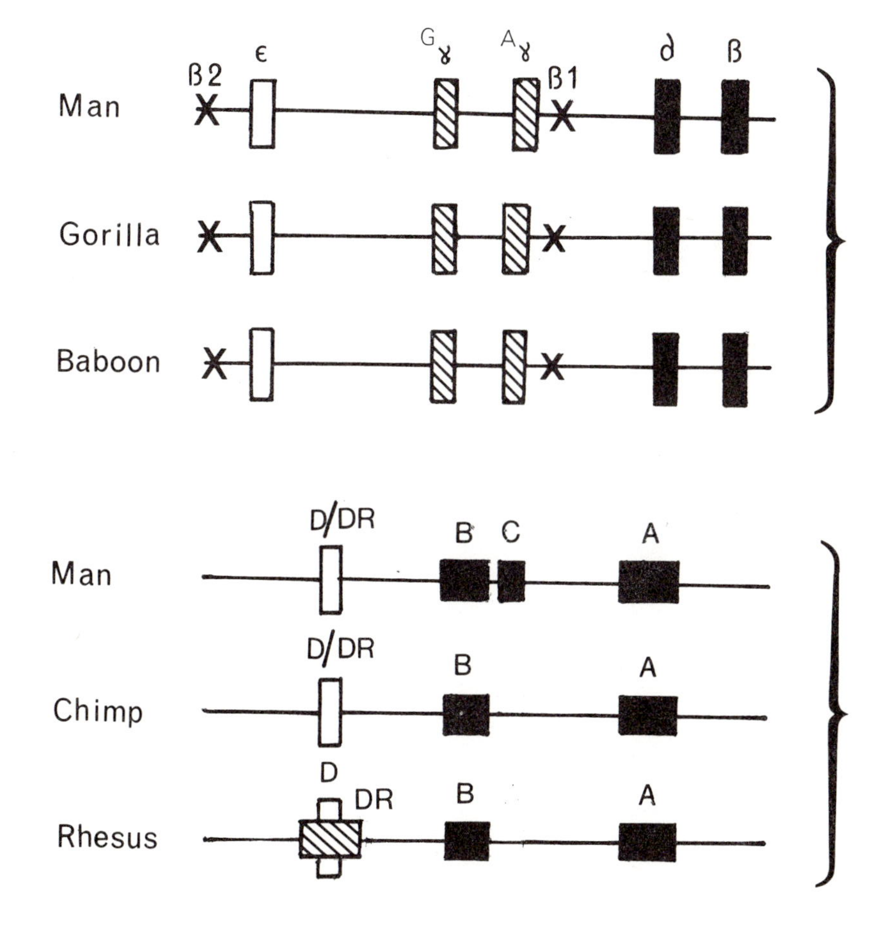

comparative anatomy of the genome as it gave the first real chance to sample the genes of related species. Electrophoresis of 23 enzymes in human, chimp, gorilla and other primates shows that, once again, there has been rather little evolutionary divergence among them (Bruce and Ayala, 1979).

Electrophoresis is not sensitive enough to detect all amino acid substitutions. The complete sequences of several proteins are now known. The best studied molecule is haemoglobin (Goodman et al, 1983). In man, adult haemoglobin consists of two alpha chains, each with 141 amino acids, and two beta chains with 146. There have been no sequence changes between pygmy chimp, chimp and man; and only two changes - one in each chain - between this group and gorilla. Partial or complete sequences of other proteins (such as cytochromes, fibrinopeptides, and carbonic anhydrase) show that again there is remarkable structural similarity between man and his relatives with a divergence at only 3 of 963 amino acid sites in these proteins between man and chimpanzee (Goodman et al, 1982).

Comparisons of DNA sequence

The study of gene products has now been extended by work on the structure of DNA itself. This depends on new techniques in molecular biology, such as our ability to cut DNA at specific sites using restriction enzymes, to produce copies of particular genes by identifying their messenger RNAs and using these to generate artificial complementary DNA sequences, to obtain multiple copies of genes by cloning them into bacteria, and to sequence DNA in various ways.

Much of this work involves primate globins (Jeffreys, 1982). There are 8 active globin genes, arranged into two 'gene families' of related sequence, which code for alpha and beta globins, foetal globins, myoglobins and the like. Each gene family contains some "pseudogenes" which have lost their coding ability, together with a number of intervening sequences which interrupt the coding segments of the active genes. The various members of each cluster arose by duplication from a common ancestor, followed by divergence. Divergence has taken place about ten times more rapidly at those DNA sites which do not produce a change in amino acid sequence than at those which do, so that most of the functional substitutions have been removed by selection. The primate globin gene clusters contain about 70,000 base pairs, only one sixth of which is known to have a coding function. What the rest might do is quite unknown.

There is a large amount of genetic diversity in this part of the genome: in the intervening sequences, one base pair in 1000 may be polymorphic (Higgs et al, 1981). Nevertheless, the general arrangement of the beta-globin genes (which are better known than are the alpha-globins) is very similar in man, gorilla and baboon (Fig. 1; Jeffreys, 1982). Sequence divergence is also limited, with only a 3% change between man and gorilla (Jeffreys and Barrie, 1981). Sequence conservation is found elsewhere in the primate genome. For example, the interferon gene family - a series of genes coding for proteins which attack viruses and influence the immune system - shows close sequence homology between gorilla and man (Wilson et al, 1983). A similar picture will no doubt emerge when further genes are examined.

The new DNA technology gives evidence not only on enormous numbers of new variants but also on the *order* in which these are arranged along segments of chromosome short enough to be safe from disruption by crossing over. This is very useful in studying the comparative anatomy of primate DNA: data from single sites are in one sense no more than the individual letters of names in a family tree, while information on haplotypes (as the ordered sets of closely linked variants are known) presents the surnames themselves. A test of a sample of the 20,000 nucleotides in the primate serum albumin locus with 27 restriction enzymes identified 8 polymorphic sites. There are 256 possible combinations of these variants but in fact only 7 haplotypes were found among the 160 human chromosomes studied. Chimps have a haplotype similar in nucleotide identity and order to the commonest human haplotype at 6 of the 8 variant sites; gorillas at 5 of the 8 and baboon at only one. The arrangement, as well as the identity, of nucleotides in this region has therefore been conserved during primate evolution (Murray et al, 1984).

The value of closely linked DNA sequences can clearly be seen in the mitochondrial genome. This is present in thousands of copies in each cell. Each genome is a closed circle of DNA made up of about 16,500 base pairs, which contains 37 coding genes (and almost no non-coding regions) in man (Anderson et al, 1981). The complete sequence of human mitochondrial DNA (mtDNA) is known, and the position of variants can be identified by comparing them with this standard. Mitochondrial DNA segregates during mitosis as well as at meiosis, and, as the sperm's mitochondria do not enter the fertilised egg, it is maternally inherited. Mitochondrial DNA also has less efficient means of repairing mutations than does nuclear DNA. All this means that it evolves rapidly; some mitochondrial genes have diverged between mice and men at a rate 100 times greater than have their nuclear equivalents (Barton and Jones, 1983).

The mitochondrial genomes of man and his close relatives differ to a greater extent than do most segments of nuclear DNA. A restriction enzyme survey of 50 variable sites in man, chimpanzee, pygmy chimpanzee, gorilla, orang utan and gibbon (Ferris et al, 1981) revealed differences at 121 positions. The most striking divergence involves the gorilla, which has a deletion of 95 base pairs; but even man and the chimpanzee differ at 44 mtDNA sites. In one 896 base pair length of mtDNA (Brown et al, 1982) man and chimp differ in about 8% of sequences, while man and gorilla differ by 10%. One disadvantage of such a rapidly evolving character in the analysis of patterns of anatomical relatedness is that changes in the mitochondrial genome become 'saturated' as mutations in one direction are reversed by back mutations. The mouse, for example, differs from man in only about 30% of its mtDNA sequence. Nevertheless, the technical advantages of studying mtDNA, and the associated availability of sequence information in many species and individuals, means that it will continue to be a useful anatomical character in the study of primate evolution.

DNA Hybridisation

It will be an enormous task to dissect the 30 billion sequences found in the nucleus with a precision close to that already possible for mtDNA. DNA hybridisation is an alternative method of comparing the gross anatomy of primate genomes (Sibley & Ahlquist, 1984). When DNA is 'melted' by heating it to boiling point, the strands separate, and will reassociate on cooling. If strands having different sequences are melted together, the rate of reassociation into hybrid molecules depends on the degree to which they share base sequences. This technique of measuring structural homology has been used in many animals, including some in which dates of divergence can be inferred from the geological record (as is the case, for example, for the separation of the African ostrich and the South American rhea with the opening of the Atlantic Ocean 80 my ago). It provides measures of relative divergence which can, in principle at least, be used to estimate when groups of organisms last shared a common ancestor. DNA hybridisation figures for man and chimpanzee suggest that they diverged about 7.5 my ago; a date rather earlier than that derived from direct comparison of individual nuclear or mitochondrial genes but one which agrees with them in suggesting that Homo sapiens shares a large part of his molecular anatomy with his living relatives.

Whatever technique is used, the verdict of molecular biology is unanimous: the emergence of man was accompanied by rather little structural change in DNA. In this sense, the evidence for close relatedness among the hominids is overwhelming. Molecular biology is also useful in studying the evolutionary changes which have taken place within the human species since its origin.

MOLECULAR DIVERGENCE AMONG HUMAN POPULATIONS

Many speculations about hominid evolution depend on comparing the genetic constitution of living primates. The human population can itself be divided into identifiable groups, and the geographic distribution and genetic differentiation of modern human 'races' can, in the same way, be used to speculate about man's evolutionary history.

Human populations show extensive geographic structuring for many of their genetic polymorphisms, some of which results from natural selection, while most have no obvious selective correlates. The frequencies of skin colour alleles are correlated with sunshine, probably because of selection on vitamin balance. White skins transmit ten times as much solar UV as do black and hence produce vitamin D at a rate up to forty times greater than can blacks (Clemens et al, 1982). A decrease in skin pigment was therefore favoured by selection as man migrated from the tropics. Dark skins persist in equatorial populations because of the destruction by penetrating sunlight of circulating vitamins: ultraviolet can destroy up to half of the blood's vitamins B12, E and riboflavins in whites (Branda and Eaton, 1978).

There is extensive geographic change in the frequency of cell surface antigens. The Duffy blood group, for example, has an allele which reaches fixation in the Pacific Islands, but is absent from South American Indians.

Histocompatibility antigens also show geographic structuring, with the frequency of particular antigens varying from 0 - 15% among populations (Thomson, 1983). Some cell surface polymorphisms are associated with susceptibility to diseases. For example, homozygotes for the Fy-allele in the Duffy system are resistant to *Plasmodium vivax* malaria (perhaps because their cell surface lacks the parasite's attachment site) and the frequency of this allele is correlated with the distribution of the parasite (Luzzatto, 1979). The extent of association of particular cell surface cues with individual diseases varies greatly; the increase in ankylosing spondylitis among individuals of one antigenic type ranges from 36 fold among American Blacks to 324 fold among Japanese. Disease associations may even involve different antigens in different populations, suggesting that selection is acting not on individual cell surface antigens, but on closely linked and undetected polymorphic loci which themselves show geographical structuring (Thomson, 1983).

Some protein polymorphisms also show geographical change. A number of haemoglobin variants (some, such as HbS and HbE, involving single amino acid substitutions; and others, such as the thalassaemias, changes in whole sections of the molecule) are found at high frequency in certain populations. This pattern arises from resistance to malaria conferred by the destruction of parasitised cells because of reduced stability of the mutant haemoglobin molecule, or the variant cells' increased sensitivity to parasite metabolites (Friedman and Trager, 1981). Other polymorphisms - such as the lactase variant which is common only in populations which have a history of milk drinking (Johnson *et al*, 1981) - are associated with diet. However, most human enzyme variants cannot be related to a selective agent, and geographical changes in their frequency probably result from random events during man's evolutionary history.

Random processes are also likely to determine geographical changes in most DNA sequence variants, particularly in those which have no coding effects (Shows *et al*, 1982). Geographical trends in nuclear DNA sequences become most obvious when haplotypes (combinations of bases over short chromosomal distances) are mapped. For example, only 2 of the 7 known haplotypes of serum albumin DNA are found in all human populations, and the remainder exist only on single continents (Murray *et al*, 1984). The same is true in the beta-globin gene family; only a few of forty or so haplotypes are widespread and the others are confined to local populations (Pagnier *et al*, 1984; Antonarakis *et al*, 1984).

Geographical differentiation of this kind is most manifest in mitochondrial DNA. The dispersal rate of maternally inherited genes is that of females (rather than the average of the two sexes), and their effective population size is lower; patterns of distribution of mitochondrial genes in many animals are therefore more subdivided than are those of the nucleus (e.g. Lansman *et al*, 1983). A survey of 112 human mitochondrial genomes from Australia, Africa, Asia and Europe, using 12 restriction enzymes, revealed 163 variable sites (Cann *et al*, 1984), most of which were found in only a single population.

Spatial changes in gene frequency can be used to identify patterns of relatedness among evolving groups. The most striking point which emerges from studies of human geographic variation is that, in spite of marked local trends in the frequency of many alleles, *Homo sapiens* is in general a geographically uniform species, as clines at different loci a are not concordant. A measure of the proportion of genes shared by two randomly chosen individuals (based on 18 polymorphic enzymes and blood-groups from 180 populations from each of six "racial" groups defined on the basis of visible characters such as skin colour) shows that the largest component of man's overall genetic diversity - about 84% of the total - comes from the differences which exist between individuals from the same tribe or nationality. About six per cent arises from differences between nationalities; and about ten per cent from the genetic divergence which exists between human 'races' (Latter, 1980). Mankind is not divided into a series of genetically distinct units, and there is hence little evidence that, as is sometimes suggested, modern racial divergence is a relic of human evolution from a series of different pre-hominid ancestors. The difficulty of making any objective genetic definition of human races may mean that this term has outlived its usefulness in physical anthropology.

Trees of evolutionary relatedness among human populations (Nei and Roychoudhury, 1982) show, not surprisingly, that adjacent populations are usually genetically similar (although they may be linguistically very different: thus, the Basques are not well differentiated from other Spaniards) and that there is no close correlation between morphological and molecular evolution. For example, the aboriginal populations of North and South America are genetically distant from each other, although they are morphologically similar, and the physically rather different populations of Europe and northern India appear to be closely related when enzymes and blood groups are analysed. The amount of interpopulation differentiation has been greater in American Indians, Australians and Papuans than elsewhere, perhaps because of their small populations and the associated increase in random differentiation. European populations are rather homogeneous, and the average genetic distance between the English and the Italians is only about a tenth of that between Chinese and Japanese. This might indicate that modern man invaded Europe relatively recently and as a single group. The genetic evidence accords with palaeontological information to suggest that this invasion coincided with the diffusion of early farmers from the East (Ammerman and Cavalli-Sforza, 1984).

The distribution of human enzyme and cell surface variants has been used to attempt to identify the point of origin of modern man, and the course of his spread over the world, by asking which human populations are genetically central, and which are most divergent from others (Piazza *et al*, 1981). The picture which emerges is fairly consistent over loci; south-central and eastern Asia have intermediate values of gene frequency, while the populations of Africa, Australia and America tend towards extremes. This might mean that the migrations which peopled the modern world started in Asia, possibly passing through the fertile crescent of the Near East 10,000 years ago, and reaching North America rather later. Indeed, modern man may have arisen in Asia, although there still remains controversy as to his point of origin.

Much less is known about geographical patterns of nuclear DNA sequence variation than about the geography of gene products. Such information would be extremely useful as combinations of closely linked variants preserve much more of their history than do freely recombining genes such as enzyme or blood group alleles. The potential of such sequence data for tracing migrations is seen in recent work on the origin of sickle cell haemoglobin alleles. The distribution of nucleotide substitutions in the beta-globin region was studied in 170 separate chromosomes bearing this mutation from American and Jamaican Blacks. Sixteen different haplotypes - combinations of tightly linked DNA variants - were associated with the various sickle cell alleles. However, three of these accounted for 151 of the chromosomes. This pattern of association of different sickle cell alleles with a limited number of genetically distinct chromosomes probably results from a few independent mutations of HbA to HbS in different populations (perhaps with the assistance of some exchange of alleles among chromosomes by crossing over)(Antonarakis et al, 1984). Different haplotypes are associated with sickle cell alleles from Benin, Senegal and the Central African Republic, again suggesting an independent origin of the mutation in each of these regions. The Benin haplotype is found in Algerian sickle cell patients, suggesting that Benin may have been the point of origin of the northward migration of Black populations at the time of the fall of the Roman Empire (Pagnier et al, 1984).

The mitochondrial haplotype consists of 16,569 base pairs, and should in principle be very valuable in studying the evolutionary relatedness of human populations. A restriction enzyme survey of mtDNA from 200 individuals, originating from five widely separated human populations, produced 35 different haplotypes (Johnson et al, 1983). Only one of these types appears in all ethnic groups, and as this is also present in other primates it is probably ancestral. The other thirty or so haplotypes (most of which are found in only one ethnic group) have been derived since the expansion of man from his origin. Trees of relatedness among these mitochondrial genotypes do not accord particularly well with the evidence from palaeontology or from the distribution of nuclear genes. Mitochondrially central populations are somewhere near those of modern Africa, although within Africa there is considerable divergence between South African Bantu and Bushmen. This might indicate that Bushmen have undergone unusually rapid divergence of their mitochondrial genome, or perhaps that they split off from the Homo sapiens lineage considerably before the time of origin of other Africans. Australian Aboriginals also appear to be unusually distant from nearby Pacific populations in their mitochondrial constitution (Cann et al, 1982).

It is rather disturbing that patterns of human phylogeny deduced from mtDNA do not accord particularly well with those which come from studies of nuclear genes. Recent work on the transfer of cytoplasmic genes across species barriers suggests that mitochondrial phylogenies may not be reliable indicators of overall evolutionary change. Two closely related species of mice, Mus musculus and Mus domesticus meet in southern Denmark. Here they hybridise in a narrow zone, and exchange nuclear genes for a few kilometres into each other's range. However, mtDNA from domesticus penetrates into musculus throughout much of Scandinavia, so that in this region there are mice which have the nuclear genome of one species, but the mitochondrial genomes of another (Ferris et al, 1983). Evolutionary trees based on mitochondrial genes hence give a misleading picture of genetic change in the nucleus. Such independent movements of cytoplasmic and nuclear genes may explain the anomalies in the human mitochondrial pedigree: it might be, for example, that isolated populations such as Bushmen or Australian Aborigines were not invaded by the mitochondrial genomes which have spread through other human populations since the emergence of modern man.

The genetic evidence on human origins suggests that modern man originated from a single population rather than from a series of simultaneously evolving prehominid ancestors, and that his emergence was not accompanied by a drastic reorganisation in molecular anatomy. Use of the 'molecular clock' on the observed differences between modern human populations suggests that the racial differences between the world's human groups may have begun to appear at about the time that man began to migrate from his centre of origin.

HUMAN EVOLUTION IN CONTEXT

The studies of the genetics and evolution of man are greatly constrained by the intrinsic problems of working with human populations, and it may be instructive to compare patterns of molecular change in the primates with those in other species more amenable to genetic analysis. It is already clear that there is often rather poor agreement between the structural changes uncovered by comparative anatomy, or by molecular biology, and those which lead to speciation in various evolving groups. Among land snails, for example, some genera have undergone extensive evolution at the loci which produce reproductive isolation with almost no divergence in loci coding for anatomical characters such as structural enzymes. In Polynesian snails of the genus Partula, distinct species living on islands thousands of kilometres apart have undergone very little molecular divergence (Johnson et al, 1977). In contrast, some populations of the European snail which interbreed freely in the laboratory, and which in the 'wild' are separated by only 20 kilometres, are fixed for alternative alleles at a number of enzyme loci (Ochman et al, 1983). In these cases molecular anatomy is a misleading indicator of taxonomic relatedness, a point which may be important when considering arguments about primate evolution based on structural change.

The evolutionary history of Homo sapiens has some parallels with that of the equally famous genetical organism, Drosophila melanogaster. Man took about 3000 generations to people the world; D. melanogaster was largely confined to the forests of west Africa until a few hundred years - and a few thousand fruit fly generations - ago. Since its expansion, D. melanogaster has undergone considerable differentiation in morphology, chromosomal constitution, soluble enzymes, and DNA sequences. Like man, this species is not divided into genetically distinct units, and like him there is no evidence that crosses between individuals from different parts of the world are less successful than are crosses from within the same population. However, D. melanogaster is much more easily studied by the traditional methods of genetics than is man, and it is here that we begin to see the limitations of comparative anatomy (at whatever level of sophistication) as an indicator of the genetic changes involved in evolution.

D. melanogaster is one of a group of morphologically very similar species, some of which show close homologies in chromosomal and molecular structure. Fertile hybrids between many of them can be produced in the laboratory, and manipulations of genetically marked chromosomes allow measurement of the decrease in fitness of hybrids possessing various combinations of parental genes. It is therefore possible to estimate the number of loci involved in the maintenance of each species as a genetic entity (eg Pontecorvo, 1943; Coyne, 1984). In every case the answer is clear: many genes are associated with species divergence, and the number detected is limited only by the acuity of the experiment. This is true in species which are structurally very close to each other, as well as those which show greater molecular divergence (see Barton and Charlesworth, 1984 for a fuller discussion of this point). There is hence rather little concordance between the extent of evolutionary change perceived by molecular anatomists and that which is obtained by students of the genes actually involved in the origin of species. The two approaches are asking different questions about the process of evolution, and it is not easy to see how questions about the numbers of loci involved in the reproductive isolation of modern man from his ancestors and relatives will be answered. Indeed, we now realise that some major genetic changes during the evolutionary history of particular species can only be detected by the traditional methods of Mendelian genetics. Crosses between D. melanogaster collected in the wild 50 years ago (and maintained in the laboratory since then) and modern wild flies show that there has been a marked genetic change within this evolving lineage in the past 500 generations; the offspring of such crosses show 'hybrid dysgenesis', a syndrome of increased mutation rate and decreased fitness which is due to a change in the interactions between nuclear and cytoplasmic genes (Green, 1980). This important evolutionary event has left almost no trace in the physical structure of this species, and there is as yet no way of identifying whether such events might have occurred in primate evolution.

SUMMARY AND CONCLUSIONS

The study of the comparative anatomy of primate and human genomes has told us a great deal about some of the genetic events in man's evolutionary history. However, comparison of our knowledge of primate evolution with that of other groups makes it clear that these studies leave a great deal unanswered. We are left with the uncomfortable possibility that the new information on molecular anatomy may mean that we still know rather little about the genetic changes involved in the origin of man.

REFERENCES

Key references

Ammerman, A.J. & Cavalli-Sforza, L.L. (1984). The Neolithic Transition and the Genetics of Populations in Europe. Princeton, N.J.: Princeton Univ. Press.

Balner, H. (1981). The major histocompatibility complex of primates, evolutionary aspects and comparative histogenetics. Phil. Trans. Roy. Soc. Lond. Ser. B. *292*, 109-20.

Cann, R.L., Brown, W.M. & Wilson, A.C. (1984). Polymorphic sites and the mechanism of evolution in human mitochondrial DNA. Genetics *106*, 479-99.

Latter, B.D.H. (1980). Genetic differences within and between populations of the major human subgroups. Amer. Nat. *116*, 220-37.

Piazza, A., Menozzi, P. & Cavalli-Sforza, L.L. (1981). Synthetic gene frequency maps of man and selective effects of climate. Proc. Natl. Acad. Sci. US. *78*, 2638-42.

Thomson, G. (1983). The human histocompatibility system: anthropological considerations. Amer. J. Phys. Anthropol. *62*, 81-9.

Yunis, J.J. & Prakash, O. (1982). The origin of man: a chromosomal pictorial legacy. Science *215*, 1525-30.

Main references

Ammerman, A.J. & Cavalli-Sforza, L.L. (1984). The Neolithic Transition and the Genetics of Populations in Europe. Princeton, N.J.: Princeton Univ. Press.

Anderson, S., Bankier, A.T., Barrell, B.G., DeBruijn, M.H.L., Coulson, A.R., Drouin, J., Eperon, I.C., Nierlich, D.P., Roe, B.A., Sanger, F., Schreier, P.H., Smith, A.J.H., Staden, R. & Young, I.G. (1981). Sequence and organization of the human mitochondrial genome. Nature *290*, 457-65.

Antonarakis, S.E., Boehm, C.D., Serjeant, G.R., Theisen, C.E., Dover, G.J. & Kazazian, H.H. (1984). Origin of the beta-S globin gene in Blacks: The contribution of recurrent mutation or gene conversion or both. Proc. Natl. Acad. Sci. US. *81*, 853-6.

Balner, H. (1981). The major histocompatibility complex of primates: evolutionary aspects and comparative histogenetics. Phil. Trans. Roy. Soc. Lond. Ser. B. *292*, 109-20.

Barton, N.H. & Charlesworth, B. (1984). Genetic revolutions, founder effects and speciation. Ann. Rev. Ecol. Syst. *15*, 133-64.

Barton, N.H. & Jones, J.S. (1983). Mitochondrial DNA: new clues about evolution. Nature *306*, 317-8.

Branda, R.F. & Eaton, J.W. (1978). Skin colour and nutrient photolysis: an evolutionary hypothesis. Science *201*, 625-6.

Brown, W.M., Prager, E.M., Wang, A. & Wilson, A.C. (1982). Mitochondrial DNA sequences of primates: tempo and mode of evolution. J. Mol. Evol. *18*, 225-39.

Bruce, E.J. & Ayala, F.J. (1979). Phylogenetic relationships between man and the apes: electrophoretic evidence. Evolution *33*, 1040-56.

Cann, R.L., Brown, W.M. & Wilson, A.C. (1982). Evolution of human mitochondrial DNA: a preliminary report. *In* Human Genetics, part A: The Unfolding Genome. Ed. B. Bonne-Tamir,, P. Cohen & R.N. Goodman. pp. 157-65. New York: Alan R. Liss Inc.

Cann, R.L., Brown, W.M. & Wilson, A.C. (1984). Polymorphic sites and the mechanism of evolution in human mitochondrial DNA. Genetics *106*, 479-99.

Clemens, T.L., Henderson, S.L., Adams, J.S. & Holick, M.F. (1982). Increased skin pigment reduces the capacity of skin to synthesise Vitamin D3. The Lancet 1982 *(1)*, 74-6.

Coyne, J.A. (1984). Genetic basis of male sterility between two closely related species of Drosophila. Proc. Natl. Acad. Sci. US. *81*, 4444-7.

Ferris, S.D., Sage, R.D., Huang, C.M., Nielsen, J.T., Ritte, U. & Wilson, A.C. (1983). Flow of mitochondrial DNA across a species boundary. Proc. Natl. Acad. Sci. US. *80* 2290-4.

Ferris, S.D., Wilson, A.C. & Brown, W.M. (1981). Evolutionary tree for apes and humans based on cleavage maps of mitochondrial DNA. Proc. Natl. Acad. Sci. US. *78*, 2432-6.

Friedman, M.J. & Trager, W. (1981). The biochemistry of resistance to malaria. Sci. Amer. *244*, 154-64.

Goodman, M., Braunitzer, G., Stangl, A. & Shrank, B. (1983). Evidence on human origins from haemoglobin of African apes. Nature *303*, 546-8.

Goodman, M., Romero-Herrera, A.E., Dene, H., Czelusniak, J., & Tahian, R.E. (1982). Amino acid sequence evidence on the phylogeny of primates and other eutherians. *In* Macromolecular Sequences in Systematic and Evolutionary Biology, Ed. M. Goodman. pp. 115-91. New York: Plenum Press.

Green, M.M. (1980). Transposable elements in Drosophila and other diptera. Ann. Rev. Genet. *14*, 109-20.

Higgs, D.R., Goodbourn, S.E.Y., Wainscoat, J.S., Clegg, J.B. & Weatherall, D.J. (1981). Highly variable regions of DNA flank the human alpha-globin genes. Nucl. Acid. Res. *9*, 4213-24.

Jeffreys, A.J. (1982). Evolution of globin genes. *In* Genome Evolution, Ed. G.A. Dover & R.B. Flavell, pp.157-76. London: Academic Press.

Jeffreys, A.J. & Barrie, P.A. (1981). Sequence variation and evolution of nuclear DNA in man and the primates. Phil. Trans. Roy. Soc. London. Ser. B. *292*, 133-42.

Johnson, M.J., Wallace, D.C., Ferris, S.D., Rattazzi, M.C. & Cavalli-Sforza, L.L. (1983). Radiation of human mitochondrial DNA types analyzed by restriction endonuclease cleavage patterns. J. Mol. Evol. *19*, 255-71.

Johnson, M.S., Clarke, B.C. & Murray, J.J. (1977). Genetic variation and reproductive isolation in Partula. Evolution *31*, 116-26.

Johnson, R.C., Cole, R.E. & Ahearn, F.M. (1981). Genetic interpretation of racial/ethnic differences in lactose absorption and tolerance: a review. Human Biol. *53*, 1-14.

Lansman, R.A., Avise, J.C., Aquadro, C.F., Shapira, J.F. & Daniel, S.W. (1983). Extensive genetic variation in mitochondrial DNA among geographic populations of the deer mouse, Peromyscus maniculatus. Evolution *37*, 1-16.

Latter, B.D.H. (1980). Genetic differences within and between populations of the major human subgroups. Amer. Nat. *116*, 220-37.

Luzzatto, L. (1979). Genetics of red cells and susceptibility to malaria. Blood *54*, 961-76.

Murray, J.C., Mills, K.A., Demopulos, C.M., Hornung, S. & Motulsky, A.G. (1984). Linkage disequilibrium and evolutionary relationships of DNA variants (restriction fragment length polymorphisms) at the serum albumin locus. Proc. Natl. Acad. Sci. US. *81*, 3486-90.

Nei, M. & Roychoudhury, A.K. (1982). Genetic relationship and evolution of human races. Evol. Biol. *14*, 1-59.

Ochman, H., Jones, J.S. & Selander, R.K. (1983). Molecular area effects in Cepaea. Proc. Natl. Acad. Sci. US. *80*, 4189-93.

Page, D.C., Harper, M.E., Love, J. & Botstein, D. (1984). Occurrence of a transposition from the X-chromosome long arm to the Y-chromosome short arm during human evolution. Nature *311*, 119-23.

Pagnier, J., Mears, J.G., Dunda-Belkhodja, O., Schaefer-Rego, K.E., Beldjord, C., Nagel, R.L. & Labie, D. (1984). Evidence for the multicentric origin of the sickle cell hemoglobin gene in Africa. Proc. Natl. Acad. Sci. US. *81*, 1771-73.

Piazza, A., Menozzi, P. & Cavalli-Sforza, L.L. (1981). Synthetic gene frequency maps of man and selective effects of climate. Proc. Natl. Acad. Sci. US. *78*, 2638-42.

Pontecorvo, G. (1943). Viability interactions between chromosomes of Drosophila melanogaster and D. simulans. J. Genet. *45*, 51-66.

Roderick, T.B., Lalley, P.A., Davisson, M.T., O'Brien, S.J., Womack, J.E., Creau-Goldberg, N., Echard, G. & Moore, K.L. (1984). Report of the Committee on Comparative Mapping. Cytogenet. Cell Genet. *37*, 312-39.

Shows, T.B., Sakaguchi, A.Y. & Naylor, S.L. (1982). Mapping the human genome: cloned genes, DNA polymorphisms, and inherited disease. Human Genetics *12*, 241-52.

Sibley, C.G. & Ahlquist, J.E. (1984). The phylogeny of the hominoid primates, as indicated by DNA-DNA hybridisation. J. Mol. Evol. *20*, 2-15.

Socha, W.W. & Ruffie, J. (1983). Blood Groups of Primates: Theory, Practice and Evolutionary Meaning. New York: Alan R. Liss.

Thomson, G. (1983). The human histocompatibility system: anthropological considerations. Amer. J. Phys. Anthropol. *62*, 81-9.

Wilson, V., Jeffreys, A.J., Barrie, P.A., Boseley, P.G., Slocombe, P.M., Easton, A. & Burke, D.C. (1983). A comparison of vertebrate interferon gene families detected by hybridisation with human interferon DNA. J. Mol. Biol. *166*, 457-75.

Yunis, J.J. & Prakash, O. (1982). The origin of man: a chromosomal pictorial legacy. Science *215*, 1525-30.

18 THE ORIGIN OF HOMO SAPIENS: THE FOSSIL EVIDENCE

David Pilbeam,
Department of Anthropology,
Peabody Museum,
Harvard University,
Cambridge, MA 02138 U.S.A.

INTRODUCTION

I write on this topic very much as an outsider, but one who has long had an active interest in the origins of modern humans. My natural inclination to make this a general survey is strengthened by the recent appearance of several excellent and exhaustive reviews of the evidence, which fortunately I do not have to repeat (for example, Klein, 1983a, b; Ronen, 1982; Trinkaus, 1983a, b; Smith & Spencer, 1984). We palaeoanthropologists sometimes forget our prime reason for studying human evolution: namely to understand the critical behavioural stages in our history and the evolutionary transitions which lie between them. To this end, we try to build up a realistic behaviour sequence, behaviour being reflected in the morphological and, where present, archaeological records. Two activities are involved here. Firstly, comes the phylogenetic reconstruction of evolutionary sequences. Secondly, there is the behavioural reconstruction of past species as much as possible as though they were alive today, which means approaching the past as though it were a kind of present, asking essentially the same questions of it as we do in trying to understand the behavioural-ecological dynamics of the living world (Pilbeam, 1984a).

FOSSIL RECORD

As the hominid fossil record expands, some of the critical steps in human evolution become clearer. The most obvious ones are firstly, the evolution of large brained hominoids; secondly, the origin of bipedal hominids; thirdly, the appearance of larger brained Homo and the establishment of the 'archaic' Homo pattern and, fourthly, the appearance of modern Homo sapiens (Pilbeam, 1984b). This last step, the evolution of modern humans, is certainly as dramatic and important as any other in human prehistory, although it has not lately attracted the amount of attention - at least public - as some of the earlier phases. This is a pity for it is certainly the only event that is remotely well documented. It took place in less than 100 tyr, and probably a shorter time than that and both the 'before' (ancestral) and 'after' (descendant) stages are well sampled. In addition, paraphrasing Pogo, the descendant is us. For parts of the ancestral stage, in particular from the circum-Mediterranean region, we have an excellent fossil and archaeological record (Smith and Spencer, 1984). The neanderthals from this region are well known and we have a reasonable understanding of their stature, robusticity, head and body anatomy, sexual dimorphism, and general morphological variation (Trinkaus, 1983a, b). The fossil record from elsewhere, although poor, is still good enough to give us some idea about regional variation within what is probably a single later Pleistocene species. Until that record improves it will be unclear to what extent any particular neanderthal feature is, or is not, present in other archaic populations. Finally, the archaeological record is excellent, both for this 'archaic' human phase and for the early modern humans (Gowlett, 1984) and lively progress is being made toward understanding both the anatomical and the artefactual records (White, 1982; Binford, 1982; Mellars, 1982; Binford, 1984).

Compare this last great evolutionary transition with some of the earlier ones, and consider particularly information density (which means standardizing the fossil data base both for time segment length and for geographical distribution). Relevant to understanding the origins of modern humans, I estimate that we have at least an order of magnitude more information, perhaps two, than for the next best understood transition. Remember, for hominid origins we have _no_ (recognized) fossil record; if (and only if) early australopithecines are counted the descendants in that transition then we still have only half a record. For the origins of _Homo_ there is agreement about neither the exact ancestral nor descendant populations and sampling and radiometric control are poor. In addition both anatomical generalizations and behavioural inferences are inevitably much less precise because we are dealing with species which are even less like us than 'archaic' _H. sapiens_.

The transition to modern _Homo sapiens_ was complete in most, perhaps all, inhabited parts of the Old World by a little over 30 tyr ago (Smith and Spencer, 1984; Gowlett, 1984). The exact time of the completion almost certainly varied locally, perhaps by several thousand years. The beginning of the transition is less clear. It began no more than 100 tyr earlier and probably much more recently. We need much more geochronological work on this critical time period; unfortunately, many of the key specimens may now be undatable. The precise pattern of the transition from 'archaic' _Homo sapiens_ to modern humans is also a problem.

It is very unlikely that an evolutionary transition occurred in western Europe (the best sampled and dated area). However, western Europe is only a small part of the Old World land mass so this does not much reduce our area of ignorance. Modern humans were present in southwest Asia earlier than in west Europe, and perhaps in Africa earlier than in southwest Asia (Smith and Spencer, 1984). However, dates on almost all specimens have been, or could be, questioned, and enormous parts of the Old World (south Asia, southeast Asia) are effectively unknown for this range of time. I suspect that, if what actually happened was 'knowable', we would see that modern humans evolved first in a localized area, then spread by replacing and hybridizing with archaic populations. Given sufficient resolution, this would mean that modern and archaic humans (different species?) would have coexisted. I realize the alarm this will generate in some quarters, but consider the following. Perhaps it is only a problem because we are working in such fine grain. An equivalent time period of a few tens of thousands of years would simply be not observable in the case of the evolution, approximately 1.7 million years ago, of _H. erectus_ from _H. habilis_, or some other _Homo_. W.W. Howells' classic paper (Howells,

1976) thus still stands in its basic conclusions: neither the palaeontological, nor the genetical, nor the archaeological records as they now stand can tell us exactly 'when, where, or how'.

It is worth reflecting briefly on the kinds of evolutionary questions that can, and cannot, be realistically answered, and adjust our language and our efforts to reflect this. Remember Lewontin's caveats of a decade ago about the difficulties of achieving dynamically sufficient descriptions of evolutionary change, even in the genetic present (Lewontin, 1974); that is, a description for which all relevant parameters are known, and in which their interactions have been worked out. Just to characterize this hominid transition adequately in terms of anatomical features is at present not possible (let alone in any other detail) and such a project would require much more information than we now have. At the very least we would need well-sampled, well-dated sequences spanning the late Pleistocene in upwards of a dozen, representative, parts of the Old World.

A view of late Pleistocene hominid evolution which has gained considerable ground in the last two or three decades, and which was given a major push in 1964 by Brace's now classic paper (Brace, 1964), sees behavioural change as steadily accelerating once modern brain size was established several hundred thousand years ago (Tobias, 1981). Clear anatomical and archaeological boundaries are hard to draw between modern and archaic humans, and the latter are solidly placed in Homo sapiens. The transition to anatomically modern humans is thought to have occurred widely throughout the inhabited Old World. In contrast is a view that has both old and new inspirations: this holds that were was little change over long time periods which were punctuated by rapid bursts of change emanating from highly localized populations. Africa leads the 'favourite-area' poll at present, although perhaps only because data from Asia are so sparse.

Putting on (or hiding behind) my outsider's hat and echoing Howells again, I see no way at present that either alternative can be categorically ruled out. If they are converted from assertions into hypotheses to be tested by local palaeontological and archaeological records, current data are inadequate to settle the issue. Similarly, despite the desirability of pursuing a cladistic strategy here (see Santa Luca, 1978), in dealing with such anatomically and genetically similar populations I see no way in which any morphological differences between these archaic and modern populations can be used to rule out relatedness.

The record is relatively excellent, but let us not forget it is only relative excellence. With this later hominid record, we tend to change mental gears without noticing, and slip into patterns of thinking and talking that are essentially microevolutionary, rather than broadly macroevolutionary. The solution is to focus for the moment on potentially more tractable questions, and get more material. Just as a trend towards a more functional analysis has transformed the early phases of hominid studies, we also now have a more biomechanical and functional approach towards the later stages. In the last few years we are seeing very creative attempts to render these archaic ancestors of ours more lifelike, and to use the anatomical and archaeological records to establish behavioural contrasts between modern humans and their archaic predecessors.

I shall advocate a very general ('macro') framework that sees 'archaic' Homo as being broadly ancestral to modern humans, and which views the circum-Mediterranean neanderthals as very similar to 'archaic' Homo populations found elsewhere between about 300 and about 40 tyr ago. Behaviour, inferred from both morphological and archaeological patterns, seems to change relatively little within this period. Around 40 tyr ago, perhaps earlier, it begins to change rapidly, then settles into a second 'modern' pattern in which morphological change is minimal, while behaviour changes radically, almost certainly because it is controlled and organized in a new, cultural, way. The contrast between the two archaeological patterns may well reflect differences in the amount and efficiency of information storage and transmission within each system.

The perspective one adopts on this 'before' phase is important. Although it may eventually turn out that 'archaic' Homo behaviour was, in its essentials, modern, I think that is unlikely, and it will be safer to adopt a perspective that assumes initially that it was not so (Pilbeam, 1984a). Let us, instead, approach the anatomical and archaeological data expecting it to tell us that the total behavioural system of 'archaic' Homo was different from that of modern humans - perhaps radically so. We should try to imagine that we can watch these pre-humans as though they were alive today, and ask the same questions that a behavioural ecologist would in a field study of human, or any mammal, populations; what were the ecological context, the body size and extent of dimorphism, the positional behaviour and ranging patterns, diet and subsistence behaviour, life history patterns, intra- and intergroup relationships, equipment use and skills and communication systems (Hinde, 1983)? Obviously for fossil species we shall only be able to 'know', or record, some of these items.

The salient anatomical features of 'archaic' H. sapiens - best represented by neanderthals - can be quickly summarized. Crania are relatively long, low, and large with big faces (Stringer and Trinkaus, 1981). Brain volumes are within the range of modern humans. Basicrania are, on average, different from modern humans. Teeth are larger than in modern humans, especially so the anterior teeth, which show heavy wear. The postcranial skeleton is very robust, with pronounced muscle markings; distal limb segments are short in some populations. Stature and body weight estimates fall within the ranges of modern populations, although at least some, and probably all, archaic populations were relatively muscular and powerfully built. There are minor differences in phalangeal proportions between archaic and modern hands, and also in the anatomy of the pelvis particularly the length of the pubis (Smith and Spencer, 1984; Trinkaus, 1983a, b; 1984; Lieberman, 1984).

It should be emphasized that most of these features are known only for some archaic populations, and may or may not be generalizable to their unknown contemporaries. Nonetheless, one is not being too daring in pointing to major contrasts between archaic and modern humans. What do these anatomical differences mean both physiologically and behaviourally? We have a substantial range of suggestions to choose from (Trinkaus,

1983a, b; Smith and Spencer, 1984; Lieberman, 1984; Carrier, 1984), and what follows are my own preferences of what seem the most plausible proposals.

The changes in tooth size and wear, in facial size and prognathism, and perhaps some of the cranial shape changes, could be related to shifts in the use of faces and teeth in non-masticatory behaviour: making tools, holding objects in food preparation, and so on. The suggestion that basicranial flexion involves changes in phonetic abilities has been controversial and generally disregarded, nonetheless, in my view, it remains one that is still worth considering. The postcranial changes imply, among other things, that levels of activity and habitual stress became reduced in modern humans. One plausible explanation is that 'archaic' Homo robusticity was at least partially related to endurance running, which formed part of a particular adaptive strategy for running down small game (Trinkaus in Carrier, 1984). Changes in limb proportions might be related to improved thermal control (through cultural adaptations). Modifications in hand anatomy, especially thumb proportions, possibly imply changes in manipulative abilities, and reduction in pelvic dimensions might be related to shortened gestation time and the birth of less mature newborn infants.

Clearly, all of these suggestions, and many others, need to be carefully reviewed and where possible experimentally or comparatively evaluated. For example, the inference that high activity levels are linked to skeletal robusticity can be tested through inter-species allometric analyses (Carrier, 1984). Whatever the exact nature of the behavioural differences between modern humans and their ancestors, and of the transition between them, there is a plausible case to be made for the argument that the biobehavioural gap was wide, that 'archaic' human behaviour was different from the behaviour of anatomically modern groups, and that we see in the 'archaics' the final representatives of a very long phase of human evolution, during which only limited changes took place.

I am even less qualified to review the archaeological record, but, nonetheless, I shall do so with enthusiasm. It is important for non-archaeologists to remember that the archaeological record is a better source of information about past behaviour than is the anatomical evidence. There is, of course, considerable diversity of opinion about the artefactual record, but a consensus is emerging that emphasizes both the differences between 'archaic' and 'modern' behavioural patterns, and the fact that 'archaic' patterns were extremely long lasting (Gowlett, 1984; White, 1982; Binford, 1982, 1984; Klein, 1983a, b; Mellars, 1982). In their regional and temporal homogeneity, tool kits are impressively unchanging between roughly 200 tyr or more and about 40 tyr ago. We can label these patterns 'palaeocultural', implying a different system of biobehavioural control than the 'cultural' systems of modern humans (see Jelinek, 1977). If we could indeed sit and watch neanderthals, as though we were at Gombe or Dobe, would we find them different from us? If so, how different?

'Archaic' Homo were probably scavengers and hunters of mainly small game (see Potts, 1984; Shipman and Rose, 1983; Binford, 1984), perhaps incapable of killing at a distance (no bows and arrows, or spear throwers), or of elaborately planned hunts of large mammals. They may also have had limited food storage and preparation abilities. Sharing may have been less developed than in modern humans, perhaps much less, and it is possible that not all critical social and subsistence behaviours were yet focused in a single main place or base camp as they are in modern humans. Possibly patterns of interactions within sexes and between males and females were also different. These and other contrasts would imply the biobehavioural systems were different from modern humans. Around the time of appearance of anatomically modern Homo sapiens archaeological and therefore behavioural patterns change rapidly: soon thereafter we have evidence that behavioural contrasts and change are mediated culturally. The precise patterns and correlation of morphological and archaeological change are not yet fully clear. The amount of information stored in, and flowing through, the social system expanded dramatically, presumably reflecting the final evolution of fully modern human language capabilities. The full modern human spectrum from hunting and gathering to urban living in complex states becomes possible, but all these cultural conditions are based on the same basic biology.

I repeat that the 'tempo and mode' of the behavioural-morphological transition to modern humans is unknown, and the precise coupling of behaviour and morphology is unclear. But a case can be made for change being relatively very rapid, following a period of much slower evolution, pointing to language as a potentially important factor in the complex sequence of shifts that ultimately produced a hominid that was energetically more efficient, and behaviourally much more flexible; the same hominid has been able to 'beat' nature in the sense of transcending critical ecological and reproductive constraints. Who finally 'beats' whom is, of course, a question that Lenin might usefully have asked.

Two of our most important palaeoanthropological challenges are thus to understand the 'archaic' human past as though it were a kind of present, albeit a rather different one, and to try to unravel this last great evolutionary transition. The data are good and can be improved. These enduring and successful later Pleistocene hominids were our immediate ancestors and their temporal proximity and increased density make them at once both more compelling and more tractable than our more distant relatives, and they deserve a major research effort that is coordinated, international, and well focussed.

Some arguments probably cannot be settled given the quality of the current data; these include the tempo, mode, or nature of the transition and the precise cladistic relationships of the 'archaic' groups and modern humans. We should seek more data. Some issues concerning behaviour can potentially be settled, and I urge my colleagues to focus on them. I am struck by the way in which the interpretive pendulum is swinging today in palaeoanthropology, with all stages in human evolution being looked at with fresh historicist eyes and being interpreted as significantly less human than we had previously thought (Pilbeam, 1984a). Even so, it is surprising to see such recent ancestors being potentially so unlike us in many and significant ways.

SUMMARY AND CONCLUSIONS

The evolutionary transition to modern humans is one of the most dramatic of all steps in hominid evolution, and it is the only one that is reasonably well documented both anatomically and archaeologically. The transition took less than 100 thousand years (tyr), perhaps substantially less, and both the 'before' and 'after' stages are well sampled. At least for the neanderthal type of 'archaic' sapiens, skeletal part representation is excellent so that we have reasonable knowledge of intra- and inter-sexual variation. Other archaic groups are becoming better known.

The transition to modern humans was essentially complete in most parts of the Old World by just over 30tyr ago, with local completion apparently being heterochronous. The start of the transition is less clear: it began no more than 100 tyr earlier, and perhaps more recently than that. The pattern of the transition is as equivocal as its timing and at present it is not possible to decide among several alternatives. The very abundant archaeological record is of little help in resolving these problems. It is likely that at least an order of magnitude more information will be necessary to settle this microevolutionary problem.

Attention is increasingly being paid rather to the more tractable question of the behavioural meaning of the anatomical and archaeological changes from 'archaic' to 'modern' sapiens. Modern patterns evolved relatively rapidly following a much longer period of slow change and stasis. Possibly, changes in information storage and transmission played important roles. Since this last great hominid transition is the best known of all, it deserves close study as a transition.

A case can be made that the nomen H. sapiens should apply only to hominids for which modern behavioural patterns can reasonably be inferred; another name would then be needed for 'archaic' H. sapiens.

REFERENCES

Key references

Binford, L. (1984). Faunal remains from Klasies River Mouth. New York: Academic Press.

Pilbeam, D. (1984a). Reflections on early human ancestors. J. Anth. Res., *40*, 14-22.

Ronen, A., ed. (1982). The transition from Lower to Middle Palaeolithic and the origin of modern man. Oxford: BAR International Series 151.

White, R. (1982). Rethinking the Middle/Upper Paleolithic transition. Curr. Anth., *23*, 169-92.

Main references

Binford, L. (1982). Reply to White. Curr. Anth., *23*, 177-81.

Binford, L. (1984). Faunal remains from Klasies River Mouth. New York: Academic Press.

Brace, C.L. (1964). The fate of the classic Neanderthals: a consideration of hominid catastrophism. Curr. Anth., *5*, 1-43.
Carrier, D. (1984). The energetic paradox of human running and hominid evolution. Curr. Anth., *25*, 483-96.
Gowlett, J. (1984). Ascent to civilization. New York: Knopf.
Hinde, R. (1983). Primate social relationships. Sunderland, Mass.: Sinauer.
Howells, W.W. (1976). Explaining modern man: evolutionists versus migrationists. J. Hum. Evol., *5*, 477-95.
Jelinek, A. (1977). The lower Paleolithic: current evidence and interpretations. Ann. Rev. Anthropol., *6*, 11-32.
Klein, R. (1983a). What do we know about Neanderthals and Cro-Magnon man? Amer. Schol., *52*, 386-92.
Klein, R. (1983b). The stone age prehistory of southern Africa. Ann. Rev. Anthropol., *12*, 25-48.
Lewontin, R. (1974). The genetic basis of evolutionary change. New York: Columbia University Press.
Lieberman, P. (1984). The biology and evolution of language. Cambridge, Mass.: Harvard University Press.
Mellars, P. (1982). On the Middle/Upper Paleolithic transition: a reply to White. Curr. Anth., *23*, 238-40.
Pilbeam, D. (1984a). Reflections on early human ancestors. J. Anth. Res., *40*, 14-22.
Pilbeam, D. (1984b). The descent of hominoids and hominids. Sci. Amer., *250*, 84-96.
Potts, R. (1984). Hominid hunters? Problems of identifying the earliest hunter/gatherers. *In* Hominid evolution and community ecology, ed. R. Foley, pp. 129-66. New York: Academic Press.
Ronen, A., ed. (1982). The transition from Lower to Middle Palaeolithic and the origin of modern man. Oxford: BAR International Series 151.
Santa Luca, A. (1978). A re-examination of presumed neanderthal-like fossils. J. Hum. Evol., *7*, 619-36.
Shipman, P. & Rose, J. (1983). Early hominid hunting, butchering and carcass-processing behaviors; approaches to the fossil record. J. Anthropol. Res., *2*, 57-98.
Smith, F. & Spencer, F., eds. (1984). The origins of modern humans. New York: Liss.
Stringer, C. & Trinkaus, E. (1981). The Shanidar neanderthal crania. *In* Aspects of human evolution, ed. C.B. Stringer, pp.129-66. London: Taylor & Francis.
Tobias, P.V. (1981). The emergence of man in Africa and beyond. Phil. Trans. Soc. Lond. B, *292*, 43-56.
Trinkaus, E. (1983a). The Mousterian legacy. Oxford: BAR International Series 164.
Trinkaus, E. (1983b). The Shanidar neanderthals. New York: Academic.
Trinkaus, E. (1984). Neanderthal pubic morphology and gestation length. Curr. Anth., *25*, 509-14.
White, R. (1982). Rethinking the Middle/Upper Paleolithic transition. Curr. Anth., *23*, 169-92.

AUTHOR INDEX

SUBJECT INDEX